Universitext

Universitext

Universitext is a series of textbooks that presents material from a wide variety of mathematical disciplines at master's level and beyond. The books, often well class-tested by their author, may have an informal, personal even experimental approach to their subject matter. Some of the most successful and established books in the series have evolved through several editions, always following the evolution of teaching curricula, to very polished texts.

Thus as research topics trickle down into graduate-level teaching, first textbooks written for new, cutting-edge courses may make their way into *Universitext*.

For further volumes:
http://www.springer.com/series/223

R. Balakrishnan • K. Ranganathan

A Textbook of Graph Theory

Second Edition

Springer

R. Balakrishnan
Department of Mathematics
Bharathidasan University
Tiruchirappalli, India

K. Ranganathan
Deceased

ISSN 0172-5939   ISSN 2191-6675 (electronic)
ISBN 978-1-4614-4528-9   ISBN 978-1-4614-4529-6 (eBook)
DOI 10.1007/978-1-4614-4529-6
Springer New York Heidelberg Dordrecht London

Library of Congress Control Number: 2012946176

Mathematics Subject Classification: 05Cxx

Printed on acid-free paper

Springer is part of Springer Science+Business Media (www.springer.com)

Preface to the Second Edition

As I set out to prepare this Second Edition, I realized that I missed very much my coauthor K. Ranganathan, who had an untimely death in 2002; but then his guiding spirit was always there to get me going.

This Second Edition is a revised and enlarged edition with two new chapters—one on domination in graphs (Chap. 10) and another on spectral properties of graphs (Chap. 11)—and an enlarged chapter on graph coloring (Chap. 7). Chapter 10 presents the basic properties of the domination number of a graph and also deals with Vizing's conjecture on the domination number of the Cartesian product of two graphs. Chapter 11 contains several results on the eigenvalues of graphs and includes a section on the Ramanujan graphs and another on the energy of graphs. The new additions in Chap. 7 include the introduction of b-coloring in graphs and an extension of the discussion of the Myceilskian of a graph over what was given in the First Edition. The sections of Chap. 10 of the First Edition that contained some applications of graph theory have been shifted in the Second Edition to the relevant chapters: "The Connector Problems" to Chap. 4, "The Timetable Problem" to Chap. 5 and the "Application to Social Psychology" to Chap. 1.

There are many who helped me to bring out this Second Edition. First and foremost, I owe my thanks to my former colleague S. Baskaran, who class-tested most of the First Edition, pointed out errors, and came up with many useful suggestions. My thanks are also due to Ashwin Ganesan, S. Francis Raj, P. Paulraja, and N. Sridharan, who read portions of the book; R. Sampathkumar, who proofread most of this edition; and A. Anuradha, who fixed all the figures and consolidated the entire material. Typesetting in LaTeX was done by Mohammed Parvees and R. Sampathkumar, and it is my pleasure to thank them.

I welcome any comments, suggestions, and corrections from readers. They can be sent to me at the email address: mathbala@sify.com.

It was a pleasure working with Springer New York, especially Kaitlin Leach, who was in charge of publishing this edition.

Tiruchirappalli, Tamil Nadu, India   R. Balakrishnan

Preface to the First Edition

Graph theory has witnessed an unprecedented growth in the 20th century. The best barometer to indicate this growth is the explosion in the number of pages that section 05: Combinatorics (in which the major share is taken by graph theory) occupies in the *Mathematical Reviews*. One of the main reasons for this growth is the applicability of graph theory in many other disciplines, such as physics, chemistry, psychology, and sociology. Yet another reason is that some of the problems in theoretical computer science that deal with complexity can be transformed into graph-theoretical problems.

This book aims to provide a good background in the basic topics of graph theory. It does not presuppose deep knowledge of any branch of mathematics. As a basic text in graph theory, it contains, for the first time, Dirac's theorem on k-connected graphs (with adequate hints), Harary–Nash–Williams' theorem on the hamiltonicity of line graphs, Toida–McKee's characterization of Eulerian graphs, the Tutte matrix of a graph, David Sumner's result on claw-free graphs, Fournier's proof of Kuratowski's theorem on planar graphs, the proof of the nonhamiltonicity of the Tutte graph on 46 vertices, and a concrete application of triangulated graphs.

An ambitious teacher can cover the entire book in a one-year (equivalent to two semesters) master's course in mathematics or computer science. However, a teacher who wants to proceed at leisurely pace can omit the sections that are starred. Exercises that are starred are nonroutine.

The book can also be adapted for an undergraduate course in graph theory by selecting the following sections: 1.1–1.6, 2.1–2.3, 3.1–3.4, 4.1–4.5, 5.1–5.4, 5.5 (omitting consequences of Hall's theorem), 5.5 (omitting the Tutte matrix), 6.1–6.3, 7.1, 7.2, 7.5 (omitting Vizing's theorem), 7.8, 8.1–8.4, and Chap. 10.

Several people have helped us by reviewing the manuscript in parts and offering constructive suggestions: S. Arumugam, S. A. Choudum, P. K. Jha, P. Paulraja, G. Ramachandran, S. Ramachandran, G. Ravindra, E. Sampathkumar, and R. Sampathkumar. We thank all of them most profusely for their kindness in sparing for our sake a portion of their precious time. Our special thanks are due to P. Paulraja and R. Sampathkumar, who have been a constant source of inspiration to us ever since we started working on this book rather seriously. We also thank D. Kannan,

Department of Mathematics, University of Georgia, for reading the manuscript and suggesting some stylistic changes.

We also take this opportunity to thank the authorities of our institutions, Annamalai University, Annamalai Nagar, and National College, Tiruchirappalli, for their kind encouragement. Finally, we thank the University Grants Commission, Government of India, for its financial support for writing this book.

Our numbering scheme for theorems and exercises is as follows. Each exercise bears two numbers, whereas each theorem, lemma, and so forth bears three numbers. Therefore, Exercise 3.4 is the fourth exercise of Sect. 3 of a particular chapter, and Theorem 6.6.1 is the first result of Sect. 6 of Chap. 6.

Tiruchirappalli, Tamil Nadu, India   R. Balakrishnan
                                                             K. Ranganathan

Contents

Chapter 1
Basic Results

1.1 Introduction

Graphs serve as mathematical models to analyze many concrete real-world problems successfully. Certain problems in physics, chemistry, communication science, computer technology, genetics, psychology, sociology, and linguistics can be formulated as problems in graph theory. Also, many branches of mathematics, such as group theory, matrix theory, probability, and topology, have close connections with graph theory.

Some puzzles and several problems of a practical nature have been instrumental in the development of various topics in graph theory. The famous Königsberg bridge problem has been the inspiration for the development of Eulerian graph theory. The challenging Hamiltonian graph theory has been developed from the "Around the World" game of Sir William Hamilton. The theory of acyclic graphs was developed for solving problems of electrical networks, and the study of "trees" was developed for enumerating isomers of organic compounds. The well-known four-color problem formed the very basis for the development of planarity in graph theory and combinatorial topology. Problems of linear programming and operations research (such as maritime traffic problems) can be tackled by the theory of flows in networks. Kirkman's schoolgirl problem and scheduling problems are examples of problems that can be solved by graph colorings. The study of simplicial complexes can be associated with the study of graph theory. Many more such problems can be added to this list.

1.2 Basic Concepts

Consider a road network of a town consisting of streets and street intersections. Figure 1.1a represents the road network of a city. Figure 1.1b denotes the corresponding graph of this network, where the street intersections are represented by

R. Balakrishnan and K. Ranganathan, *A Textbook of Graph Theory,*
Universitext, DOI 10.1007/978-1-4614-4529-6_1,
© Springer Science+Business Media New York 2012

Fig. 1.1 (**a**) A road network
and (**b**) the graph
corresponding to the road
network in (**a**)

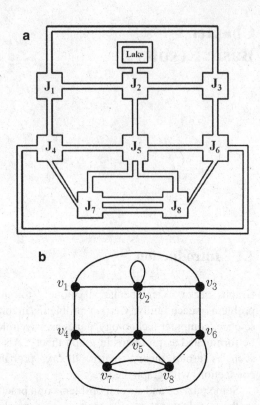

points, and the street joining a pair of intersections is represented by an arc (not
necessarily a straight line). The road network in Fig. 1.1 is a typical example of a
graph in which intersections and streets are, respectively, the "vertices" and "edges"
of the graph. (Note that in the road network in Fig. 1.1a, there are two streets joining
the intersections J_7 and J_8, and there is a loop street starting and ending at J_2.)

We now present a formal definition of a graph.

Definition 1.2.1. A *graph* is an ordered triple $G = (V(G), E(G), I_G)$, where
$V(G)$ is a nonempty set, $E(G)$ is a set disjoint from $V(G)$, and I_G is an "incidence"
relation that associates with each element of $E(G)$ an unordered pair of elements
(same or distinct) of $V(G)$. Elements of $V(G)$ are called the *vertices* (or *nodes* or
points) of G, and elements of $E(G)$ are called the *edges* (or *lines*) of G. $V(G)$ and
$E(G)$ are the *vertex set* and *edge set* of G, respectively. If, for the edge e of G,
$I_G(e) = \{u, v\}$, we write $I_G(e) = uv$.

Example 1.2.2. If $V(G) = \{v_1, v_2, v_3, v_4, v_5\}$, $E(G) = \{e_1, e_2, e_3, e_4, e_5, e_6\}$, and
I_G is given by $I_G(e_1) = \{v_1, v_5\}$, $I_G(e_2) = \{v_2, v_3\}$, $I_G(e_3) = \{v_2, v_4\}$, $I_G(e_4) = \{v_2, v_5\}$, $I_G(e_5) = \{v_2, v_5\}$, $I_G(e_6) = \{v_3, v_3\}$, then $(V(G), E(G), I_G)$ is a graph.

Diagrammatic Representation of a Graph 1.2.3. Each graph can be represented
by a diagram in the plane. In this diagram, each vertex of the graph is represented

Fig. 1.2 Graph
$(V(G), E(G), I_G)$ described
in Example 1.2.2

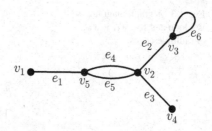

by a point, with distinct vertices being represented by distinct points. Each edge is
represented by a simple "Jordan" arc joining two (not necessarily distinct) vertices.
The diagrammatic representation of a graph aids in visualizing many concepts
related to graphs and the systems of which they are models. In a diagrammatic
representation of a graph, it is possible that two edges intersect at a point that is not
necessarily a vertex of the graph.

Definition 1.2.4. If $I_G(e) = \{u, v\}$, then the vertices u and v are called the *end
vertices* or *ends* of the edge e. Each edge is said to join its ends; in this case, we
say that e is *incident* with each one of its ends. Also, the vertices u and v are then
incident with e. A set of two or more edges of a graph G is called a set of *multiple*
or *parallel edges* if they have the same pair of distinct ends. If e is an edge with end
vertices u and v, we write $e = uv$. An edge for which the two ends are the same is
called a *loop* at the common vertex. A vertex u is a *neighbor* of v in G, if uv is an
edge of G, and $u \neq v$. The set of all neighbors of v is the *open neighborhood* of
v or the *neighbor set* of v, and is denoted by $N(v)$; the set $N[v] = N(v) \cup \{v\}$ is
the *closed neighborhood* of v in G. When G needs to be made explicit, these open
and closed neighborhoods are denoted by $N_G(v)$ and $N_G[v]$, respectively. Vertices
u and v are *adjacent* to each other in G if and only if there is an edge of G with u
and v as its ends. Two distinct edges e and f are said to be *adjacent* if and only if
they have a common end vertex. A graph is *simple* if it has no loops and no multiple
edges. Thus, for a simple graph G, the incidence function I_G is one-to-one. Hence,
an edge of a simple graph is identified with the pair of its ends. A simple graph
therefore may be considered as an ordered pair $(V(G), E(G))$, where $V(G)$ is a
nonempty set and $E(G)$ is a set of unordered pairs of elements of $V(G)$ (each edge
of the graph being identified with the pair of its ends).

Example 1.2.5. In the graph of Fig. 1.2, edge $e_3 = v_2v_4$, edges e_4 and e_5 form
multiple edges, e_6 is a loop at v_3, $N(v_2) = \{v_3, v_4, v_5\}$, $N(v_3) = \{v_2\}$, $N[v_2] =
\{v_2, v_3, v_4, v_5\}$, and $N[v_2] = N(v_2) \cup \{v_2\}$. Further, v_2 and v_5 are adjacent vertices
and e_3 and e_4 are adjacent edges.

Definition 1.2.6. A graph is called *finite* if both $V(G)$ and $E(G)$ are finite. A graph
that is not finite is called an *infinite* graph. Unless otherwise stated, all graphs
considered in this text are finite. Throughout this book, we denote by $n(G)$ and
$m(G)$ the number of vertices and edges of the graph G, respectively. The number
$n(G)$ is called the *order* of G and $m(G)$ is the *size* of G. When explicit reference to

Fig. 1.3 A graph diagram; e_1
is a loop and $\{e_2, e_3\}$ is a set
of multiple edges

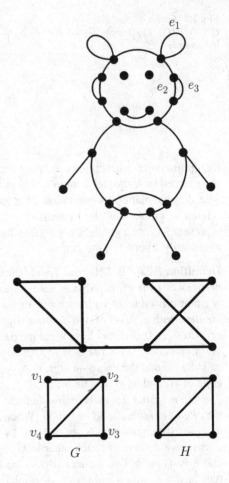

Fig. 1.4 A simple graph

Fig. 1.5 A labeled graph G
and an unlabeled graph H

the graph G is not needed, $V(G)$, $E(G)$, $n(G)$, and $m(G)$ will be denoted simply
by V, E, n, and m, respectively.

Figure 1.3 is a graph with loops and multiple edges, while Fig. 1.4 represents a
simple graph.

Remark 1.2.7. The representation of graphs on other surfaces such as a sphere, a
torus, or a Möbius band could also be considered. Often a diagram of a graph is
identified with the graph itself.

Definition 1.2.8. A graph is said to be *labeled* if its n vertices are distinguished
from one another by labels such as $v_1, v_2, \ldots, v_n$ (see Fig. 1.5).

Note that there are three different labeled simple graphs on three vertices each
having two edges, whereas there is only one unlabeled simple graph of the same
order and size (see Fig. 1.6).

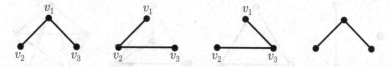

Fig. 1.6 Labeled and unlabeled simple graphs on three vertices

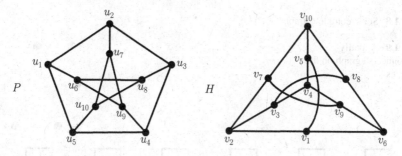

Fig. 1.7 Isomorphic graphs

Isomorphism of Graphs 1.2.9. A graph isomorphism, which we now define, is a concept similar to isomorphism in algebraic structures. Let $G = (V(G), E(G), I_G)$ and $H = (V(H), E(H), I_H)$ be two graphs. A *graph isomorphism* from G to H is a pair (ϕ, θ), where $\phi : V(G) \to V(H)$ and $\theta : E(G) \to E(H)$ are bijections with the property that $I_G(e) = \{u, v\}$ if and only if $I_H(\theta(e)) = \{\phi(u), \phi(v)\}$. If (ϕ, θ) is a graph isomorphism, the pair of inverse mappings (ϕ^{-1}, θ^{-1}) is also a graph isomorphism. Note that the bijection ϕ satisfies the condition that *u and v are end vertices of an edge e of G if and only if $\phi(u)$ and $\phi(v)$ are end vertices of the edge $\theta(e)$ in H*. It is clear that isomorphism is an equivalence relation on the set of all graphs. Isomorphism between graphs is denoted by the symbol $\simeq$ (as in algebraic structures).

Simple Graphs and Isomorphisms 1.2.10. If graphs G and H are simple, any bijection $\phi : V(G) \to V(H)$ such that u and v are adjacent in G if and only if $\phi(u)$ and $\phi(v)$ are adjacent in H induces a bijection $\theta : E(G) \to E(H)$ satisfying the condition that $I_G(e) = \{u, v\}$ if and only if $I_H(\theta(e)) = \{\phi(u), \phi(v)\}$. Hence, ϕ itself is referred to as an isomorphism in the case of simple graphs G and H. Thus, if G and H are simple graphs, an isomorphism from G to H is a bijection $\phi : V(G) \to V(H)$ such that u and v are adjacent in G if and only if $\phi(v)$ and $\phi(v)$ are adjacent in H. Figure 1.7 exhibits two isomorphic graphs P and H, where P is the well-known Petersen graph. We observe that P is a simple graph.

Exercise 2.1. Let G and H be simple graphs and let $\phi : V(G) \to V(H)$ be a bijection such that $uv \in E(G)$ implies that $\phi(u)\phi(v) \in E(H)$. Show by means of an example that ϕ need not be an isomorphism from G to H.

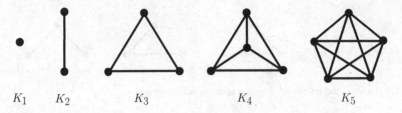

K_1 K_2 K_3 K_4 K_5

Fig. 1.8 Some complete graphs

Fig. 1.9 A totally
disconnected graph on five
vertices

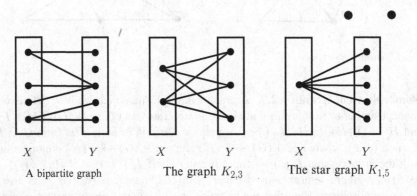

X Y X Y X Y

A bipartite graph   The graph $K_{2,3}$ The star graph $K_{1,5}$

Fig. 1.10 Bipartite graphs

Definition 1.2.11. A simple graph G is said to be *complete* if every pair of distinct vertices of G are adjacent in G. Any two complete graphs each on a set of n vertices are isomorphic; each such graph is denoted by K_n (Fig. 1.8).

A simple graph with n vertices can have at most $\binom{n}{2} = \frac{n(n-1)}{2}$ edges. The complete graph K_n has the maximum number of edges among all simple graphs with n vertices. At the other extreme, a graph may possess no edge at all. Such a graph is called a *totally disconnected graph* (see Fig. 1.9). Thus, *for a simple graph G with n vertices, we have* $0 \leq m(G) \leq \frac{n(n-1)}{2}$.

Definition 1.2.12. A graph is *trivial* if its vertex set is a singleton and it contains no edges. A graph is *bipartite* if its vertex set can be partitioned into two nonempty subsets X and Y such that each edge of G has one end in X and the other in Y. The pair (X, Y) is called a *bipartition* of the bipartite graph. The bipartite graph G with bipartition (X, Y) is denoted by $G(X, Y)$. A simple bipartite graph $G(X, Y)$ is *complete* if each vertex of X is adjacent to all the vertices of Y. If $G(X, Y)$ is complete with $|X| = p$ and $|Y| = q$, then $G(X, Y)$ is denoted by $K_{p,q}$. A complete bipartite graph of the form $K_{1,q}$ is called a *star* (see Fig. 1.10).

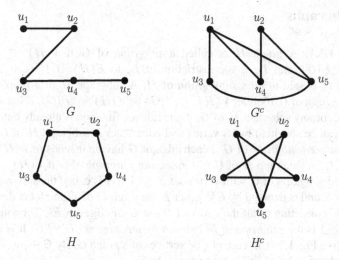
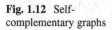

G $\qquad$ G^c

H $\qquad$ H^c

Fig. 1.11 Two simple graphs and their complements

Fig. 1.12 Self-complementary graphs

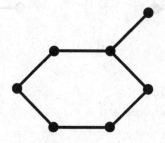

Definition 1.2.13. Let G be a simple graph. Then the *complement* G^c of G is defined by taking $V(G^c) = V(G)$ and making two vertices u and v adjacent in G^c if and only if they are nonadjacent in G (see Fig. 1.11). It is clear that G^c is also a simple graph and that $(G^c)^c = G$.

If $|V(G)| = n$, then clearly, $|E(G)| + |E(G^c)| = |E(K_n)| = \frac{n(n-1)}{2}$.

Definition 1.2.14. A simple graph G is called *self-complementary* if $G \simeq G^c$.

For example, the graphs shown in Fig. 1.12 are self-complementary.

Exercise 2.2. Find the complement of the following simple graph:

1.3   Subgraphs

Definition 1.3.1. A graph H is called a *subgraph* of G if $V(H) \subseteq V(G)$, $E(H) \subseteq E(G)$, and I_H is the restriction of I_G to $E(H)$. If H is a subgraph of G, then G is said to be a *supergraph* of H. A subgraph H of a graph G is a *proper subgraph* of G if either $V(H) \neq V(G)$ or $E(H) \neq E(G)$. (Hence, when G is given, for any subgraph H of G, the incidence function is already determined so that H can be specified by its vertex and edge sets.) A subgraph H of G is said to be an *induced subgraph* of G if each edge of G having its ends in $V(H)$ is also an edge of H. A subgraph H of G is a *spanning subgraph* of G if $V(H) = V(G)$. The induced subgraph of G with vertex set $S \subseteq V(G)$ is called the *subgraph of G induced by S* and is denoted by $G[S]$. Let E' be a subset of E and let S denote the subset of V consisting of all the end vertices in G of edges in E'. Then the graph $(S, E', I_G|_{E'})$ is the *subgraph of G induced by the edge set E' of G*. It is denoted by $G[E']$ (see Fig. 1.13). Let u and v be vertices of a graph G. By $G + uv$, we mean the graph obtained by adding a new edge uv to G.

Definition 1.3.2. A *clique* of G is a complete subgraph of G. A clique of G is a *maximal clique* of G if it is not properly contained in another clique of G (see Fig. 1.13).

Definition 1.3.3. *Deletion of vertices and edges in a graph:* Let G be a graph, S a proper subset of the vertex set V, and E' a subset of E. The subgraph $G[V \setminus S]$ is said to be obtained from G by the *deletion* of S. This subgraph is denoted by $G - S$. If $S = \{v\}$, $G - S$ is simply denoted by $G - v$. The spanning subgraph of G with the edge set $E \setminus E'$ is the subgraph obtained from G by deleting the edge subset E'. This subgraph is denoted by $G - E'$. Whenever $E' = \{e\}$, $G - E'$ is

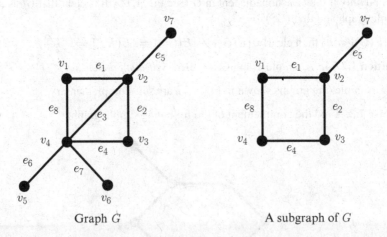

Graph G A subgraph of G

Fig. 1.13 Various subgraphs and cliques of G

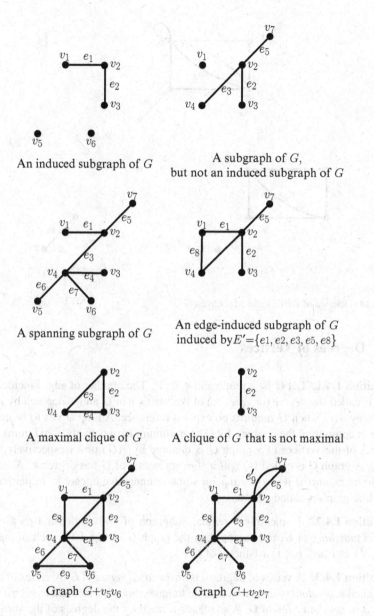

Fig. 1.13 (continued)

simply denoted by $G - e$. Note that when a vertex is deleted from G, all the edges incident to it are also deleted from G, whereas the deletion of an edge from G does not affect the vertices of G (see Fig. 1.14).

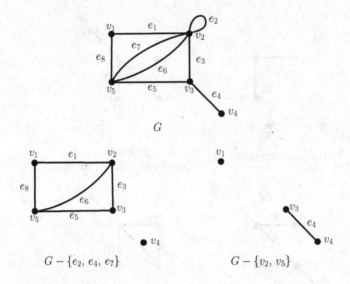

Fig. 1.14 Deletion of vertices and edges from G

1.4   Degrees of Vertices

Definition 1.4.1. Let G be a graph and $v \in V$. The number of edges incident at v in G is called the *degree* (or *valency*) of the vertex v in G and is denoted by $d_G(v)$, or simply $d(v)$ when G requires no explicit reference. A loop at v is to be counted twice in computing the degree of v. The minimum (respectively, maximum) of the degrees of the vertices of a graph G is denoted by $\delta(G)$ or δ (respectively, $\Delta(G)$ or Δ). A graph G is called k-*regular* if every vertex of G has degree k. A graph is said to be *regular* if it is k-regular for some nonnegative integer k. In particular, a 3-regular graph is called a *cubic graph.*

Definition 1.4.2. A spanning 1-regular subgraph of G is called a 1-*factor* or a *perfect matching* of G. For example, in the graph G of Fig. 1.15, each of the pairs $\{ab, cd\}$ and $\{ad, bc\}$ is a 1-factor of G.

Definition 1.4.3. A vertex of degree 0 is an *isolated vertex* of G. A vertex of degree 1 is called a *pendant vertex* of G, and the unique edge of G incident to such a vertex of G is a *pendant edge* of G. A sequence formed by the degrees of the vertices of G, when the vertices are taken in the same order, is called a *degree sequence* of G. It is customary to give this sequence in the nonincreasing or nondecreasing order, in which case the sequence is unique.

In the graph G of Fig. 1.16, the numbers within the parentheses indicate the degrees of the corresponding vertices. In G, v_7 is an isolated vertex, v_6 is a pendant vertex, and $v_5 v_6$ is a pendant edge. The degree sequence of G is $(0, 1, 2, 2, 4, 4, 5)$.

Fig. 1.15 Graph with
1-factors

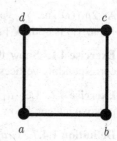

Fig. 1.16 Degrees of vertices
of graph G

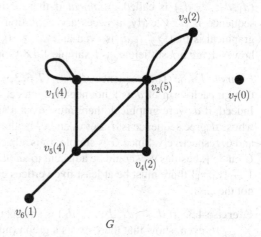

The very first theorem of graph theory was due to Leonhard Euler (1707–1783). This theorem connects the degrees of the vertices and the number of edges of a graph.

Theorem 1.4.4 (Euler). *The sum of the degrees of the vertices of a graph is equal to twice the number of its edges.*

Proof. If $e = uv$ is an edge of G, e is counted once while counting the degrees of each of u and v (even when $u = v$). Hence, each edge contributes 2 to the sum of the degrees of the vertices. Thus, the m edges of G contribute $2m$ to the degree sum. $\square$

Remark 1.4.5. If $d = (d_1, d_2, \ldots, d_n)$ is the degree sequence of G, then the above theorem gives the equation $\sum_{i=1}^{n} d_i = 2m$, where n and m are the order and size of G, respectively.

Corollary 1.4.6. *In any graph G, the number of vertices of odd degree is even.*

Proof. Let V_1 and V_2 be the subsets of vertices of G with odd and even degrees, respectively. By Theorem 1.4.4,

$$2m(G) = \sum_{v \in V} d_G(v) = \sum_{v \in V_1} d_G(v) + \sum_{v \in V_2} d_G(v).$$

As $2m(G)$ and $\sum_{v \in V_2} d_G(v)$ are even, $\sum_{v \in V_1} d_G(v)$ is even. Since for each $v \in V_1$, $d_G(v)$ is odd, $|V_1|$ must be even.   □

Exercise 4.1. Show that if G and H are isomorphic graphs, then each pair of corresponding vertices of G and H has the same degree.

Exercise 4.2. Let $(d_1, d_2, \ldots, d_n)$ be the degree sequence of a graph and r be any positive integer. Show that $\sum_{i=1}^{n} d_i^r$ is even.

Definition 1.4.7. *Graphical sequences*: A sequence of nonnegative integers $d = (d_1, d_2, \ldots, d_n)$ is called *graphical* if there exists a simple graph whose degree sequence is d. Clearly, a necessary condition for $d = (d_1, d_2, \ldots, d_n)$ to be graphical is that $\sum_{i=1}^{n} d_i$ is even and $d_i \geq 0$, $1 \leq i \leq n$. These conditions, however, are not sufficient, as Example 1.4.8 shows.

Example 1.4.8. The sequence $d = (7, 6, 3, 3, 2, 1, 1, 1)$ is not graphical even though each term of d is a nonnegative integer and the sum of the terms is even. Indeed, if d were graphical, there must exist a simple graph G with eight vertices whose degree sequence is d. Let v_0 and v_1 be the vertices of G whose degrees are 7 and 6, respectively. Since G is simple, v_0 is adjacent to all the remaining vertices of G, and v_1, besides v_0, should be adjacent to another five vertices. This means that in $V - \{v_0, v_1\}$ there must be at least five vertices each of degree at least 2; but this is not the case.   □

Exercise 4.3. If $d = (d_1, d_2, \ldots, d_n)$ is any sequence of nonnegative integers with $\sum_{i=1}^{n} d_i$ even, show that there exists a graph (not necessarily simple) with d as its degree sequence.

We present a simple application whose proof just depends on the degree sequence of a graph.

Application 1.4.9. *In any group of n persons ($n \geq 2$), there are at least two with the same number of friends.*

Proof. Denote the n persons by $v_1, v_2, \ldots, v_n$. Let G be the simple graph with vertex set $V = \{v_1, v_2, \ldots, v_n\}$ in which v_i and v_j are adjacent if and only if the corresponding persons are friends. Then the number of friends of v_i is just the degree of v_i in G. Hence, to solve the problem, we must prove that there are two vertices in G with the same degree. If this were not the case, the degrees of the vertices of G must be $0, 1, 2, \ldots, (n - 1)$ in some order. However, a vertex of degree $(n - 1)$ must be adjacent to all the other vertices of G, and consequently there cannot be a vertex of degree 0 in G. This contradiction shows that the degrees of the vertices of G cannot all be distinct, and hence at least two of them should have the same degree.   □

Exercise 4.4. Let G be a graph with n vertices and m edges. Assume that each vertex of G is of degree either k or $k + 1$. Show that the number of vertices of degree k in G is $(k + 1)n - 2m$.

1.5 Paths and Connectedness

Definition 1.5.1. A *walk* in a graph G is an alternating sequence W : $v_0 e_1 v_1 e_2 v_2$ $\ldots e_p v_p$ of vertices and edges beginning and ending with vertices in which v_{i-1} and v_i are the ends of e_i; v_0 is the *origin* and v_p is the *terminus* of W. The walk W is said to join v_0 and v_p; it is also referred to as a v_0-v_p walk. If the graph is simple, a walk is determined by the sequence of its vertices. The walk is *closed* if $v_0 = v_p$ and is *open* otherwise. A walk is called a *trail* if all the edges appearing in the walk are distinct. It is called a *path* if all the vertices are distinct. Thus, a path in G is automatically a trail in G. When writing a path, we usually omit the edges. A *cycle* is a closed trail in which the vertices are all distinct. The *length* of a walk is the number of edges in it. A walk of length 0 consists of just a single vertex.

Example 1.5.2. In the graph of Fig. 1.17, $v_5 e_7 v_1 e_1 v_2 e_4 v_4 e_5 v_1 e_7 v_5 e_9 v_6$ is a walk but not a trail (as edge e_7 is repeated) $v_1 e_1 v_2 e_2 v_3 e_3 v_2 e_1 v_1$ is a closed walk; $v_1 e_1 v_2 e_4 v_4 e_5 v_1 e_7 v_5$ is a trail; $v_6 e_8 v_1 e_1 v_2 e_2 v_3$ is a path and $v_1 e_1 v_2 e_4 v_4 e_6 v_5 e_7 v_1$ is a cycle. Also, $v_6 v_1 v_2 v_3$ is a path, and $v_1 v_2 v_4 v_5 v_6 v_1$ is a cycle in this graph. Very often a cycle is enclosed by ordinary parentheses.

Definition 1.5.3. A cycle of length k is denoted by C_k. Further, P_k denotes a path on k vertices. In particular, C_3 is often referred to as a *triangle*, C_4 as a *square*, and C_5 as a *pentagon*. If $P = v_0 e_1 v_1 e_2 v_2 \ldots e_k v_k$ is a path, then $P^{-1} = v_k e_k v_{k-1} e_{k-1} v_{k-2} \ldots v_1 e_1 v_0$ is also a path and P^{-1} is called the *inverse* of the path P. The subsequence $v_i e_{i+1} v_{i+1} \ldots e_j v_j$ of P is called the v_i-v_j *section* of P.

Definition 1.5.4. Let G be a graph. Two vertices u and v of G are said to be *connected* if there is a u-v path in G. The relation "connected" is an equivalence relation on $V(G)$. Let $V_1, V_2, \ldots, V_\omega$ be the equivalence classes. The subgraphs $G[V_1], G[V_2], \ldots, G[V_\omega]$ are called the *components* of G. If $\omega = 1$, the graph G is *connected*; otherwise, the graph G is *disconnected* with $\omega \geq 2$ components (see Fig. 1.18).

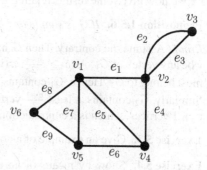

Fig. 1.17 Graph illustrating walks, trails, paths, and cycles

Fig. 1.18 A graph G with
three components

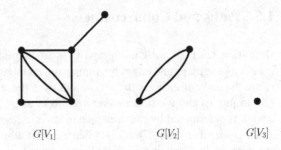

$\qquad G[V_1] \qquad\qquad\qquad\qquad G[V_2] \qquad\qquad G[V_3]$

Definition 1.5.5. The components of G are clearly the maximal connected sub-graphs of G. We denote the number of components of G by $\omega(G)$. Let u and v be two vertices of G. If u and v are in the same component of G, we define $d(u, v)$ to be the length of a shortest u-v path in G; otherwise, we define $d(u, v)$ to be ∞. If G is a connected graph, then d is a distance function or metric on $V(G)$; that is, $d(u, v)$ satisfies the following conditions:

(i) $d(u, v) \geq 0$, and $d(u, v) = 0$ if and only if $u = v$.
(ii) $d(u, v) = d(v, u)$.
(iii) $d(u, v) \leq d(u, w) + d(w, v)$, for every w in $V(G)$.

Exercise 5.1. Prove that the function d defined above is indeed a metric on $V(G)$.

Exercise 5.2. In the following graph, find a closed trail of length 7 that is not a cycle:

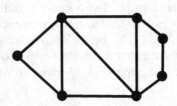

We now give some results relating to connectedness of graphs.

Proposition 1.5.6. *If G is simple and $\delta \geq \frac{n-1}{2}$, then G is connected.*

Proof. Assume the contrary. Then G has at least two components, say G_1, G_2. Let v be any vertex of G_1. As $\delta \geq \frac{n-1}{2}$, $d(v) \geq \frac{n-1}{2}$. All the vertices adjacent to v in G must belong to G_1. Hence, G_1 contains at least $d(v) + 1 \geq \frac{n-1}{2} + 1 = \frac{n+1}{2}$ vertices. Similarly, G_2 contains at least $\frac{n+1}{2}$ vertices. Therefore G has at least $\frac{n+1}{2} + \frac{n+1}{2} = n + 1$ vertices, which is a contradiction. $\qquad\square$

Exercise 5.3. Give an example of a nonsimple disconnected graph with $\delta \geq \frac{n-1}{2}$.

Exercise 5.4. Show by means of an example that the condition $\delta \geq \frac{n-2}{2}$ for a simple graph G need not imply that G is connected.

Exercise 5.5. In a group of six people, prove that there must be three people who are mutually acquainted or three people who are mutually nonacquainted.

Our next result shows that of the two graphs G and G^c, at least one of them must be connected.

Theorem 1.5.7. *If a simple graph G is not connected, then G^c is connected.*

Proof. Let u and v be any two vertices of G^c (and therefore of G). If u and v belong to different components of G, then obviously u and v are nonadjacent in G and so they are adjacent in G^c. Thus u and v are connected in G^c. In case u and v belong to the same component of G, take a vertex w of G not belonging to this component of G. Then uw and vw are not edges of G and hence they are edges of G^c. Then uwv is a u-v path in G^c. Thus G^c is connected. $\qquad\square$

Exercise 5.6. Show that if G is a self-complementary graph of order n, then $n \equiv 0$ or 1 (mod 4).

Exercise 5.7. Show that if a self-complementary graph contains a pendant vertex, then it must have at least another pendant vertex.

The next theorem gives an upper bound on the number of edges in a simple graph.

Theorem 1.5.8. *The number of edges of a simple graph of order n having ω components cannot exceed $\frac{(n-\omega)(n-\omega+1)}{2}$.*

Proof. Let $G_1, G_2, \ldots, G_\omega$ be the components of a simple graph G and let n_i be the number of vertices of G_i, $1 \le i \le \omega$. Then $m(G_i) \le \frac{n_i(n_i-1)}{2}$, and hence $m(G) \le \sum_{i=1}^{\omega} \frac{n_i(n_i-1)}{2}$. Since $n_i \ge 1$ for each i, $1 \le i \le \omega$, $n_i = n - (n_1 + \ldots + n_{i-1} + n_{i+1} + \ldots + n_\omega) \le n - \omega + 1$. Hence, $\sum_{i=1}^{\omega} \frac{n_i(n_i-1)}{2} \le \sum_{i=1}^{\omega} \frac{(n-\omega+1)(n_i-1)}{2} = \frac{(n-\omega+1)}{2} \sum_{i=1}^{\omega}(n_i - 1) = \frac{(n-\omega+1)}{2}\left[\left(\sum_{i=1}^{\omega} n_i\right) - \omega\right] = \frac{(n-\omega+1)(n-\omega)}{2}$. $\qquad\square$

Definition 1.5.9. A graph G is called *locally connected* if, for every vertex v of G, the subgraph $N_G(v)$ induced by the neighbor set of v in G is connected.

A cycle is *odd* or *even* depending on whether its length is odd or even. We now characterize bipartite graphs.

Theorem 1.5.10. *A graph is bipartite if and only if it contains no odd cycles.*

Proof. Suppose that G is a bipartite graph with the bipartition (X, Y). Let $C = v_1e_1v_2e_2v_3 \ldots v_ke_kv_1$ be a cycle in G. Without loss of generality, we can suppose that $v_1 \in X$. As v_2 is adjacent to v_1, $v_2 \in Y$. Similarly, v_3 belongs to X, v_4 to Y, and so on. Thus, $v_i \in X$ or Y according as i is odd or even, $1 \le i \le k$. Since v_kv_1 is an edge of G and $v_1 \in X$, $v_k \in Y$. Accordingly, k is even and C is an even cycle.

Conversely, let us suppose that G contains no odd cycles. We first assume that G is connected. Let u be a vertex of G. Define $X = \{v \in V \mid d(u, v)$ is even$\}$ and $Y = \{v \in V \mid d(u, v)$ is odd$\}$. We will prove that (X, Y) is a bipartition of G. To prove this we have only to show that no two vertices of X as well as no two

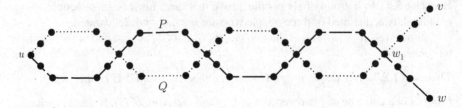

Fig. 1.19 Graph for proof of Theorem 1.5.10

vertices of Y are adjacent in G. Let v, w be two vertices of X. Then $p = d(u, v)$ and $q = d(u, w)$ are even. Further, as $d(u, u) = 0$, $u \in X$. Let P be a u-v shortest path of length p and Q, a u-w shortest path of length q. (See Fig. 1.19.) Let w_1 be a vertex common to P and Q such that the w_1-v section of P and the w_1-w section of Q contain no vertices common to P and Q. Then the u-w_1 sections of both P and Q have the same length.

Hence, the lengths of the w_1-v section of P and the w_1-w section of Q are both even or both odd. Now if $e = vw$ is an edge of G, then the w_1-v section of P followed by the edge vw and the w-w_1 section of the w-u path Q^{-1} is an odd cycle in G, contradicting the hypothesis. This contradiction proves that no two vertices of X are adjacent in G. Similarly, no two vertices of Y are adjacent in G. This proves the result when G is connected.

If G is not connected, let $G_1, G_2, \ldots, G_\omega$ be the components of G. By hypothesis, no component of G contains an odd cycle. Hence, by the previous paragraph, each component $G_i, 1 \le i \le \omega$, is bipartite. Let (X_i, Y_i) be the bipartition of G_i. Then (X, Y), where $X = \bigcup_{i=1}^{\omega} X_i$ and $Y = \bigcup_{i=1}^{\omega} Y_i$, is a bipartition of G, and G is a bipartite graph.   □

Exercise 5.8. Prove that a simple nontrivial graph G is connected if and only if for any partition of V into two nonempty subsets V_1 and V_2, there is an edge joining a vertex of V_1 to a vertex of V_2.

Example 1.5.11. Prove that in a connected graph G with at least three vertices, any two longest paths have a vertex in common.

Proof. Suppose $P = u_1 u_2 \ldots u_k$ and $Q = v_1 v_2 \ldots v_k$ are two longest paths in G having no vertex in common. As G is connected, there exists a u_1-v_1 path P' in G. Certainly there exist vertices u_r and v_s of P', $1 \le r \le k$, $1 \le s \le k$ such that the u_r-v_s section P'' of P' has no internal vertex in common with P or Q.

Now, of the two sections u_1-u_r and u_r-u_k of P, one must have length at least $\frac{k}{2}$. Similarly, of the two sections v_1-v_s and v_s-v_k of Q, one must have length at least $\frac{k}{2}$. Let these sections be P_1 and Q_1, respectively. Then $P_1 \cup P'' \cup Q_1$ is a path of length at least $\frac{k}{2} + 1 + \frac{k}{2}$, contradicting that k is the length of a longest path in G (see Fig. 1.20).   □

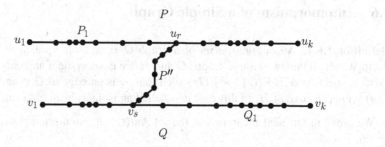

Fig. 1.20 Graph for the solution to Example 1.5.11

Exercise 5.9. Prove that in a simple graph G, the union of two distinct paths joining two distinct vertices contains a cycle.

Exercise 5.10. Show by means of an example that the union of two distinct walks joining two distinct vertices of a simple graph G need not contain a cycle.

Exercise 5.11. If a simple connected graph G is not complete, prove that there exist three vertices u, v, w of G such that uv and vw are edges of G, but uw is not an edge of G.

Exercise 5.12. (see reference: [174]) Show that a simple connected graph G is complete if and only if for some vertex v of G, $N[v] = N[u]$ for every $u \in N[v]$.

Exercise 5.13. A simple graph G is called *highly irregular* if, for each $v \in V(G)$, the degrees of the neighbors of v are all distinct. (For example, P_4 is a graph with this property.) Prove that there exist no connected highly irregular graphs of orders 3 and 5.

Exercise 5.14. The generalized Petersen graph $P(n,k)$ is defined by taking

$$V(P(n,k)) = \{a_i, b_i : 0 \le i \le n-1\}$$

and

$$E(P(n,k)) = \{a_i a_{i+1}, a_i b_i, b_i b_{i+k}, 0 \le i \le n-1\},$$

where the subscripts are integers modulo n, $n \ge 5$ and $1 \le k \le \lfloor \frac{n-1}{2} \rfloor$. Prove that if n is even and k is odd, then $P(n,k)$ is bipartite.

Example 1.5.12. If G is simple and $\delta \ge k$, then G contains a path of length at least k.

Proof. Let $P = v_0 v_1 \dots v_r$ be a longest path in G. Then the vertices adjacent to v_r can only be from among $v_0, v_1, \dots, v_{r-1}$. Hence, the length of $P = r \ge d_G(v_r) \ge \delta \ge k$. $\square$

1.6   Automorphism of a Simple Graph

Definition 1.6.1. An *automorphism* of a graph G is an isomorphism of G onto itself. We recall that two simple graphs G and H are isomorphic if and only if there exists a bijection $\phi : V(G) \to V(H)$ such that uv is an edge of G if and only if $\phi(u)\phi(v)$ is an edge of H. In this case ϕ is called an isomorphism of G onto H.

We prove in our next theorem that the set $\text{Aut}(G)$ of automorphisms of G is a group.

Theorem 1.6.2. *The set $\text{Aut}(G)$ of all automorphisms of a simple graph G is a group with respect to the composition $\circ$ of mappings as the group operation.*

Proof. We shall verify that the four axioms of a group are satisfied by the pair $(\text{Aut}(G), \circ)$.

  (i) Let ϕ_1 and ϕ_2 be bijections on $V(G)$ preserving adjacency and nonadjacency. Clearly, the mapping $\phi_1 \circ \phi_2$ is a bijection on $V(G)$. If u and v are adjacent in G, then $\phi_2(u)$ and $\phi_2(v)$ are adjacent in G. But $(\phi_1 \circ \phi_2)(u) = \phi_1(\phi_2(u))$ and $(\phi_1 \circ \phi_2)(v) = \phi_1(\phi_2(v))$. Hence, $(\phi_1 \circ \phi_2)(u)$ and $(\phi_1 \circ \phi_2)(v)$ are adjacent in G; that is, $\phi_1 \circ \phi_2$ preserves adjacency. A similar argument shows that $\phi_1 \circ \phi_2$ preserves nonadjacency. Thus, $\phi_1 \circ \phi_2$ is an automorphism of G.
 (ii) It is a well-known result that the composition of mappings of a set onto itself is associative.
(iii) The identity mapping I of $V(G)$ onto itself is an automorphism of G, and it satisfies the condition $\phi \circ I = I \circ \phi = \phi$ for every $\phi \in \text{Aut}(G)$. Hence, I is the identity element of $\text{Aut}(G)$.
 (iv) Finally, if ϕ is an automorphism of G, the inverse mapping ϕ^{-1} is also an automorphism of G (see Sect. 1.1). $\square$

Theorem 1.6.3. *For any simple graph G, $\text{Aut}(G) = \text{Aut}(G^c)$.*

Proof. Since $V(G^c) = V(G)$, every bijection on $V(G)$ is also a bijection on $V(G^c)$. As an automorphism of G preserves the adjacency and nonadjacency of vertices of G, it also preserves the adjacency and nonadjacency of vertices of G^c. Hence, every element of $\text{Aut}(G)$ is also an element of $\text{Aut}(G^c)$, and vice versa. $\square$

Exercise 6.1. Show that the automorphism group of K_n (or K_n^c) is isomorphic to the symmetric group S_n of degree n.

In contrast to the complete graphs for which the automorphism group consists of every bijection of the vertex set, there are graphs whose automorphism groups consist of just the identity permutation. Such graphs are called *identity graphs*.

Example 1.6.4. The graph G shown in Fig. 1.21 is an identity graph.

Proof. Let γ be an automorphism of G. Then γ preserves degrees; that is, $d(v) = d(\gamma(v))$ for all $v \in V(G)$ (see Exercise 4.1). Since u_4 is the only vertex of degree 3 in G, $\gamma(u_4) = u_4$. Now u_1, u_2, and u_7 are the vertices of degree 1 in G. Hence, $\gamma(u_1) \in \{u_1, u_2, u_7\}$. Also, since u_1 is adjacent to u_4, $\gamma(u_1)$ is adjacent to

Fig. 1.21 An identity graph

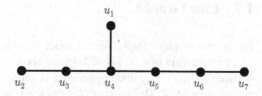

$\gamma(u_4) = u_4$. Hence, the only possibility is $\gamma(u_1) = u_1$. Now, u_3, u_5, and u_6 are the vertices of degree 2 in G. Hence, $\gamma(u_3) \in \{u_3, u_5, u_6\}$. Also, as u_3 is adjacent to u_4, $\gamma(u_3)$ is adjacent to $\gamma(u_4) = u_4$. Hence, $\gamma(u_3)$ is u_3 or u_5. Again, $\gamma(u_3)$ is adjacent to a vertex of degree 1. This forces $\gamma(u_3) \neq u_5$ since u_5 is not adjacent to a vertex of degree 1. Consequently, $\gamma(u_3) = u_3$. This again forces $\gamma(u_2) = u_2$. Having proved $\gamma(u_2) = u_2$, we must have $\gamma(u_7) = u_7$ on degree consideration. Using similar arguments, one easily proves that $\gamma(u_5) = u_5$ and $\gamma(u_6) = u_6$. Consequently, $\gamma = I$, which implies that $\mathrm{Aut}(G) = \{I\}$.   □

Exercise 6.2. Let G be a simple connected graph with n vertices such that $\mathrm{Aut}(G) \simeq S_n$. Show that G is the complete graph K_n.

Exercise 6.3. For $n > 1$, give

(a) A simple connected graph G with $G \neq K_n$ and $\mathrm{Aut}(G) \simeq S_n$,
(b) A simple disconnected graph G with $G \neq K_n^c$ and $\mathrm{Aut}(G) \simeq S_n$.

Exercise 6.4. Find the automorphism groups of the following graphs:

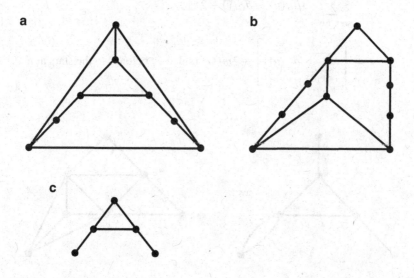

Exercise 6.5. Let G be a simple graph and $\gamma \in \mathrm{Aut}(G)$. Prove that $\gamma(N(v)) = N(\gamma(v))$ and $\gamma(N[v]) = N[\gamma(v)]$ for every $v \in V$.

1.7 Line Graphs

Let G be a loopless graph. We construct a graph $L(G)$ in the following way:

The vertex set of $L(G)$ is in 1-1 correspondence with the edge set of G and two vertices of $L(G)$ are joined by an edge if and only if the corresponding edges of G are adjacent in G. The graph $L(G)$ (which is always a simple graph) is called the *line graph* or the *edge graph* of G.

Figure 1.22 shows a graph and its line graph in which v_i of $L(G)$ corresponds to the edge e_i of G for each i. Isolated vertices of G do not have any bearing on $L(G)$, and hence we assume in this section that G has no isolated vertices. We also assume that G has no loops. (See Exercise 7.4.)

Some simple properties of the line graph $L(G)$ of a graph G follow:

1. G is connected if and only if $L(G)$ is connected.
2. If H is a subgraph of G, then $L(H)$ is a subgraph of $L(G)$.
3. The edges incident at a vertex of G give rise to a maximal complete subgraph of $L(G)$.
4. If $e = uv$ is an edge of a simple graph G, the degree of e in $L(G)$ is the same as the number of edges of G adjacent to e in G. This number is $d_G(u) + d_G(v) - 2$. Hence, $d_{L(G)}(e) = d_G(u) + d_G(v) - 2$.
5. Finally, if G is a simple graph,

$$\sum_{e \in V(L(G))} d_{L(G)}(e)$$

$$= \sum_{uv \in E(G)} (d_G(u) + d_G(v) - 2)$$

$$= \left[\sum_{u \in V(G)} d_G(u)^2 \right] - 2m(G) \text{ (since } uv \text{ belongs to the stars at } u$$

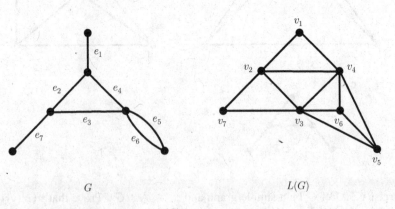

$$G \qquad\qquad\qquad L(G)$$

Fig. 1.22 A graph G and its line graph $L(G)$

and v; see Definitions 1.2.12)

$$= \left[\sum_{i=1}^{n} d_i^2 \right] - 2m,$$

where $(d_1, d_2, \ldots, d_n)$ is the degree sequence of G, and $m = m(G)$. By Euler's theorem (Theorem 1.4.4), it follows that the number of edges of $L(G)$ is given by

$$m(L(G)) = \frac{1}{2} \left[\sum_{i=1}^{n} d_i^2 \right] - m.$$

Exercise 7.1. Show that the line graph of the star $K_{1,n}$ is the complete graph K_n.

Exercise 7.2. Show that $L(C_n) \simeq C_n, n \geq 3$.

Theorem 1.7.1. *The line graph of a simple graph G is a path if and only if G is a path.*

Proof. Let G be the path P_n on n vertices. Then clearly, $L(G)$ is the path P_{n-1} on $n - 1$ vertices.

Conversely, let $L(G)$ be a path. Then no vertex of G can have degree greater than 2 because if G has a vertex v of degree greater than 2, the edges incident to v would form a complete subgraph of $L(G)$ with at least three vertices. Hence, G must be either a cycle or a path. But G cannot be a cycle, because the line graph of a cycle is again a cycle. $\square$

Exercise 7.3. Let $H = L^2(G)$ be defined as $L(L(G))$. Find $m(H)$ if G is the graph of Fig. 1.22.

Exercise 7.4. Give an example of a graph G to show that the relation $d_{L(G)}(uv) = d_G(u) + d_G(v) - 2$ may not be valid if G has a loop.

As shown in Exercise 7.2, $C_n \simeq L(C_n)$; in fact, C_n is the only simple graph with this property, as Exercise 7.5 shows.

Exercise 7.5. Prove that a simple connected graph G is isomorphic to its line graph if and only if it is a cycle.

Exercise 7.6. Disprove by a counterexample: If the graph H is a spanning subgraph of a graph G, then $L(H)$ is a spanning subgraph of $L(G)$.

Theorem 1.7.2. *If the simple graphs G_1 and G_2 are isomorphic, then $L(G_1)$ and $L(G_2)$ are isomorphic.*

Proof. Let (ϕ, θ) be an isomorphism of G_1 onto G_2. Then θ is a bijection of $E(G_1)$ onto $E(G_2)$. We show that θ is an isomorphism of $L(G_1)$ to $L(G_2)$. We prove this by showing that θ preserves adjacency and nonadjacency. Let e_i and e_j be two adjacent vertices of $L(G_1)$. Then there exists a vertex v of G_1 incident with both e_i and e_j, and so $\phi(v)$ is a vertex incident with both $\theta(e_i)$ and $\theta(e_j)$. Hence, $\theta(e_i)$ and $\theta(e_j)$ are adjacent vertices in $L(G_2)$.

Fig. 1.23 Nonisomorphic
graphs on four vertices or less

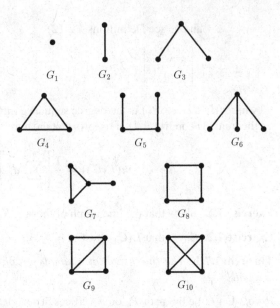

Now, let $\theta(e_i)$ and $\theta(e_j)$ be adjacent vertices in $L(G_2)$. This means that they are adjacent edges in G_2 and hence there exists a vertex v' of G_2 incident to both $\theta(e_i)$ and $\theta(e_j)$ in G_2. Then $\phi^{-1}(v')$ is a vertex of G_1 incident to both e_i and e_j, so that e_i and e_j are adjacent vertices of $L(G_1)$.

Thus, e_i and e_j are adjacent vertices of $L(G_1)$ if and only if $\theta(e_i)$ and $\theta(e_j)$ are adjacent vertices of $L(G_2)$. Hence, θ is an isomorphism of $L(G_1)$ onto $L(G_2)$. (Recall that a line graph is always a simple graph.)   □

Remark 1.7.3. The converse of Theorem 1.7.2 is not true. Consider the graphs $K_{1,3}$ and K_3. Their line graphs are K_3. But $K_{1,3}$ is not isomorphic to K_3 since there is a vertex of degree 3 in $K_{1,3}$, whereas there is no such vertex in K_3.

Theorem 1.7.4* shows that the above two graphs are the only two exceptional simple graphs of this type.

Theorem 1.7.4* (H. Whitney).   *Let G and G' be simple connected graphs with isomorphic line graphs. Then G and G' are isomorphic unless one of them is $K_{1,3}$ and the other is K_3.*

Proof. First, suppose that $n(G)$ and $n(G')$ are less than or equal to 4. A necessary condition for $L(G)$ and $L(G')$ to be isomorphic is that $m(G) = m(G')$. The only nonisomorphic connected graphs on at most four vertices are those shown in Fig. 1.23.

In Fig. 1.23, graphs G_4, G_5, and G_6 are the three graphs having three edges each. We have already seen that G_4 and G_6 have isomorphic line graphs, namely, K_3. The line graph of G_5 is a path of length 2, and hence $L(G_5)$ cannot be isomorphic to $L(G_4)$ or $L(G_6)$. Further, G_7 and G_8 are the only two graphs in the list having four edges each.

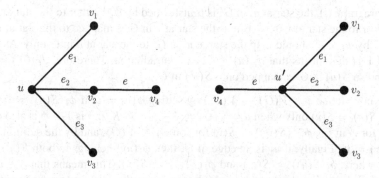

Fig. 1.24 Graphs with five vertices and edge e adjacent to one or all three other edges

Now $L(G_8) \simeq G_8$, and $L(G_7)$ is isomorphic to G_9. Thus, the line graphs of G_7 and G_8 are not isomorphic. No two of the remaining graphs have the same number of edges. Hence the only nonisomorphic graphs with at most four vertices having isomorphic line graphs are G_4 and G_6.

We now suppose that either G or G', say G, has at least five vertices and that $L(G)$ and $L(G')$ are isomorphic under an isomorphism ϕ_1. ϕ_1 is a bijection from the edge set of G onto the edge set of G'.

We now prove that ϕ_1 transforms a $K_{1,3}$ subgraph of G onto a $K_{1,3}$ subgraph of G'. Let $e_1 = uv_1$, $e_2 = uv_2$, and $e_3 = uv_3$ be the edges of a $K_{1,3}$ subgraph of G. As G has at least five vertices and is connected, there exists an edge e adjacent to only one or all the three edges e_1, e_2, and e_3, as illustrated in Fig. 1.24.

Now $\phi_1(e_1)$, $\phi_1(e_2)$, and $\phi_1(e_3)$ form either a $K_{1,3}$ subgraph or a triangle in G'. If $\phi_1(e_1)$, $\phi_1(e_2)$, and $\phi_1(e_3)$ form a triangle in G', $\phi_1(e)$ can be adjacent to precisely two of $\phi_1(e_1)$, $\phi_1(e_2)$, and $\phi_1(e_3)$ (since $L(G')$ is simple), whereas $\phi_1(e)$ must be adjacent to only one or all the three. This contradiction shows that $\{\phi_1(e_1), \phi_1(e_2), \phi_1(e_3)\}$ is not a triangle in G' and therefore forms a star at a vertex v' of G'.

It is clear that a similar result holds for ϕ_1^{-1} as well, since it is an isomorphism of $L(G')$ onto $L(G)$.

Let $S(u)$ denote the star subgraph of G formed by the edges of G incident at a vertex u of G. We shall prove that ϕ_1 maps $S(u)$ onto the star subgraph $S(u')$ of G'.

(i) First, suppose that the degree of u is at least 2. Let f_1 and f_2 be any two edges incident at u. The edges $\phi_1(f_1)$ and $\phi_1(f_2)$ of G' have an end vertex u' in common. If f is any other edge of G incident with u, then $\phi_1(f)$ is incident with u', and conversely, for every edge f' of G' incident with u', $\phi_1^{-1}(f')$ is incident with u. Thus, $S(u)$ in G is mapped to $S(u')$ in G'.

(ii) Let the degree of u in G be 1 and $e = uv$ be the unique edge incident with u. As G is connected and $n(G) \geq 5$, degree of v must be at least 2 in G, and therefore, by (i), $S(v)$ is mapped to a star $S(v')$ in G'. Also, $\phi_1(uv) = u'v'$ for some $u' \in V(G')$. Now, if the degree of u' in G' is greater than 1, by paragraph (i), the star at u' in G' is transformed by ϕ_1^{-1} either to the star at u in G or to the star at v in G. But as the star at v in G is mapped to the star at v' in G' by ϕ_1, ϕ_1^{-1} should map the star at u' in G' to the star at u in G only. As ϕ_1^{-1} is 1-1, this means that $d_G(u) \geq 2$, a contradiction. Therefore, $d_{G'}(u') = 1$, and so $S(u)$ in G is mapped onto $S(u')$ in G'.

We now define $\phi : V(G) \to V(G')$ by setting $\phi(u) = u'$ if $\phi_1(S(u)) = S(u')$. Since $S(u) = S(v)$ only when $u = v$ $(G \neq K_2, G' \neq K_2)$, ϕ is 1-1. ϕ is also onto since, for v' in G', $\phi_1^{-1}(S(v')) = S(v)$ for some $v \in V(G)$, and by the definition of ϕ, $\phi(v) = v'$. Finally, if uv is an edge of G, then $\phi_1(uv)$ belongs to both $S(u')$ and $S(v')$, where $\phi_1(S(u)) = S(u')$ and $\phi_1(S(v)) = S(v')$. This means that $u'v'$ is an edge of G'. But $u' = \phi(u)$ and $v' = \phi(v)$. Consequently, $\phi(u)\phi(v)$ is an edge of G'. If u and v are nonadjacent in G, $\phi(u)\phi(v)$ must be nonadjacent in G'. Otherwise, $\phi(u)\phi(v)$ belongs to both $S(\phi(u))$ and $S(\phi(v))$, and hence $\phi_1^{-1}(\phi(u)\phi(v)) = uv \in E(G)$, a contradiction. Thus, G and G' are isomorphic under ϕ. □

Definition 1.7.5. A graph H is called a *forbidden subgraph* for a property P of graphs if it satisfies the following condition: If a graph G has property P, then G cannot contain an induced subgraph isomorphic to H.

Beineke [17] obtained a forbidden-subgraph criterion for a graph to be a line graph. In fact, he showed that a graph G is a line graph if and only if the nine graphs of Fig. 1.25 are forbidden subgraphs for G. However, for the sake of later reference, we prove only the following result.

Theorem 1.7.6. *If G is a line graph, then $K_{1,3}$ is a forbidden subgraph of G.*

Proof. Suppose that G is the line graph of graph H and that G contains a $K_{1,3}$ as an induced subgraph. If v is the vertex of degree 3 in $K_{1,3}$ and v_1, v_2, and v_3 are the neighbors of v in this $K_{1,3}$, then the edge e corresponding to v in H is adjacent to the three edges e_1, e_2, and e_3 corresponding to the vertices v_1, v_2, and v_3. Hence, one of the end vertices of e must be the end vertex of at least two of e_1, e_2, and e_3 in H, and hence v together with two of v_1, v_2, and v_3 form a triangle in G. This means that the $K_{1,3}$ subgraph of G considered above is not an induced subgraph of G, a contradiction. □

1.8 Operations on Graphs

In mathematics, one always tries to get new structures from given ones. This also applies to the realm of graphs where one can generate many new graphs from a given set of graphs. In this section we consider some of the methods of generating new graphs from a given pair of graphs.

Let $G_1 = (V_1, E_1)$ and $G_2 = (V_2, E_2)$ be two simple graphs.

Definition 1.8.1. *Union of two graphs:* The graph $G = (V, E)$, where $V = V_1 \cup V_2$ and $E = E_1 \cup E_2$, is called the union of G_1 and G_2 and is denoted by $G_1 \cup G_2$.

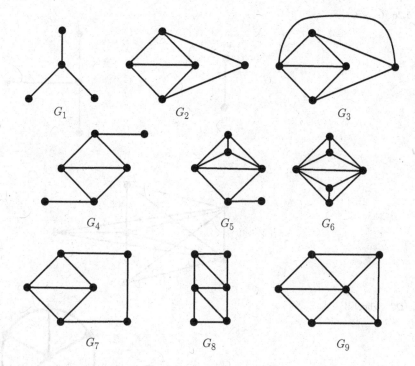

Fig. 1.25 Nine graphs of Bieneke [17]

When G_1 and G_2 are vertex disjoint, $G_1 \cup G_2$ is denoted by $G_1 + G_2$ and is called the *sum* of the graphs G_1 and G_2.

The finite union of graphs is defined by means of associativity; in particular, if $G_1, G_2, \ldots, G_r$ are pairwise vertex-disjoint graphs, each of which is isomorphic to G, then $G_1 + G_2 + \ldots + G_r$ is denoted by rG.

Definition 1.8.2. *Intersection of two graphs*: If $V_1 \cap V_2 \neq \emptyset$, the graph $G = (V, E)$, where $V = V_1 \cap V_2$ and $E = E_1 \cap E_2$ is the *intersection* of G_1 and G_2 and is written as $G_1 \cap G_2$.

Definition 1.8.3. *Join of two graphs*: Let G_1 and G_2 be two *vertex-disjoint* graphs. Then the *join* $G_1 \vee G_2$ of G_1 and G_2 is the supergraph of $G_1 + G_2$ in which each vertex of G_1 is also adjacent to every vertex of G_2.

Figure 1.26 illustrates the graph $G_1 \vee G_2$. If $G_1 = K_1$ and $G_2 = C_n$, then $G_1 \vee G_2$ is called the *wheel* W_n. W_5 is shown in Fig. 1.27.

It is worthwhile to note that $K_{m,n} = K_m^c \vee K_n^c$ and $K_n = K_1 \vee K_{n-1}$.

It follows from the above definitions that

(i) $n(G_1 \cup G_2) = n(G_1) + n(G_2) - n(G_1 \cap G_2), m(G_1 \cup G_2) = m(G_1) + m(G_2) - m(G_1 \cap G_2)$.

(ii) $n(G_1 + G_2) = n(G_1) + n(G_2), m(G_1 + G_2) = m(G_1) + m(G_2)$ and

(iii) $n(G_1 \vee G_2) = n(G_1) + n(G_2), m(G_1 \vee G_2) = m(G_1) + m(G_2) + n(G_1)n(G_2)$.

Fig. 1.26 $G_1 \vee G_2$

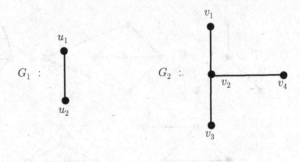

G_1 :

u_1
u_2

G_2 :

v_1
v_2 v_4
v_3

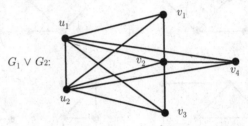

$G_1 \vee G_2$:

Fig. 1.27 Wheel W_5

1.9   Graph Products

We now define graph products. Denote a general graph product of two *simple* graphs by $G * H$. We define the product in such a way that $G * H$ is also simple. Given graphs G_1 and G_2 with vertex sets V_1 and V_2 respectively, any product graph $G_1 * G_2$ has as its vertex set the Cartesian product $V(G_1) \times V(G_2)$. For any two vertices (u_1, u_2), (v_1, v_2) of $G_1 * G_2$, consider the following possibilities:

(i) u_1 adjacent to v_1 in G_1 or u_1 nonadjacent to v_1 in G_1.
(ii) u_2 adjacent to v_2 in G_2 or u_2 nonadjacent to v_2 in G_2.
(iii) $u_1 = v_1$ and/or $u_2 = v_2$.

We use, with respect to any graph, the symbols E, N, and $=$ to denote adjacency (edge), nonadjacency (no edge), and equality of vertices, respectively. We then have the following structure table S for $G_1 * G_2$, where the rows of S correspond to G_1 and the columns to G_2 and

$$S \; : \; = \begin{array}{c} \\ E \\ E \\ N \end{array} \overset{\begin{array}{ccc} E & = & N \end{array}}{\left[\begin{array}{ccc} \circ & \circ & \circ \\ \circ & = & \circ \\ \circ & \circ & \circ \end{array} \right]},$$

where each $\circ$ in the double array S is E or N according to whether a general vertex (u_1, u_2) of $G_1 * G_2$ is adjacent or nonadjacent to a general vertex (v_1, v_2) of $G_1 * G_2$. Since each $\circ$ can take two options, there are in all $2^8 = 256$ graph products $G_1 * G_2$ that can be defined using G_1 and G_2.

If $S = \begin{bmatrix} a_{11} & a_{12} & a_{13} \\ a_{21} & = & a_{23} \\ a_{31} & a_{32} & a_{33} \end{bmatrix}$, then the edge-nonedge entry of S will correspond to the nonedge-edge entry of the structure matrix of $G_2 * G_1$. Hence, the product $*$ is commutative, that is, G_1 and G_2 commute under $*$ if and only if the double array S is symmetric. Hence, if the product is commutative, it is enough if we know the five circled positions in $\begin{bmatrix} \circ & \circ & \circ \\ & = & \circ \\ & & \circ \end{bmatrix}$ to determine S completely. Therefore, there are in all $2^5 = 32$ commuting products.

We now give the matrix S for the Cartesian, direct, composition, and strong products.

Definition 1.9.1. *Cartesian product*, $G_1 \square G_2$.

$$S \; : \; = \begin{array}{c} \\ E \\ E \\ N \end{array} \overset{\begin{array}{ccc} E & = & N \end{array}}{\left[\begin{array}{ccc} N & E & N \\ E & = & N \\ N & N & N \end{array} \right]}.$$

Hence, (u_1, u_2) and (v_1, v_2) are adjacent in $G_1 \square G_2$ if and only if either $u_1 = v_1$ and u_2 is adjacent to v_2 in G_2, or u_1 is adjacent to v_1 in G_1 and $u_2 = v_2$.

Definition 1.9.2. *Direct* (or *tensor* or *Kronecker*) *product*, $G_1 \times G_2$.

$$S \; : \; = \begin{array}{c} \\ E \\ E \\ N \end{array} \overset{\begin{array}{ccc} E & - & N \end{array}}{\left[\begin{array}{ccc} E & N & N \\ N & = & N \\ N & N & N \end{array} \right]}.$$

Hence, (u_1, u_2) is adjacent to (v_1, v_2) in $G_1 \times G_2$ if and only if u_1 is adjacent to v_1 in G_1 and u_2 is adjacent to v_2 in G_2.

Definition 1.9.3. *Composition* (or *wreath* or *lexicographic*) *product* $G_1[G_2]$.

Fig. 1.28 $G_3 = C_4 \square P_3$

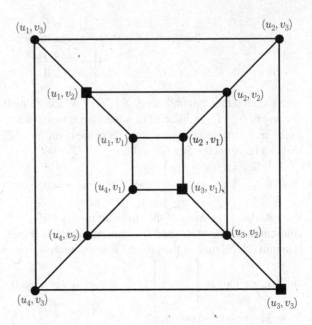

$$S : = \begin{array}{c} \\ E \\ E \\ N \end{array} \begin{array}{c} E \quad = \quad N \\ \left[\begin{array}{ccc} E & E & E \\ E & = & N \\ N & N & N \end{array} \right]. \end{array}$$

Therefore, (u_1, u_2) is adjacent to (v_1, v_2) in $G_1[G_2]$ if and only if u_1 is adjacent to v_1 in G_1 or $u_1 = v_1$, and u_2 is adjacent to v_2 in G_2.

Definition 1.9.4. *Strong* (or *normal*) *product* $G_1 \boxtimes G_2$. By definition, $G_1 \boxtimes G_2 = (G_1 \square G_2) \cup (G_1 \times G_2)$. Hence, its structure matrix is given by

$$S : = \begin{array}{c} \\ E \\ E \\ N \end{array} \begin{array}{c} E \quad = \quad N \\ \left[\begin{array}{ccc} E & E & N \\ E & = & N \\ N & N & N \end{array} \right] \end{array}$$

Remarks 1.9.5. 1. All the four products defined above are associative.
2. With the exception of composition, the other three are commutative.

Example 1.9.6. Example of Cartesian product: Let $G_1 = C_4 = (u_1 u_2 u_3 u_4)$ and $G_2 = P_3 = (v_1 v_2 v_3)$. Then $G_1 \square G_2$ is the graph G_3 given in Fig. 1.28.

Definition 1.9.7. Let (x, y) be any vertex of $G_1 \square G_2$. The G_2-*fiber* (also called G_2-*layer*) at x in $G_1 \square G_2$, denoted by $(G_2)_x$, is the subgraph of $G_1 \square G_2$ induced by the set of all vertices $\{(x, z) : z \in V(G_2)\}$. Similarly, the G_1-*fiber* at y, in symbol $(G_1)_y$, is the subgraph of $G_1 \square G_2$ induced by $\{(z, y) : z \in V(G_1)\}$.

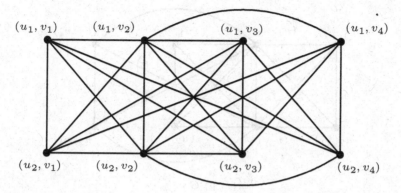

Fig. 1.29 $G_1[G_2]$

Fig. 1.30 $G_2[G_1]$

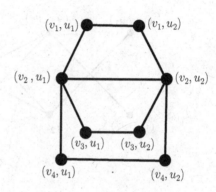

For instance, the outer square of the graph G_3 in Fig. 1.28 is the G_1-fiber $(G_1)_{v_3}$ at v_3 of G_2 while the path $(u_1, v_1)(u_1, v_2)(u_1, v_3)$ is the G_2-fiber $(G_2)_{u_1}$ at u_1. Hence, fixing a vertex of G_1, we get a G_2-fiber, and fixing a vertex of G_2, we get a G_1-fiber. Moreover, it is clear that $(G_1)_y \cap (G_2)_x = $ the vertex (x, y) of $G_1 \square G_2$.

Remark 1.9.8. It is easy to see that $n(G_1 \square G_2) = n(G_1) n(G_2)$ and $m(G_1 \square G_2) = n(G_1) m(G_2) + n(G_2) m(G_1)$.

Example 1.9.9. If G_1 and G_2 are graphs of Fig. 1.26, $G_1[G_2]$ and $G_2[G_1]$ are shown in Figs. 1.29 and 1.30, respectively.

Example 1.9.10. If G_1 and G_2 are graphs of Fig. 1.26, $G_1 \boxtimes G_2$ and $G_1 \times G_2$ are shown in Fig. 1.31.

Exercise 9.1. Show by means of an example that $G_1[G_2]$ need not be isomorphic to $G_2[G_1]$.

Exercise 9.2. Prove that $G_1 \square G_2 \simeq G_2 \square G_1$.

Exercise 9.3. Prove that $G_1 \boxtimes G_2$ is isomorphic to $G_2 \boxtimes G_1$.

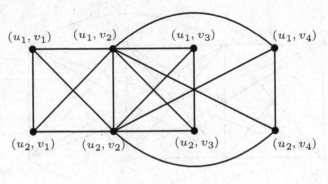

$$G_1 \boxtimes G_2$$

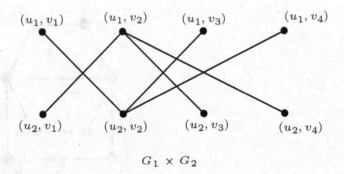

$$G_1 \times G_2$$

Fig. 1.31 $G_1 \boxtimes G_2$ and $G_1 \times G_2$

Exercise 9.4. Prove that

(a) $n(G_1[G_2]) = n(G_2[G_1]) = n(G_1 \boxtimes G_2) = n(G_1)n(G_2)$.
(b) $m(G_1[G_2]) = n(G_1)m(G_2) + n(G_2)^2 m(G_1)$.
(c) $m(G_1 \boxtimes G_2) = n(G_1)m(G_2) + m(G_1)n(G_2) + 2m(G_1)m(G_2)$.
(d) $m(G_1 \times G_2) = 2m(G_1)m(G_2)$.

Exercise 9.5. Prove: $L(K_{m,n}) \simeq K_m \square K_n$.

We now introduce the powers of a simple graph G.

Definition 1.9.11. The *k th power* G^k of G has $V(G^k) = V(G)$, where u and v are adjacent in G^k whenever $d_G(u, v) \le k$.

Two graphs H_1 and H_2 and their squares H_1^2 and H_2^2 are displayed in Fig. 1.32. By the definition of G^k, G is a *spanning subgraph* of G^k, $k \ge 1$.

Exercise 9.6. Show that if G is a connected graph with at least two edges, then each edge of G^2 belongs to a triangle.

Exercise 9.7. If $d_G(u, v) = p$, determine $d_{G^k}(u, v)$.

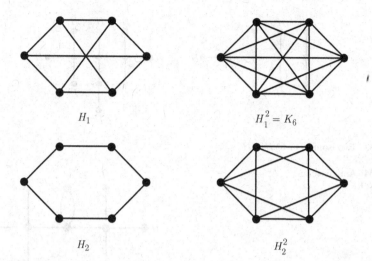

H_1

$H_1^2 = K_6$

H_2

H_2^2

Fig. 1.32 Two examples for the square of a graph

1.10 An Application to Chemistry

The earliest application of graph theory to chemistry was found by Cayley [33] in 1857 when he enumerated the number of rooted trees with a given number of vertices. A chemical molecule can be represented by a graph by taking each atom of the molecule as a vertex of the graph and making the edges of the graph represent atomic bonds between the end atoms. Thus, the degree of each vertex of such a graph gives the valence of the corresponding atom. Molecules that have the same chemical formula but have different structural properties are called isomers. In terms of graphs, isomers are two nonisomorphic graphs having the same degree sequence. The two molecular graphs (actually trees; see Chap. 4) of Fig. 1.33 represent two isomers of the molecule C_3H_7OH (propanol). The graph of Fig. 1.34 represents aminoacetone C_3H_7NO. This has a multiple bond represented by a pair of multiple edges between C and O.

The paraffins have the molecular formula C_kH_{2k+2}, having $3k + 2$ atoms (vertices), of which k are carbon atoms and the remaining $2k + 2$ are hydrogen atoms. They all have $3k + 1$ bonds (edges). Cayley used enumeration techniques of graph theory (see reference [95]) to count the number of isomers of C_kH_{2k+2}. His formula shows that for the paraffin $C_{13}H_{28}$, there are 802 different isomers.

1.11 Application to Social Psychology

The study of applications of graph theory to social psychology was initiated by Cartwright, Harary, and Norman [97] when they were at the Research Center for Group Dynamics at the University of Michigan during the 1960s.

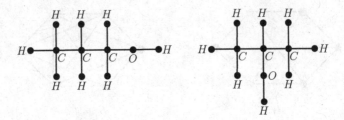

Fig. 1.33 Two isomers of C_3H_7OH

Fig. 1.34 Aminoacetone
C_3H_7NO

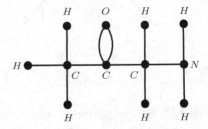

Group dynamics is the study of social relationships between people within a particular group. The graphs that are commonly used to study these relationships are signed graphs. A *signed graph* is a graph G with sign $+$ or $-$ attached to each of its edges. An edge of G is *positive* (respectively, *negative*) if the sign attached to it is $+$ (respectively, $-$). A positive sign between two persons u and v would mean that u and v are "related"; that is, they share the same social trait under consideration. A negative sign would indicate the opposite. The social trait may be "same political ideology," "friendship," "likes certain social customs," and so on.

A group of people with such relations between them is called a *social system*. A social system is called *balanced* if any two of its people have a positive relation between them, or if it is possible to divide the group into two subgroups so that any two persons in the same subgroup have a positive relation between them while two persons of different subgroups have a negative relation between them. This of course means that if both u and v have negative relation to w, then u and v must have positive relation between them.

In consonance with a balanced social system, a balanced signed graph G is defined as a graph in which the vertex set V can be partitioned into two subsets V_i, $i = 1, 2$, one of which may be empty, so that any edge in each $G[V_i]$ is positive, while any edge between V_1 and V_2 is negative.

For example, the signed graph G_1 of Fig. 1.35a is balanced (take $V_1 = \{u_1, u_2\}$ and $V_2 = \{u_3, u_4, u_5\}$). However, the signed graph G_2 of Fig. 1.35b is not balanced. (If G_2 were balanced and $v_4 \in$ (say) V_1, then $v_5 \in V_2$ and $v_0 \in V_1$, and therefore v_0v_5 should be a negative edge, which is not the case.)

A *path* or *cycle* in a signed graph is *positive* if it has an even number of $-$ signs; otherwise, the path is negative. The following characterization of balanced signed graphs is due to Harary [97].

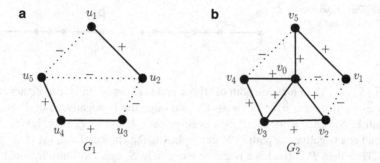

Fig. 1.35 (a) Balanced and (b) unbalanced graph

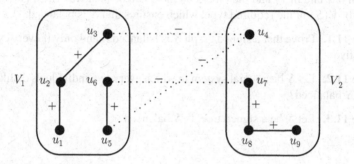

Fig. 1.36 A negative $u_1 - u_9$ path

Theorem 1.11.1 (Harary [97]). *A signed graph S is balanced if and only if the paths between any two vertices of S either are all positive paths or are all negative paths.*

Proof. Let S be a balanced signed graph with (V_1, V_2) being the partition of the vertex set $V(S)$ of S. Let $P : u = u_1 u_2 \ldots u_n = v$ be a path in S. Without loss of generality, we may assume that $u_1 \in V_1$. Then, as we traverse along P from u_1, we will continue to remain in V_1 until we traverse along a negative edge. Recall that a negative edge joins a vertex of V_1 to a vertex of V_2. Hence, if P contains an odd number of negative edges, that is, if P is negative, then $v \in V_2$, whereas if P is positive, $v \in V_1$ (see Fig. 1.36). It is clear that every u-v path in S must have the same sign as P.

Conversely, assume that S is a signed graph with the property that between any two vertices of S the paths are either all positive or all negative. We prove that S is balanced. We may assume that S is a connected graph. Otherwise, we can prove the result for the components, and if (V_i, V_i'), $1 \leq i \leq \omega$, are the partitions of the vertex subsets of the (signed) components, then $[\cup_{i=1}^{\omega} V_i, \cup_{i=1}^{\omega} V_i']$ is a partition of $V(S)$ of the requisite type.

So we assume that S is connected. Let v be any vertex of S. Denote by V_1 the set of all vertices u of S that are connected to v by positive paths of S, and let

Fig. 1.37 Path P with (**a**) $uw \in P$ and (**b**) $uw \notin P$

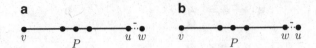

$V_2 = V(S) \backslash V_1$. Then no edge both of whose end vertices are in V_1 can be negative. Suppose, for instance, $u \in V_1$, $w \in V_1$, and edge uw is negative. Let P be any v-w path in S. Since $w \in V_1$, P is a positive path. If $uw \in P$, (Fig. 1.37a), then $P - (uw)$ is a negative v-u path in S, contradicting the choice of $u \in V_1$. If $uw \notin P$, (Fig. 1.37b), then $P \cup (wu)$ is a negative v-u path in S, again a contradiction. For a similar reason, no edge both of whose end vertices are in V_2 can be negative, and no edge with one end in V_1 and the other end in V_2 can be positive. Thus, (V_1, V_2) is a partition of $V(S)$ of the required type, which ensures that S is balanced. $\qquad \square$

Exercise 11.1. Prove that a signed graph S is balanced if and only if every cycle of S is positive.

Exercise 11.2. Let S be a social system in which every two individuals dislike each other. Is S balanced?

Exercise 11.3. Let S be a signed tree. Is S balanced?

1.12 Miscellaneous Exercises

12.1. Show that the sequences (a) (7, 6, 5, 4, 3, 3, 2) and (b) (6, 6, 5, 4, 3, 3, 1) are not graphical.

12.2. Give an example of a degree sequence that is realizable as the degree sequence by only a disconnected graph.

12.3. Show that for a simple bipartite graph, $m \le \frac{n^2}{4}$.

12.4. If δ and Δ are respectively the minimum and maximum of the degrees of a graph G, show that $\delta \le \frac{2m}{n} \le \Delta$.

12.5. For every $n \ge 3$, construct a 3-regular simple graph on $2n$ vertices containing no triangles.

12.6. If a bipartite graph $G(X, Y)$ is regular, show that $|X| = |Y|$.

12.7. Show that in a simple graph, any closed walk of odd length contains a cycle.

12.8. Give an example of a disconnected simple graph having ω components, n vertices, and $\frac{(n-\omega)(n-\omega+1)}{2}$ edges.

12.9. Prove or disprove: If H is a subgraph of G, then

(a) $\delta(H) \le \delta(G)$
(b) $\Delta(H) \le \Delta(G)$.

12.10. If $m > \frac{(n-1)(n-2)}{2}$, then show that G is connected.

12.11. If $\delta \ge 2$, then show that G contains a cycle.

12.12. If $\delta(G) \geq 3k - 1$ for a graph G, prove that G contains k edge-disjoint cycles.

12.13. If a simple graph has two pendant vertices, prove that G^c has at most two pendant vertices. Give an example of a graph G for which both G and G^c have exactly two pendant vertices.

12.14. Show that $\mathrm{Aut}(C_n)$ is D_n, the dihedral group of order $2n$.

12.15. Determine $\mathrm{Aut}(K_{p,q})$, $p \neq q$.

12.16. Determine the order of the automorphism group of
(a) $K_4 - e$; (b) P_n.

12.17. Show that the line graph of K_n is regular of degree $2n - 4$. Draw the line graph of K_4.

12.18. Show that the line graph of $K_{p,q}$ is regular of degree $p + q - 2$. Draw the line graph of $K_{2,3}$.

12.19. Show that a connected graph G is complete bipartite if and only if no induced subgraph of G is a K_3 or P_4.

12.20. If G is connected and $\mathrm{diam}(G) \geq 3$, where $\mathrm{diam}(G) = \max\{d(u,v) : u, v \in V(G)\}$, show that G^c is connected and $\mathrm{diam}(G^c) \leq 3$.

12.21. Show that the complement of a simple connected graph G is connected if and only if G has contains no spanning complete bipartite subgraph.

Notes

Graph theory, which had arisen out of puzzles solved for the sake of curiosity, has now grown into a major discipline in mathematics with problems permeating into almost all subjects—physics, chemistry, engineering, psychology, computer science, and more! It is customary to assume that graph theory originated with Leonhard Euler (1707–1783), who formulated the first few theorems in the subject. The subject, which was lying almost dormant for more than 100 years after Euler's death, suddenly started exploding at the turn of the 20th century, and today it has branched off in various directions—coloring problems, Ramsey theory, hypergraph theory, Eulerian and Hamiltonian graphs, decomposition and factorization theory, directed graphs, just to name a few. Some of the standard texts in graph theory are Refs. [14, 16, 27, 28, 34, 41, 51, 77, 80, 93, 155, 192]. A good account of enumeration theory of graphs is given in Ref. [95]. Further, a comprehensive account of applications of graph theory to chemistry is given in Refs. [176, 177].

Theorem 1.7.4* is due to H. Whitney [193], and the proof given in this chapter is due to Jüng [118].

Chapter 2
Directed Graphs

2.1 Introduction

Directed graphs arise in a natural way in many applications of graph theory. The
street map of a city, an abstract representation of computer programs, and network
flows can be represented only by directed graphs rather than by graphs. Directed
graphs are also used in the study of sequential machines and system analysis in
control theory.

2.2 Basic Concepts

Definition 2.2.1. A *directed graph* D is an ordered triple $(V(D), A(D), I_D)$, where
$V(D)$ is a nonempty set called the set of *vertices* of D; $A(D)$ is a set disjoint from
$V(D)$, called the set of *arcs* of D; and I_D is an *incidence map* that associates with
each arc of D an ordered pair of vertices of D. If a is an arc of D, and $I_D(a) =
(u, v)$, u is called the *tail* of a, and v is the *head* of a. The arc a is said to join v with
u. u and v are called the *ends* of a. A directed graph is also called a *digraph*.

With each digraph D, we can associate a graph G (written $G(D)$ when reference
to D is needed) on the same vertex set as follows: Corresponding to each arc
of D, there is an edge of G with the same ends. This graph G is called the
underlying graph of the digraph D. Thus, every digraph D defines a unique
(up to isomorphism) graph G. Conversely, given any graph G, we can obtain a
digraph from G by specifying for each edge of G an order of its ends. Such a
specification is called an *orientation* of G.

Just as with graphs, digraphs have a diagrammatic representation. A digraph is
represented by a diagram of its underlying graph together with arrows on its edges,
the arrow pointing toward the head of the corresponding arc. A digraph and its
underlying graph are shown in Fig. 2.1.

R. Balakrishnan and K. Ranganathan, *A Textbook of Graph Theory*,
Universitext, DOI 10.1007/978-1-4614-4529-6_2,
© Springer Science+Business Media New York 2012

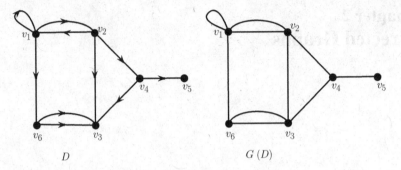

Fig. 2.1 Digraph D and its underlying graph $G(D)$

Many of the concepts and terminology for graphs are also valid for digraphs. However, there are many concepts of digraphs involving the notion of orientation that apply only to digraphs.

Definition 2.2.2. If $a = (u, v)$ is an arc of D, a is said to be *incident out of u* and *incident into v*. v is called an *outneighbor* of u, and u is called an *inneighbor* of v. $N_D^+(u)$ denotes the set of outneighbors of u in D. Similarly, $N_D^-(u)$ denotes the set of inneighbors of u in D. When no explicit reference to D is needed, we denote these sets by $N^+(u)$ and $N^-(u)$, respectively. An arc a is *incident with u* if it is either incident into or incident out of u. An arc having the same ends is called a *loop* of D. The number of arcs incident out of a vertex v is the *outdegree* of v and is denoted by $d_D^+(v)$ or $d^+(v)$. The number of arcs incident into v is its *indegree* and is denoted by $d_D^-(v)$ or $d^-(v)$.

For the digraph D of Fig. 2.2, we have $d^+(v_1) = 3$, $d^+(v_2) = 3$, $d^+(v_3) = 0$, $d^+(v_4) = 2$, $d^+(v_5) = 0$, $d^+(v_6) = 2$, $d^-(v_1) = 2$, $d^-(v_2) = 1$, $d^-(v_3) = 4$, $d^-(v_4) = 1$, $d^-(v_5) = 1$, and $d^-(v_6) = 1$. (The loop at v_1 contributes 1 each to $d^+(v_1)$ and $d^-(v_1)$.)

The *degree* $d_D(v)$ of a vertex v of a digraph D is the degree of v in $G(D)$. Thus, $d(v) = d^+(v) + d^-(v)$. As each arc of a digraph contributes 1 to the sum of the outdegrees and 1 to the sum of indegrees, we have

$$\sum_{v \in V(D)} d^+(v) = \sum_{v \in V(D)} d^-(v) = m(D),$$

where $m(D)$ is the number of arcs of D.

A vertex of D is *isolated* if its degree is 0; it is *pendant* if its degree is 1. Thus, for a pendant vertex v, either $d^+(v) = 1$ and $d^-(v) = 0$, or $d^+(v) = 0$ and $d^-(v) = 1$.

Definitions 2.2.3. 1. A digraph D' is a *subdigraph* of a digraph D if $V(D') \subseteq V(D)$, $A(D') \subseteq A(D)$, and $I_{D'}$ is the restriction of I_D to $A(D')$.

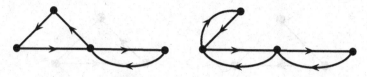

Fig. 2.2   A strong digraph (*left*) and a symmetric digraph (*right*)

2. A *directed walk* joining the vertex v_0 to the vertex v_k in D is an alternating sequence $W = v_0 a_1 v_1 a_2 v_2 \ldots a_k v_k$, $1 \leq i \leq k$, with a_i incident out of v_{i-1} and incident into v_i. *Directed trails, directed paths, directed cycles*, and *induced subdigraphs* are defined analogously as for graphs.

3. A vertex v is *reachable* from a vertex u of D if there is a directed path in D from u to v.

4. Two vertices of D are *diconnected* if each is reachable from the other in D. Clearly, diconnection is an equivalence relation on the vertex set of D, and if the equivalence classes are $V_1, V_2, \ldots, V_\omega$, the subdigraphs of D induced by $V_1, V_2, \ldots, V_\omega$ are called the *dicomponents* of D.

5. A digraph is *diconnected* (also called *strongly-connected*) if it has exactly one dicomponent. A diconnected digraph is also called a *strong digraph*.

6. A digraph is *strict* if its underlying graph is simple. A digraph D is *symmetric* if, whenever (u, v) is an arc of D, then (v, u) is also an arc of D (see Fig. 2.2).

Exercise 2.1.   How many orientations does a simple graph of m edges have?

Exercise 2.2.   Let D be a digraph with no directed cycle. Prove that there exists a vertex whose indegree is 0. Deduce that there is an ordering $v_1, v_2, \ldots, v_n$ of V such that, for $2 \leq i \leq n$, every arc of D with terminal vertex v_i has its initial vertex in $\{v_1, v_2, \ldots, v_{i-1}\}$.

2.3   Tournaments

A digraph D is a *tournament* if its underlying graph is a complete graph. Thus, in a tournament, for every pair of distinct vertices u and v, either (u, v) or (v, u), but not both, is an arc of D. Figures 2.3a, b display all tournaments on three and four vertices, respectively.

The word "tournament" derives its name from the usual round-robin tournament. Suppose there are n players in a tournament and that every player is to play against every other player. The results of such a tournament can be represented by a tournament on n vertices, where the vertices represent the n players and an arc (u, v) represents the victory of player u over player v.

Suppose the players of a tournament have to be ranked. The corresponding digraph T, a tournament, could be used for such a ranking. The ranking of the vertices of T is as follows: One way of doing it is by looking at the sequence of outdegrees of T. This is because $d_T^+(v)$ stands for the number of players defeated by

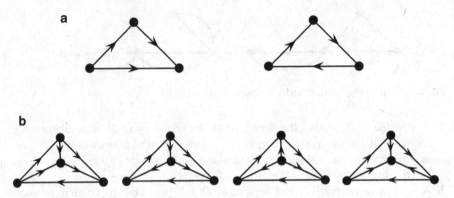

Fig. 2.3 Tournaments on (**a**) three and (**b**) four vertices

the player v. Another way of doing it is by finding a *directed Hamilton path*, that is, a spanning directed path in T. One could rank the players as per the sequence of this path so that each player defeats his or her successor. We now prove the existence of a directed Hamilton path in any tournament.

Theorem 2.3.1 (Rédei [165]). *Every tournament contains a directed Hamilton path.*

Proof. (By induction on the number of vertices n of the tournament.) The result can be directly verified for all tournaments having two or three vertices. Hence, suppose that the result is true for all tournaments on $n \geq 3$ vertices. Let T be a tournament on $n + 1$ vertices $v_1, v_2, \ldots, v_{n+1}$. Now, delete v_{n+1} from T. The resulting subdigraph T' of T is a tournament on n vertices and hence by the induction hypothesis contains a directed Hamilton path. Assume that the Hamilton path is $v_1 v_2 \ldots v_n$, relabeling the vertices, if necessary.

If the arc joining v_1 and v_{n+1} has v_{n+1} as its tail, then $v_{n+1} v_1 v_2 \ldots v_n$ is a directed Hamilton path in T and the result stands proved (see Fig. 2.4a).

If the arc joining v_n and v_{n+1} is directed from v_n to v_{n+1}, then $v_1 v_2 \ldots v_n v_{n+1}$ is a directed Hamilton path in T (see Fig. 2.4b).

Now suppose that none of (v_{n+1}, v_1) and (v_n, v_{n+1}) is an arc of T. Hence, (v_1, v_{n+1}) and (v_{n+1}, v_n) are arcs of T—the first arc incident into v_{n+1} and the second arc incident out of v_{n+1}. Thus, as we pass on from v_1 to v_n, we encounter a reversal of the orientation of edges incident with v_{n+1}. Let v_i, $2 \leq i \leq n$, be the first vertex where this reversal takes place, so that (v_{i-1}, v_{n+1}) and (v_{n+1}, v_i) are arcs of T. Then $v_1 v_2 \ldots v_{i-1} v_{n+1} v_i v_{i+1} \ldots v_n$ is a directed Hamilton path of T (see Fig. 2.4c).   □

Theorem 2.3.2 (Moon [141, 143]). *Every vertex of a diconnected tournament T on n vertices with $n \geq 3$ is contained in a directed k-cycle, $3 \leq k \leq n$. (T is then said to be vertex-pancyclic.)*

Proof. Let T be a diconnected tournament with $n \geq 3$ and u, a vertex of T. Let $S = N^+(u)$, the set of all outneighbors of u in T, and $S' = N^-(u)$, the set of all

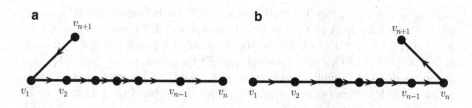

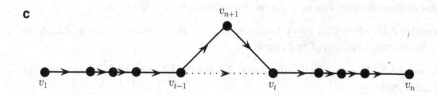

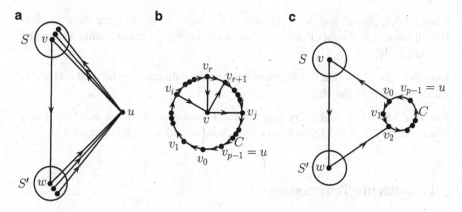

Fig. 2.4 Digraphs for proof of Theorem 2.3.1

Fig. 2.5 Digraphs for proof of Theorem 2.3.2

inneighbors of u in T. As T is diconnected, none of S and S' is empty. If $[S, S']$ denotes the set of all arcs of T having their tails in S and heads in S', then $[S, S']$ is also nonempty for the same reason. If (v, w) is an arc of $[S, S']$, then (u, v, w, u) is a directed 3-cycle in T containing u. (see Fig. 2.5a.)

Suppose that u belongs to directed cycles of T of all lengths k, $3 \le k \le p$, where $p < n$. We shall prove that there is a directed $(p + 1)$-cycle of T containing u.

Let C : $(v_0, v_1, \ldots, v_{p-1}, v_0)$ be a directed p-cycle containing u, where $v_{p-1} = u$. Suppose that v is a vertex of T not belonging to C such that for some i and j, $0 \le i, j \le p-1, i \ne j$, there exist arcs (v_i, v) and (v, v_j) of T (see Fig. 2.5b). Then there must exist arcs (v_r, v) and (v, v_{r+1}) of $A(T)$, $i \le r \le j - 1$ (suffixes taken modulo p), and hence $(v_0, v_1, \ldots, v_r, v, v_{r+1}, \ldots, v_{p-1}, v_0)$ is a directed $(p + 1)$-cycle containing u (see Fig. 2.5b).

If no such v exists, then for every vertex v of T not belonging to $V(C)$, either $(v_i, v) \in A(T)$ for every i, $0 \le i \le p - 1$, or $(v, v_i) \in A(T)$ for every i, $0 \le i \le p - 1$. Let $S = \{v \in V(T) \setminus V(C) : (v_i, v) \in A(T)$ for each i, $0 \le i \le p - 1\}$ and $S' = \{w \in V(T) \setminus V(C) : (w, v_i) \in A(T)$ for each i, $0 \le i \le p - 1\}$. The diconnectedness of T implies that none of S, S', and $[S, S']$ is empty. Let (v, w) be an arc of $[S, S']$. Then $(v_0, v, w, v_2, \ldots, v_{p-1}, v_0)$ is a directed $(p + 1)$-cycle of T containing $v_{p-1} = u$ (see Fig. 2.5c). $\qquad\qquad\square$

Remark 2.3.3. Theorem 2.3.2 shows, in particular, that every diconnected tournament is Hamiltonian; that is, it contains a directed spanning cycle.

Exercise 3.1. Show that every tournament T is diconnected or can be made into one by the reorientation of just one arc of T.

Exercise 3.2. Show that a tournament is diconnected if and only if it has a spanning directed cycle.

Exercise 3.3. Show that every tournament of order n has at most one vertex v with $d^+(v) = n - 1$.

Exercise 3.4. Show that for each positive integer $n \ge 3$, there exists a non-Hamiltonian tournament of order n (that is, a tournament not containing a spanning directed cycle).

Exercise 3.5. Show that if a tournament contains a directed cycle, then it contains a directed cycle of length 3.

Exercise 3.6. Show that every tournament T contains a vertex v such that every other vertex of T is reachable from v by a directed path of length at most 2.

2.4 k-Partite Tournaments

Definition 2.4.1. A *k-partite graph*, $k \ge 2$, is a graph G in which $V(G)$ is partitioned into k nonempty subsets $V_1, V_2, \ldots, V_k$, such that the induced subgraphs $G[V_1], G[V_2], \ldots, G[V_k]$ are all totally disconnected. It is said to be *complete* if, for $i \ne j$, each vertex of V_i is adjacent to every vertex of V_j, $1 \le i, j \le k$. A *k-partite tournament* is an oriented complete k-partite graph (see Fig. 2.6). The subsets $V_1, V_2, \ldots, V_k$ are often referred to as the *partite sets* of G.

The next three theorems are based on Goddard et al. [74]. We now give a characterization of a k-partite tournament containing a 3-cycle.

Theorem 2.4.2. *Let T be a k-partite tournament, $k \ge 3$. Then T contains a directed 3-cycle if and only if there exists a directed cycle in T that contains vertices from at least three partite sets.*

Proof. Suppose that T contains a directed 3-cycle C. Then the three vertices of C must belong to three distinct partite sets of T.

Fig. 2.6 A 3-partite
tournament

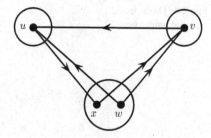

Fig. 2.7 Digraphs for proof
of Theorem 2.4.2

a

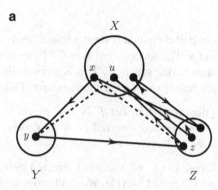

b

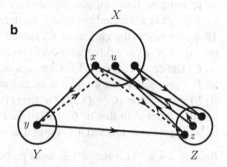

Conversely, suppose that T contains a directed cycle C that in turn contains
vertices from at least three partite sets. Assume that C has the least possible length.
Then there exist three consecutive vertices x, y, z on C that belong to distinct partite
sets of T, say X, Y, Z, respectively. We claim that C is a directed 3-cycle.

As x and z are in different partite sets of the k-partite tournament T, either
$(z, x) \in A(T)$, the arc set of T, or $(x, z) \in A(T)$. If $(z, x) \in A(T)$, then (x, y, z, x)
is a directed 3-cycle containing vertices from three partite sets of T. If $(x, z) \in A(T)$,
then consider $C' = (C - y) + (x, z)$. C' is a directed cycle of length one less than
that of C. So by assumption on C, C' contains vertices from only two partite sets,
namely, X and Z. Let u be the vertex of C immediately following z on C'. Then
$u \in X$. If $(u, y) \in A(T)$, then $C_1'' = (y, z, u, y)$ is a directed 3-cycle containing
vertices from three partite sets of T. Hence, assume that $(u, y) \notin A(T)$, and so
$(y, u) \in A(T)$ (see Fig. 2.7a, b). Now consider $C_2'' = (C - z) + (y, u)$. C_2'' is a

Fig. 2.8 Diconnected
3-partite tournament T

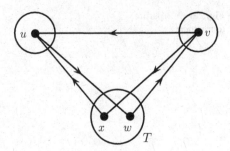

directed cycle that is shorter than C and contains at least one vertex other than x, y, and u. The successor of u in C_2'' belongs to Z, and thus C_2'' contains vertices from three partite sets of T. This is a contradiction to the choice of C. Thus, (y, u) does not belong to $A(T)$, a contradiction. This proves the result.   □

Theorem 2.4.3. *Let T be a k-partite tournament, $k \geq 3$. Then every vertex u belonging to a directed cycle in T must belong to either a directed 3-cycle or a directed 4-cycle.*

Proof. Let C be a shortest directed cycle in T that contains u. Suppose that C is not a directed 3-cycle. We shall prove that u is a vertex of a directed 4-cycle. Let u, x, y, and z be four consecutive vertices of C. If $(u, y) \in A(T)$, then $C' = (C - x) + (u, y)$ is a directed cycle in T containing u and having a length shorter than C. This contradicts the choice of C. Hence $(u, y) \notin A(T)$. Also, if $(y, u) \in A(T)$, then (u, x, y, u) is a directed 3-cycle containing u. This contradicts our assumption on C. Hence, $(y, u) \notin A(T)$. Consequently, y and u belong to the same partite set of T. This means that u and z must belong to distinct partite sets of T. If $(u, z) \in A(T)$, then $C'' = (C - \{x, y\}) + (u, z)$ is a directed cycle containing u and having length shorter than that of C. Hence $(u, z) \notin A(T)$. Therefore, $(z, u) \in A(T)$ and (u, x, y, z, u) is a directed 4-cycle containing u.   □

Remark 2.4.4. Theorem 2.3.2 states that every vertex of a diconnected tournament lies on a k-cycle for every k, $3 \leq k \leq n$. However, this property is not true for a diconnected k-partite tournament. The tournament T of Fig. 2.8 is a counterexample. T is a 3-partite tournament with $\{x, w\}$, $\{u\}$, and $\{v\}$ as partite sets, $(uwvxu)$ is a spanning directed cycle in T, and hence T is strongly connected, but x is not a vertex of any directed 3-cycle.

Definition 2.4.5. The *score* of a vertex v in a tournament T is its outdegree. (This corresponds to the number of players who are beaten by player v.) If $v_1, v_2, \ldots, v_n$ are the vertices of T and $S(v_i)$ is the score of v_i in T, then $(S(v_1), S(v_2), \ldots, S(v_n))$ is the *score vector* of T. An ordered triple (u, v, w) of vertices of T is a *transitive triple* of T if $(u, v) \in A(T)$ and $(v, w) \in A(T)$, then $(u, w) \in A(T)$.

Remarks 2.4.6. 1. If v is any vertex of a tournament T and u, w are two outneighbors of v, then $\{u, v, w\}$ determines a unique transitive triple in T. Such a transitive

Fig. 2.9 Three-partite
tournaments with eight
directed 3-cycles (T_1) and
five directed 3-cycles (T_2).
Both T_1 and T_2 have the same
score vector

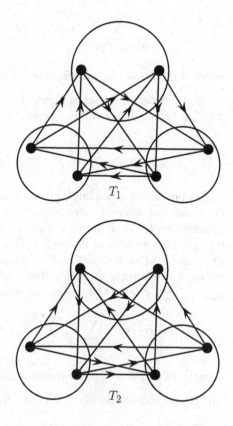

triple is said to be defined by the vertex v. Clearly, any transitive triple of T is
defined by some vertex of T. Further, the number of transitive triples defined by
v is $\binom{S(v)}{2}$.

2. The number of directed 3-cycles in a tournament T of order n is obtained by
subtracting the total number of transitive triples of vertices of T from the total
number of triples of vertices of T. Thus, the total number of directed 3-cycles in
T is equal to $\binom{n}{3} - \sum_{v \in V(T)} \binom{S(v)}{2} = \frac{n(n-1)(n-2)}{6} - \frac{1}{2} \sum_{v \in V(T)} S(v)(S(v) - 1)$.

Thus, the score vector of a tournament T determines the number of directed
3-cycles in T. But in a general k-partite tournament, the score vector need not
determine the number of directed 3-cycles. Consider the two 3-partite tournaments
T_1 and T_2 of Fig. 2.9. Both have the same score vector $(2, 2, 2, 2, 2, 2)$. But T_1 has
eight directed 3-cycles, whereas T_2 has only five directed 3-cycles.

Theorem 2.4.7 gives a formula for the number of directed 3-cycles in a k-partite
tournament.

Theorem 2.4.7. *Let T be a k-partite tournament, $k \geq 3$, having partite sets
$V_0, V_1, \ldots, V_{k-1}$. Then the number of directed 3-cycles in T is given by*

$$\sum_{0 \le i < j < \ell \le k-1} |V_i| |V_j| |V_\ell| - \sum_{v \in V(T)} \sum_{i<j} O_i(v) O_j(v),$$

where $O_i(v)$ denotes the number of outneighbors of v in V_i.

Proof. Let S denote the set of triples of vertices of T such that the three vertices of the triple belong to three different partite sets, and let $N = |S|$. Then

$$N = \sum_{0 \le i < j < \ell \le k-1} |V_i| |V_j| |V_\ell|.$$

Any orientation of a triangle gives a directed 3-cycle or a transitive triple. Hence the number of directed 3-cycles in $T = N - N_1$, where N_1 is the number of transitive triples in T. Also, a triple of vertices of T is transitive if and only if there exists a vertex of the triple having the other two vertices as outneighbors. The number of such triples of T to which a vertex v can belong and for which the other two vertices are outneighbors of v is $\sum_{i<j} O_i(v) O_j(v)$. Hence $N_1 = \sum_{v \in V(T)} \sum_{i<j} O_i(v) O_j(v)$. Thus the number of directed 3-cycles in T is given by

$$N - N_1 = \sum_{0 \le i < j < \ell \le k-1} |V_i| |V_j| |V_\ell| - \sum_{v \in V(T)} \sum_{i<j} O_i(v) O_j(v). \qquad \square$$

Remark 2.4.8. For $k = 3$, the results of Theorem 2.4.7 simplify as follows:

(i) The number of transitive triples in a 3-partite tournament equals

$$\sum_{i=0}^{2} \sum_{v \in V_i} O_{i+1}(v) O_{i+2}(v),$$

where the suffixes are taken modulo 3.

(ii) The number of directed 3-cycles in a 3-partite tournament is given by

$$|V_0| |V_1| |V_2| - \sum_{i=0}^{2} \sum_{v \in V_i} O_{i+1}(v) O_{i+2}(v),$$

where the suffixes are taken modulo 3.

Remark 2.4.9. Consider the two 3-partite tournaments of Fig. 2.10. T_1 has $|V_0||V_1||V_2|$ directed 3-cycles and has no transitive triples, whereas T_2 contains no directed 3-cycles but contains $|V_0||V_1||V_2|$ transitive triples.

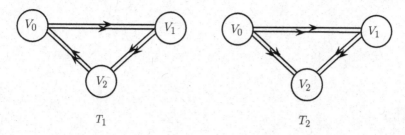

Fig. 2.10 Three-partite tournaments T_1 (with directed 3-cycles and no transitive triples) and T_2 (with transitive triples and no directed 3-cycles). *Double arrows* indicate that all arcs joining corresponding partite sets have the same orientation

2.5 Exercises

5.1. If $|V_i| = n_i$, $1 \leq i \leq k$, find the number of edges in the complete multipartite graph $G(V_1, V_2, \ldots, V_k)$. (See [27], p. 6.)

5.2. Show that if T is a strongly connected 3-partite tournament with partite sets V_0, V_1, V_2, then the maximum number of transitive triples in T is $|V_0||V_1||V_2| - 1$ unless $|V_0| = |V_1| = |V_2| = 2$, in which case T has at most $|V_0||V_1||V_2| - 2 = 6$ transitive triples.

5.3. Construct a strongly connected 3-partite tournament containing exactly six transitive triples.

5.4. Give a definition of digraph isomorphism similar to that of graph isomorphism.

5.5. Give an example of two nonisomorphic tournaments on five vertices. Justify your answer.

5.6. If u and v are distinct vertices of a tournament T such that both $d(u, v)$ and $d(v, u)$ are defined [where $d(u, v)$ denotes the length of a shortest directed (u, v)-path in T], show that $d(u, v) \neq d(v, u)$.

5.7. (A tournament T is called *transitive* if (a, b) and (b, c) are arcs of T, then (a, c) is also an arc of T.) Prove that a transitive tournament contains a Hamilton path with any preassigned orientation. [Hint: Use the fact that T has a vertex of outdegree $(n - 1)$ and a vertex of outdegree zero.]

Notes

The earliest of the books on directed graphs is by Harary, Norman, and Cartwright [97]. *Topics on Tournaments* by Moon [143] deals exclusively with tournaments. Theorems 2.4.2, 2.4.3 and 2.4.7 are based on [74].

Chapter 3
Connectivity

3.1 Introduction

The connectivity of a graph is a "measure" of its connectedness. Some connected graphs are connected rather "loosely" in the sense that the deletion of a vertex or an edge from the graph destroys the connectedness of the graph. There are graphs at the other extreme as well, such as the complete graphs K_n, $n \geq 2$, which remain connected after the removal of any k vertices, $1 \leq k \leq n - 1$.

Consider a communication network. Any such network can be represented by a graph in which the vertices correspond to communication centers and the edges represent communication channels. In the communication network of Fig. 3.1a, any disruption in the communication center v will result in a communication breakdown, whereas in the network of Fig. 3.1b, at least two communication centers have to be disrupted to cause a breakdown. It is needless to stress the importance of maintaining reliable communication networks at all times, especially during times of war, and the reliability of a communication network has a direct bearing on its connectivity.

In this chapter, we study the two graph parameters, namely, vertex connectivity and edge connectivity. We also introduce the parameter cyclical edge connectivity. We prove Menger's theorem and several of its variations. In addition, the theorem of Ford and Fulkerson on flows in networks is established.

3.2 Vertex Cuts and Edges Cuts

We now introduce the notions of vertex cuts, edge cuts, vertex connectivity, and edge connectivity.

R. Balakrishnan and K. Ranganathan, *A Textbook of Graph Theory*,
Universitext, DOI 10.1007/978-1-4614-4529-6_3,
© Springer Science+Business Media New York 2012

Fig. 3.1 Two types of
communication networks

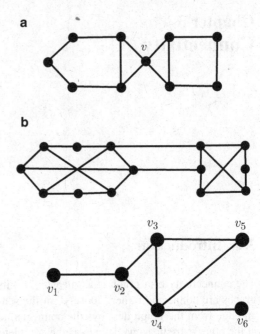

Fig. 3.2 Graph illustrating
vertex cuts and edge cuts

Definitions 3.2.1. 1. A subset V' of the vertex set $V(G)$ of a connected graph G is
   a *vertex cut* of G if $G - V'$ is disconnected; it is a *k-vertex cut* if $|V'| = k$. V'
   is then called a *separating set of vertices* of G. A vertex v of G is a *cut vertex* of
   G if $\{v\}$ is a vertex cut of G.
2. Let G be a nontrivial connected graph with vertex set $V(G)$ and let S be a
   nonempty subset of $V(G)$. For $\bar{S} = V \backslash S \neq \emptyset$, let $[S, \bar{S}]$ denote the set of
   all edges of G that have one end vertex in S and the other in $\bar{S}$. A set of edges of
   G of the form $[S, \bar{S}]$ is called an *edge cut* of G. An edge e is a *cut edge* of G if
   $\{e\}$ is an edge cut of G. An edge cut of cardinality k is called a *k-edge cut* of G.

Example 3.2.2. For the graph of Fig. 3.2, $\{v_2\}$, and $\{v_3, v_4\}$ are vertex cuts. The edge
subsets $\{v_3v_5, v_4v_5\}$, $\{v_1v_2\}$, and $\{v_4v_6\}$ are all edge cuts. Of these, v_2 is a cut vertex,
and v_1v_2 and v_4v_6 are both cut edges. For the edge cut $\{v_3v_5, v_4v_5\}$, we may take
$S = \{v_5\}$ so that $\bar{S} = \{v_1, v_2, v_3, v_4, v_6\}$.

Remarks 3.2.3. 1. If uv is an edge of an edge cut E', then all the edges having u
and v as their ends also belong to E'.
2. No loop can belong to an edge cut.

Exercise 2.1. If $\{x, y\}$ is a 2-edge cut of a graph G, show that every cycle of G that
contains x must also contain y.

Remarks 3.2.4. If G is connected and E' is a set of edges whose deletion results in
a disconnected graph, then E' contains an edge cut of G. It is clear that if e is a cut
edge of a connected graph G, then $G - e$ has exactly two components.

Remarks 3.2.5. Since the removal of a parallel edge of a connected graph does not result in a disconnected graph, such an edge cannot be a cut edge of the graph. A set of edges of a connected graph G whose deletion results in a disconnected graph is called a *separating set of edges.* In particular, any edge cut of a connected graph G is a separating set of edges of G.

We now characterize a cut vertex of G.

Theorem 3.2.6. *A vertex v of a connected graph G with at least three vertices is a cut vertex of G if and only if there exist vertices u and w of G distinct from v such that v is in every u-w path in G.*

Proof. If v is a cut vertex of G, then $G - v$ is disconnected and has at least two components, G_1 and G_2. Take $u \in V(G_1)$ and $w \in V(G_2)$. Then every u-w path in G must contain v, as otherwise u and w would belong to the same component of $G - v$.

Conversely, suppose that the condition of the theorem holds. Then the deletion of v destroys every u-w path in G, and hence u and w lie in distinct components of $G - v$. Therefore, $G - v$ is disconnected and v is a cut vertex of G. $\square$

Theorems 3.2.7 and 3.2.8 characterize a cut edge of a graph.

Theorem 3.2.7. *An edge $e = xy$ of a connected graph G is a cut edge of G if and only if e belongs to no cycle of G.*

Proof. Let e be a cut edge of G and let $[S, \bar{S}] = \{e\}$ be the partition of V defined by $G - e$ so that one of x and y belongs to S, and the other to $\bar{S}$, say, $x \in S$ and $y \in \bar{S}$. If e belongs to a cycle of G, then $[S, \bar{S}]$ must contain at least one more edge, contradicting that $\{e\} = [S, \bar{S}]$. Hence, e cannot belong to a cycle.

Conversely, assume that e is not a cut edge of G. Then $G - e$ is connected, and hence there exists an x-y path P in $G - e$. Then $P \cup \{e\}$ is a cycle in G containing e.
                $\square$

Theorem 3.2.8. *An edge $e = xy$ is a cut edge of a connected graph G if and only if there exist vertices u and v such that e belongs to every u-v path in G.*

Proof. Let $e = xy$ be a cut edge of G. Then $G - e$ has two components, say, G_1 and G_2. Let $u \in V(G_1)$ and $v \in V(G_2)$. Then, clearly, every u-v path in G contains e.

Conversely, suppose that there exist vertices u and v satisfying the condition of the theorem. Then there exists no u-v path in $G - e$ so that $G - e$ is disconnected. Hence, e is a cut edge of G. $\square$

Remark 3.2.9. There exist graphs in which every edge is a cut edge. It follows from Theorem 3.2.7 that if G is a simple connected graph with at least one edge and without cycles, then every edge of G is a cut edge of G. A similar result is not true for cut vertices. Our next result shows that not every vertex of a connected graph (with at least two vertices) can be a cut vertex of G.

Theorem 3.2.10. *A connected graph G with at least two vertices contains at least two vertices that are not cut vertices.*

Fig. 3.3 Graph for proof of
Theorem 3.2.10

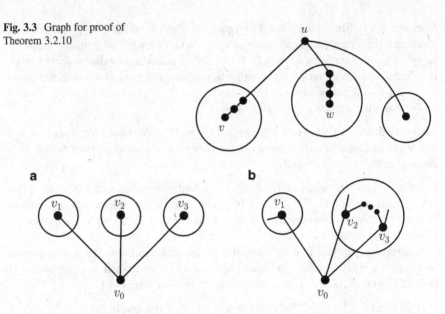

Fig. 3.4 Graph for proof of Proposition 3.2.11

Proof. First, suppose that $n(G) \geq 3$. Let u and v be vertices of G such that $d(u, v)$ is maximum. Then neither u nor v is a cut vertex of G. For if u were a cut vertex of G, $G - u$ would be disconnected, having at least two components. The vertex v belongs to one of these components. Let w be any vertex belonging to a component of $G - u$ not containing v. Then every v-w path in G must contain u (see Fig. 3.3). Consequently, $d(v, w) > d(v, u)$, contradicting the choice of u and v. Hence, u is not a cut vertex of G. Similarly, v is not a cut vertex of G.

If $n(G) = 2$, then K_2 is a spanning subgraph of G, and so no vertex of G is a cut vertex of G. This completes the proof of the theorem. □

Proposition 3.2.11. *A simple cubic (i.e., 3-regular) connected graph G has a cut vertex if and only if it has a cut edge.*

Proof. Let G have a cut vertex v_0. Let v_1, v_2, v_3 be the vertices of G that are adjacent to v_0 in G. Consider $G - v_0$, which has either two or three components. If $G - v_0$ has three components, no two of v_1, v_2, and v_3 can belong to the same component of $G - v_0$. In this case, each of v_0v_1, v_0v_2, and v_0v_3 is a cut edge of G. (See Fig. 3.4a.) In the case when $G - v_0$ has only two components, one of the vertices, say v_1, belongs to one component of $G - v_0$, and v_2 and v_3 belong to the other component. In this case, v_0v_1 is a cut edge. (See Fig. 3.4b.)

Conversely, suppose that $e = uv$ is a cut edge of G. Then the deletion of u results in the deletion of the edge uv. Since G is cubic, $G - u$ is disconnected. Accordingly, u is a cut vertex of G. □

Exercise 2.2. Find the vertex cuts and edge cuts of the graph of Fig. 3.2.

Exercise 2.3. Prove or disprove: Let G be a simple connected graph with $n(G) \geq 3$. Then G has a cut edge if and only if G has a cut vertex.

Exercise 2.4. Show that in a graph, the number of edges common to a cycle and an edge cut is even.

3.3 Connectivity and Edge Connectivity

We now introduce two parameters of a graph that in a way measure the connectedness of the graph.

Definition 3.3.1. For a nontrivial connected graph G having a pair of nonadjacent vertices, the minimum k for which there exists a k-vertex cut is called the *vertex connectivity* or simply the *connectivity* of G; it is denoted by $\kappa(G)$ or simply κ (kappa) when G is understood. If G is trivial or disconnected, $\kappa(G)$ is taken to be zero, whereas if G contains K_n as a spanning subgraph, $\kappa(G)$ is taken to be $n - 1$.

A set of vertices and/or edges of a connected graph G is said to *disconnect* G if its deletion results in a disconnected graph.

When a connected graph G (on $n \geq 3$ vertices) does not contain K_n as a spanning subgraph, κ is the connectivity of G if there exists a set of κ vertices of G whose deletion results in a disconnected subgraph of G while no set of $\kappa - 1$ (or fewer) vertices has this property.

Exercise 3.1. Prove that a simple graph G with n vertices, $n \geq 2$, is complete if and only if $\kappa(G) = n - 1$.

Definition 3.3.2. The *edge connectivity* of a connected graph G is the smallest k for which there exists a k-edge cut (i.e., an edge cut having k edges). The edge connectivity of a trivial or disconnected graph is taken to be 0. The edge connectivity of G is denoted by $\lambda(G)$. If λ is the edge connectivity of a connected graph G, there exists a set of λ edges whose deletion results in a disconnected graph, and no subset of edges of G of size less than λ has this property.

Exercise 3.2. Prove that the deletion of edges of a minimum-edge cut of a connected graph G results in a disconnected graph with exactly two components. (Note that a similar result is not true for a minimum vertex cut.)

Definition 3.3.3. A graph G is *r-connected* if $\kappa(G) \geq r$. Also, G is *r-edge connected* if $\lambda(G) \geq r$.

An r-connected (respectively, r-edge-connected) graph is also ℓ-connected (respectively, ℓ-edge connected) for each ℓ, $0 \leq \ell \leq r - 1$.

For the graph G of Fig. 3.5, $\kappa(G) = 1$ and $\lambda(G) = 2$.

We now derive inequalities connecting $\kappa(G)$, $\lambda(G)$, and $\delta(G)$.

Theorem 3.3.4. *For any loopless connected graph G, $\kappa(G) \leq \lambda(G) \leq \delta(G)$.*

Fig. 3.5 A 1-connected
graph

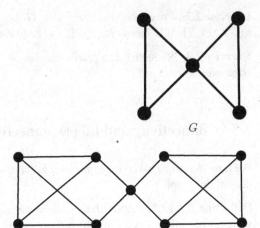

G

Fig. 3.6 Graph G with
$\kappa = 1$, $\lambda = 2$ and $\delta = 3$

G

Proof. We observe that $\kappa = 0$ if and only if $\lambda = 0$. Also, $\delta = 0$ implies that $\kappa = 0$ and $\lambda = 0$. Hence we may assume that κ, λ, and δ are all at least 1. Let $\mathscr{E}$ be an edge cut of G with λ edges. Let u and v be the end vertices of an edge of $\mathscr{E}$. For each edge of $\mathscr{E}$ that does not have both u and v as end vertices, remove an end vertex that is different from u and v. If there are t such edges, at most t vertices have been removed. If the resulting graph, say H, is disconnected, then $\kappa \leq t < \lambda$. Otherwise, there will remain a subset of edges of E having u and v as end vertices, the removal of which from H would disconnect G. Hence, in addition to the already removed vertices, the removal of one of u and v will result in either a disconnected graph or a trivial graph. In the process, a set of at most $t + 1$ vertices has been removed and $\kappa \leq t + 1 \leq \lambda$.

Finally, it is clear that $\lambda \leq \delta$. In fact, if v is a vertex of G with $d_G(v) = \delta$, then the set $[\{v\}, V \setminus \{v\}]$ of δ edges of G incident at v forms an edge cut of G. Thus, $\lambda \leq \delta$. □

It is possible that the inequalities in Theorem 3.3.4 can be strict. See the graph G of Fig. 3.6, for which $\kappa = 1$, $\lambda = 2$, and $\delta = 3$.

Exercise 3.3. Prove or disprove: If H is a subgraph of G, then

(a) $\kappa(H) \leq \kappa(G)$ and
(b) $\lambda(H) \leq \lambda(G)$.

Exercise 3.4. Determine $\lambda(K_n)$.

Exercise 3.5. Determine the connectivity and edge connectivity of the Petersen graph P. (See graph P of Fig. 1.7. Note that P is a cubic graph.)

Theorem 3.3.5 gives a class of graphs for which $\kappa = \lambda$.

Theorem 3.3.5. *The connectivity and edge connectivity of a simple cubic graph G are equal.*

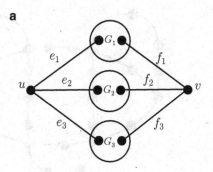

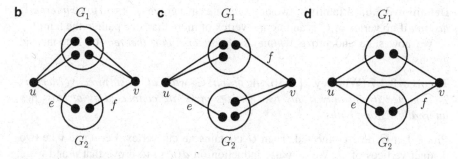

Fig. 3.7 Connected cubic graphs for proof of Theorem 3.3.5

Proof. We need only consider the case of a connected cubic graph. Again, since $\kappa \leq \lambda \leq \delta = 3$, we have only to consider the cases when $\kappa = 1, 2$, or 3. Now, Proposition 3.2.11 implies that for a simple cubic graph G, $\kappa = 1$ if and only if $\lambda = 1$.

If $\kappa = 3$, then by Theorem 3.3.4, $3 = \kappa \leq \lambda \leq \delta = 3$, and hence $\lambda = 3$.

We shall now prove that $\kappa = 2$ implies that $\lambda = 2$.

Suppose $\kappa = 2$ and $\{u, v\}$ is a 2-vertex cut of G. The deletion of $\{u, v\}$ results in a disconnected subgraph G' of G. Since each of u and v must be joined to each component of G', and since G is cubic, G' can have at most three components. If G' has three components, G_1, G_2, and G_3, and if e_i and f_i, $i = 1, 2, 3$, join, respectively, u and v with G_i, then each pair $\{e_i, f_i\}$ is an edge cut of G (see Fig. 3.7a).

If G' has only two components, G_1 and G_2, then each of u and v is joined to one of G_1 and G_2 by a single edge, say, e and f, respectively, so that $\{e, f\}$ is an edge cut of G (see Fig. 3.7b–d).

Hence, in either case there exists an edge cut consisting of two edges. As such, $\lambda \leq 2$. But by Theorem 3.3.4, $\lambda \geq \kappa = 2$. Hence $\lambda = 2$. Finally, the above arguments show that if $\lambda = 3$, then $\kappa = 3$, and if $\lambda = 2$, then $\kappa = 2$. $\square$

Exercise 3.6. Give examples of cubic graphs G_1, G_2, and G_3 with $\kappa(G_1) = 1$, $\kappa(G_2) = 2$, and $\kappa(G_3) = 3$.

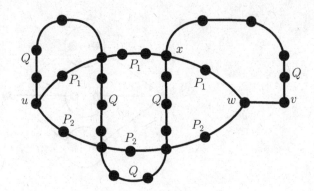

Definition 3.3.6. A family of two or more paths in a graph G is said to be *internally disjoint* if no vertex of G is an internal vertex of more than one path in the family.

We now state and prove *Whitney's characterization theorem* of 2-connected graphs.

Theorem 3.3.7 (Whitney [193]). *A graph G with at least three vertices is 2-connected if and only if any two vertices of G are connected by at least two internally disjoint paths.*

Proof. Let G be 2-connected. Then G contains no cut vertex. Let u and v be two distinct vertices of G. We now use induction on $d(u, v)$ to prove that u and v are joined by two internally disjoint paths.

If $d(u, v) = 1$, let $e = uv$. As G is 2-connected and $n(G) \geq 3$, e cannot be a cut edge of G, since if e were a cut edge, at least one of u and v must be a cut vertex. By Theorem 3.2.7, e belongs to a cycle C in G. Then $C - e$ is a u-v path in G, internally disjoint from the path uv.

Now assume that any two vertices x and y of G with $d(x, y) = k-1, k \geq 2$, are joined by two internally disjoint x-y paths in G. Let $d(u, v) = k$. Let P be a u-v path of length k and w be the vertex of G just preceding v on P. Then $d(u, w) = k - 1$. By an induction hypothesis, there are two internally disjoint u-w paths, say P_1 and P_2, in G. As G has no cut vertex, $G - w$ is connected and hence there exists a u-v path Q in $G - w$. Q is clearly a u-v path in G not containing w. Let x be the vertex of Q such that the x-v section of Q contains only the vertex x in common with $P_1 \cup P_2$ (see Fig. 3.8).

We may suppose, without loss of generality, that x belongs to P_1. Then the union of the u-x section of P_1 and x-v section of Q and $P_2 \cup (wv)$ are two internally disjoint u-v paths in G. This gives the proof in one direction.

In the other direction, assume that any two distinct vertices of G are connected by at least two internally disjoint paths. Then G is connected. Further, G cannot contain a cut vertex, since if v were a cut vertex of G, there must exist vertices u and w such that every u-w path contains v (compare with Theorem 3.2.6), contradicting the hypothesis. Hence, G is 2-connected. $\square$

Fig. 3.9 Graph for
Remark 3.3.9

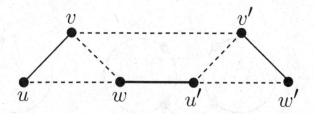

Theorem 3.3.8. *A graph G with at least three vertices is 2-connected if and only if any two vertices of G lie on a common cycle.*

Proof. Let u and v be any two vertices of a 2-connected graph G. By Theorem 3.3.7, there exist two internally disjoint paths in G joining u and v. The union of these two paths is a cycle containing u and v.

Conversely, if any two vertices u and v lie on a cycle C, then C is the union of two internally disjoint u-v paths. Again, by Theorem 3.3.7, G is 2-connected.   □

Remark 3.3.9. If G is 2-connected, if u and v are distinct vertices of G, and if P is a u-v path in G, it is not in general true that there exists another u-v path Q in G such that P and Q are internally disjoint. For example, in the 2-connected graph of Fig. 3.9, if P is the u-w' path $uwvv'u'w'$, there exists no u-w' path Q in G that is internally disjoint from P. However, there do exist two internally disjoint u-w' paths in G.

Exercise 3.7. (a) Show that a graph G with at least three vertices is 2-connected if and only if any vertex and any edge of G lie on a common cycle of G.
(b) Show that a graph G with at least three vertices is 2-connected if and only if any two edges of G lie on a common cycle.

Exercise 3.8. Prove that a graph is 2-connected if and only if for every pair of disjoint connected subgraphs G_1 and G_2, there exist two internally disjoint paths P_1 and P_2 of G between G_1 and G_2.

Exercise 3.9. *Edge form of Whitney's theorem:* Prove that a graph G with $n \geq 3$ is 2-edge connected if and only if any two distinct vertices of G are connected by at least two edge-disjoint paths in G. [Hint: Imitate the proof of Theorem 3.3.7, or pass on to $L(G)$.]

Exercise 3.10. (a) Disprove by a counterexample: If $\kappa(G) = k$, then $\kappa(L(G)) = k$.
(b) Prove: $\lambda(G) \leq \kappa(L(G))$. Give an example of a graph G for which $\lambda(G) < \kappa(L(G))$.

Theorem 3.3.10. *In a 2-connected graph G, any two longest cycles have at least two vertices in common.*

Proof. Let $C_1 = u_1u_2 \ldots u_ku_1$ and $C_2 = v_1v_2 \ldots v_kv_1$ be two longest cycles in G. If C_1 and C_2 are disjoint, there exist (since G is 2-connected) two disjoint paths,

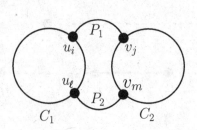

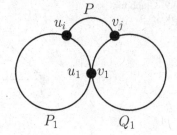

Fig. 3.10 Graphs for proof of Theorem 3.3.10

Fig. 3.11 Graph for proof of
Theorem 3.3.11

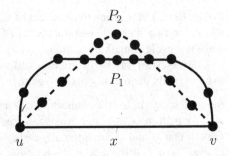

say, P_1 joining u_i and v_j and P_2 joining u_ℓ and v_p, connecting C_1 and C_2 such that $u_i \neq u_\ell$ and $v_j \neq v_p$ (see Exercise 3.8). u_i and u_ℓ divide C_1 into two subpaths. Let L_1 be the longer of these subpaths. (If both subpaths are of equal length, we take either one of them to be L_1.) Let L_2 be defined in a similar manner in C_2. Then $L_1 \cup P_1 \cup L_2 \cup P_2$ is a cycle of length greater than that of C_1 (or C_2). Hence, C_1 and C_2 cannot be disjoint. (See Fig. 3.10.)

Suppose that C_1 and C_2 have exactly one vertex, say $u_1 = v_1$, in common. Since G is 2-connected, u_1 is not a cut vertex of G, and so there exists a path P with one end vertex u_i in $C_1 - u_1$ and the other end vertex v_j in $C_2 - v_1$, which is internally disjoint from $C_1 \cup C_2$. Let P_1 denote the longer of the two u_1-u_i sections of C_1, and Q_1 denote the longer of the two v_1-v_j sections of C_2. If the two sections of C_1 or of C_2 are of equal length, take any one of them. Then $P_1 \cup P \cup Q_1$ is a cycle longer than C_1 (or C_2). But this is impossible. Thus, C_1 and C_2 must have at least two vertices in common.   □

Theorem 3.3.11 gives a simple characterization of 3-edge-connected graphs.

Theorem 3.3.11. *A connected simple graph G is 3-edge connected if and only if every edge of G is the (exact) intersection of the edge sets of two cycles of G.*

Proof. Let G be 3-edge connected and let $x = uv$ be an edge of G. Since $G - x$ is 2-edge connected, there exist two edge-disjoint u-v paths P_1 and P_2 in $G - x$ (see Exercise 3.9). Now, $P_1 \cup \{x\}$ and $P_2 \cup \{x\}$ are two cycles of G, the intersection of whose edge sets is precisely $\{x\}$ (see Fig. 3.11).

Conversely, suppose that for each edge $x = uv$ there exist two cycles C and C' such that $\{x\} = E(C) \cap E(C')$. G cannot have a cut edge since, by hypothesis, each edge belongs to two cycles and no cut edge can belong to a cycle; nor can G contain an edge cut consisting of two edges x and y, by Exercise 2.1. (Since any cycle that contains x also contains y, the intersection of any two such cycles must contain both x and y, a contradiction.) Hence, $\lambda(G) \geq 3$, and G is 3-edge connected. □

3.4 Blocks

In this section, we focus on connected graphs without cut vertices.

Definition 3.4.1. A graph G is *nonseparable* if it is nontrivial and connected and has no cut vertices. A *block of a graph* is a maximal nonseparable subgraph of G. If G has no cut vertex, G itself is a block.

In Fig. 3.12, a graph G and its blocks B_1, B_2, B_3, and B_4 are indicated. B_1, B_3, and B_4 are the *end blocks* of G (i.e., blocks having exactly one cut vertex of G). The following facts are worthy of observation.

Remarks 3.4.2. Let G be a connected graph with $n \geq 3$.

1. Each block of G with at least three vertices is a 2-connected subgraph of G.
2. Each edge of G belongs to one and only one of its blocks. Hence G is an edge-disjoint union of its blocks.
3. Any two blocks of G have at most one vertex in common. (Such a common vertex is a cut vertex of G.)
4. A vertex of G that is not a cut vertex belongs to exactly one of its blocks.
5. A vertex of G is a cut vertex of G if and only if it belongs to at least two blocks of G.

Whitney's theorem (Theorem 3.3.7) implies that a graph with at least three vertices is a block if and only if any two vertices of the graph are connected by at least two internally disjoint paths. Again by Theorem 3.3.8, we see that any two vertices of a block with at least three vertices belong to a common cycle. Thus, a block with at least three vertices contains a cycle.

Theorem 3.4.3 (Ear decomposition of a block). *If C is any cycle of a simple block G, then there exists a sequence of nonseparable subgraphs $C = B_0, B_1, \ldots, B_r = G$ such that B_{i+1} is an edge-disjoint union of B_i and a path P_i, where the only vertices common to B_i and P_i are the end vertices of P_i, $0 \leq i \leq r - 1$.*

Proof. Assume that we have already determined B_i (see Fig. 3.13). If $B_i \neq G$, there exists (as G is connected) an edge $e = uv$ not belonging to B_i but with u in B_i. If v also belongs to B_i, take $P_i = uv$ and $B_{i+1} = B_i \cup P_i$. Otherwise, $e = uv$ is an edge of G having only one of its ends, namely u, in B_i. Let u' be any other

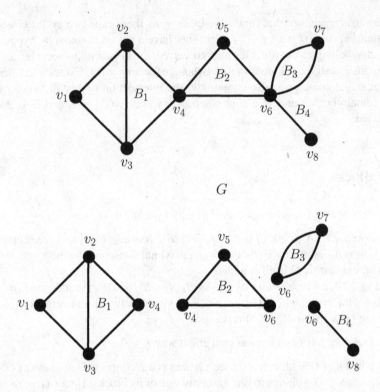

G

Fig. 3.12 A graph G and its blocks

Fig. 3.13 Graph for proof of
Theorem 3.4.3

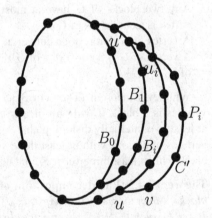

vertex of B_i. Then since G is 2-connected, e and u' belong to a common cycle C_i
(see Exercise 3.7). Let u_i be the first vertex of B_i after u in the u-u' section C' of C_i
containing v, and let P_i be the u-u_i section of C'. Define $B_{i+1} = B_i \cup P_i$. Then
B_{i+1} is nonseparable and the proof follows by induction on i.   □

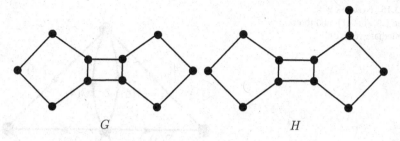

Fig. 3.14 Graphs $G(\lambda(G) = \lambda_c(G) = 2)$ and $H(\lambda(H) = 1, \lambda_c(H) = 2)$

3.5 Cyclical Edge Connectivity of a Graph

In this section we introduce the parameter "cyclical edge connectivity of a graph."
Unlike connectivity and edge connectivity, cyclical edge connectivity is not defined
for all graphs.

Definition 3.5.1. Let G be a simple connected graph containing at least two disjoint
cycles. Then the *cyclical edge connectivity* of G is defined to be the minimum
number of edges of G whose deletion results in a graph having two components,
each containing a cycle. It is denoted by $\lambda_c(G)$.

It is clear that $\lambda \leq \lambda_c$. The graphs G and H of Fig. 3.14 show that both $\lambda = \lambda_c$
and $\lambda < \lambda_c$ can happen.

Exercise 5.1. Show that the cyclical edge connectivity of the Petersen graph P is 5.

3.6 Menger's Theorem

In this section we prove different versions of the celebrated Menger's theorem,
which generalizes Whitney's theorem (Theorem 3.3.7). Menger's theorem [140]
relates the connectivity of a graph G to the number of internally disjoint paths
between pairs of vertices of G. The proofs given here make use of network analysis.
Hence we begin with the definition of a network.

Definition 3.6.1. A *network* N is a digraph D with two distinguished vertices s
and t, $(s \neq t)$, and a nonnegative integer-valued function c defined on its arc set A.
s is called the *source* and t is called the *sink* of N. The source corresponds to the
supply center and the sink corresponds to a market. Vertices of N, other than s and
t, are called the *intermediate vertices* of N. The digraph D is called the underlying
digraph of N. The function c is called the *capacity function* of N and $c(a)$, for an
arc a, denotes the *capacity* of a.

Fig. 3.15 Network with
source (*s*), sink (*t*), and three
intermediate vertices

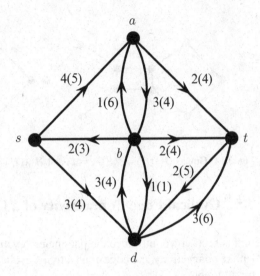

Example 3.6.2. A network N is diagrammatically represented by the underlying digraph D, labeling each arc with its capacity. Figure 3.15 is a network with source s and sink t, and three intermediate vertices. The numbers inside the brackets denote the capacities of the respective arcs.

For a real-valued function f defined on A, and $K \subseteq A$, $\sum\limits_{a \in K} f(a)$ will be denoted by $f(K)$. If K is a set of arcs of D of the form $[S, \bar{S}]$, that is, the set of arcs with heads in S and tails in $\bar{S}$, where $S \subseteq V(D)$, $\bar{S} = V(D) \setminus S$, then $f^+(S)$ and $f^-(S)$ denote $f([S, \bar{S}])$ and $f([\bar{S}, S])$, respectively. If $S = \{v\}$, then $f^+(S)$ and $f^-(S)$ are denoted by $f^+(v)$ and $f^-(v)$, respectively.

Definition 3.6.3. A *flow* in a network N is an integer-valued function f defined on $A = A(N)$ such that $0 \leq f(a) \leq c(a)$ for all $a \in A$ and $f^+(v) = f^-(v)$ for all the intermediate vertices v of N.

Remarks 3.6.4. 1. $f^+(v)$ is the flow out of v and $f^-(v)$ is the flow into v. *The condition $f^+(v) = f^-(v)$ for each intermediate vertex v then signifies that there is conservation of flow at every such vertex.*
2. If $a = (u, v)$, we denote $f(a)$ by f_{uv}. Every network N has at least one flow since the function f defined by $f(a) = 0$ for all $a \in A$ is a flow. It is called the *zero flow* in N.
3. A less trivial example of a flow in the network of Fig. 3.15 is given by f, where $f_{sa} = 4$, $f_{sd} = 3$, $f_{bs} = 2$, $f_{at} = 2$, $f_{ab} = 3$, $f_{ba} = 1$, $f_{bd} = 1$, $f_{db} = 3$, $f_{bt} = 2$, $f_{dt} = 3$, and $f_{td} = 2$.
4. If S is a subset of vertices in a network N and f is a flow in N, $f^+(S) - f^-(S)$, is called the *resultant flow out of* S and $f^-(S) - f^+(S)$, the *resultant flow into* S, relative to f.

5. The flow along any arc (u, v) is both the outflow at u along (u, v) and the inflow at v along (u, v). Hence, $\sum_{v \in V(N)} f^+(v) - \sum_{v \in V(N)} f^-(v) = 0$. This gives us

$$[f^+(s) - f^-(s)] + \sum_{\substack{v \in V(N) \\ v \neq s, t}} (f^+(v) - f^-(v)) + [f^+(t) - f^-(t)] = 0.$$

But $f^+(v) = f^-(v)$ for each $v \in V(N)$, $v \neq s, t$. Hence,

$$f^+(s) - f^-(s) = f^-(t) - f^+(t).$$

Thus, relative to any flow f, the resultant flow out of s is equal to the resultant flow into t. For a similar reason, if S is any subset of $V(N)$ containing s but not t,

$$\sum_{v \in S} f^+(v) - \sum_{v \in S} f^-(v) = f^+(s) - f^-(s). \tag{3.1}$$

This common quantity is called the *value* of f and is denoted by val f. Thus,

$$\text{val } f = f^+(s) - f^-(s) = f^-(t) - f^+(t).$$

The value of the flow f of the network of Fig. 3.15 is 5.

Definition 3.6.5. 1. A flow f in N is a *maximum flow* if there is no flow f' in N such that val $f' > $ val f.
2. A cut K in N is a set of arcs of the form $[S, \bar{S},]$ where $s \in S$ and $t \in \bar{S}$. Such a cut is said to *separate* s and t. For example, $K = \{(a, t), (b, t), (d, t)\}$ is a cut in the network of Fig. 3.15, where $S = \{s, a, b, d\}$.
3. The *capacity* of a cut K is the sum of the capacities of its arcs. We denote the capacity of K by cap K. Thus, cap $K = \sum_{a \in K} c(a)$. For the network of Fig. 3.15, cap $K = 2 + 2 + 3 = 7$.

Theorem 3.6.6 gives the relation between the value of a flow and the capacity of a cut in a network.

Theorem 3.6.6. *In any network N, the value of any flow f is less than or equal to the capacity of any cut K.*

Proof. Let $[S, \bar{S}]$ be any cut with $s \in S$ and $t \in T$. We have, by (3.1),

$$\text{val } f = f^+(s) - f^-(s)$$

$$= \sum_{v \in S} f^+(v) - \sum_{v \in S} f^-(v)$$

$$= \sum_{\substack{v \in S, u \in S, \\ (v,u) \in A(D)}} f_{vu} + \sum_{\substack{v \in S, u \in \bar{S}, \\ (v,u) \in A(D)}} f_{vu} - \sum_{\substack{u \in S, v \in S, \\ (u,v) \in A(D)}} f_{uv} - \sum_{\substack{u \in \bar{S}, v \in S, \\ (u,v) \in A(D)}} f_{uv}.$$

But

$$\sum_{\substack{v \in S, \ u \in S, \\ (v,u) \in A(D)}} f_{vu} - \sum_{\substack{u \in S, \ v \in S, \\ (u,v) \in A(D)}} f_{uv} = 0.$$

Hence

$$\text{val } f = \sum_{\substack{v \in S, \ u \in \bar{S}, \\ (v,u) \in A(D)}} f_{vu} - \sum_{\substack{u \in \bar{S}, \ v \in S, \\ (u,v) \in A(D)}} f_{uv}. \qquad (3.2)$$

Since

$$\sum_{\substack{u \in \bar{S}, \ v \in S, \\ (u,v) \in A(D)}} f_{uv} \geq 0$$

(recall that f is a nonnegative integer-valued function), we get

$$\text{val } f \leq \sum_{\substack{v \in S, \ u \in \bar{S}, \\ (v,u) \in A(D)}} f_{vu} \leq \sum_{\substack{v \in S, \ u \in \bar{S}, \\ (v,u) \in A(D)}} c(v, u) = c([S, \bar{S}]). \qquad \square$$

Note 3.6.7. Note that we have shown in (3.2) that val f is the flow out of S minus the flow into S for any $S \subset V$ with $s \in S$ and $t \in \bar{S}$.

By Theorem 3.6.6, in any network N, the value of any flow f does not exceed the capacity of any cut K. In particular, if f^* is a maximum flow in N and K^* is a minimum cut, that is, a cut with minimum capacity, then val $f^* \leq$ cap K^*.

Lemma 3.6.8. *Let f be a flow and K a cut in a network N such that val $f = $ cap K. Then f is a maximum flow and K is a minimum cut.*

Proof. Let f^* be a maximum flow and K^* be a minimum cut in N. Then we have, by Theorem 3.6.6, val $f \leq$ val $f^* \leq$ cap $K^* \leq$ cap K. But by hypothesis, val $f = $ cap K. Hence, val $f = $ val $f^* = $ cap $K^* = $ cap K. Thus, f is a maximum flow and K is a minimum cut. $\qquad \square$

Theorem 3.6.9 is the celebrated *max-flow min-cut theorem* due to Ford and Fulkerson [65], which establishes the equality of the value of a maximum flow and the minimum capacity of a cut separating s and t.

Theorem 3.6.9 (Ford and Fulkerson). *In a given network N (with source s and sink t), the maximum value of a flow is equal to the minimum value of the capacities of all the cuts in N.*

Proof. In view of Lemma 3.6.8, we need only prove that there exists a flow in N whose value is equal to $c([S, \bar{S}])$ for some cut $[S, \bar{S}]$ separating s and t in N. Let f be a maximum flow in N with val $f = w_0$. Define $S \subset N$ recursively as follows:

(a) $s \in S$, and
(b) If a vertex $u \in S$ and either $f_{uv} < c(u, v)$ or $f_{vu} > 0$, then include v in S.

Any vertex not belonging to S belongs to $\bar{S}$. We claim that t cannot belong to S; indeed, if we suppose that $t \in S$, then there exists a path P from s to t, say

$P : sv_1v_2 \ldots v_jv_{j+1} \ldots v_kt$, with its vertices in S such that for any arc of P, either $f_{v_jv_{j+1}} < c(v_j, v_{j+1})$ or $f_{v_{j+1}v_j} > 0$. Call an arc joining v_j and v_{j+1} of P a *forward arc* if it is directed from v_j to v_{j+1}; otherwise, it is a *backward arc*.

Let δ_1 be the minimum of all differences $(c(v_j, v_{j+1}) - f_{v_jv_{j+1}})$ for forward arcs, and let δ_2 be the minimum of all flows in backward arcs of P. Both δ_1 and δ_2 are positive, by the definition of S. Let $\delta = \min\{\delta_1, \delta_2\}$. Increase the flow in each forward arc of P by δ and also decrease the flow in each backward arc of P by δ. Keep the flows along the other arcs of N unaltered. Then there results a new flow whose value is $w_0 + \delta > w_0$, leading to a contradiction. [This is because among all arcs incident at s, only in the initial arc of P, the flow value is increased by δ if it is a forward arc or decreased by δ if it is a backward arc; see (5) of Remarks 3.6.4.] This contradiction shows that $t \notin S$, and therefore $t \in \bar{S}$. In other words, $[S, \bar{S}]$ is a cut separating s and t. If $v \in S$ and $u \in \bar{S}$, we have, by the definition of S, $f_{vu} = c(v, u)$ if (v, u) is an arc of N, and $f_{uv} = 0$ if (u, v) is an arc of N. Hence, as in the proof of Theorem 3.6.6,

$$w_0 = \sum_{\substack{u \in S \\ v \in \bar{S}}} f_{uv} - \sum_{\substack{v \in \bar{S} \\ u \in S}} f_{vu} = \sum_{\substack{u \in S \\ v \in \bar{S}}} f_{uv} - 0 = \sum_{\substack{u \in S \\ v \in \bar{S}}} c(u, v) = c([S, \bar{S}]). \qquad \square$$

We now use the max-flow min-cut theorem to prove a number of results due to Menger. We shall first prove a result for a network in which each arc has unit capacity.

Theorem 3.6.10. *Let N be a network with source s and sink t. Let each arc of N have unit capacity. Then,*

(a) *The value of a maximum flow in N is equal to the maximum number k of arc-disjoint directed (s, t)-paths in N, and*

(b) *The capacity of a minimum cut in N is equal to the minimum number ℓ of arcs whose deletion destroys all (s, t)-paths in N.*

Proof. Let f^* be a maximum flow in N, and let D^* denote the digraph obtained from D, the underlying digraph of N, by deleting all arcs whose flow is zero in f^*. Now, note that $0 < f^*(a) \leq c(a) = 1$ for all $a \in A(D^*)$, and therefore, $f^*(a) = 1$ for all $a \in A(D^*)$. Hence,

(i) $d^+_{D^*}(s) - d^-_{D^*}(s) = f^{*+}(s) - f^{*-}(s) = \text{val } f^* = f^*(t) - f^{*+}(t) = d^-_{D^*}(t) - d^+_{D^*}(t)$ and

(ii) $d^+_{D^*}(v) = d^-_{D^*}(v)$ for $v \in V(N) \setminus \{s, t\}$.

(i) and (ii) imply that there are val f^* arc-disjoint directed (s, t)-paths in D^* and hence also in D. Thus, val $f^* \leq k$. Now let $P_1, P_2, \ldots, P_k$ be any system of k arc-disjoint directed (s, t)-paths in N. Define a function f on $A(N)$ by

$$f(a) = \begin{cases} 1 & \text{if } a \text{ is an arc of } \bigcup_{i=1}^{k} P_i, \\ 0 & \text{otherwise.} \end{cases}$$

Then f is a flow in N with value k. Since f^* is a maximum flow, we have val $f^* \geq k$. Consequently, val $f^* = k$, proving (a).

Let $K^* = [S, \bar{S}]$ be a minimum cut in N so that $|K^*| \geq \ell$ by the definition of ℓ. Then, cap $K^* = |K^*| \geq \ell$.

Now let Z be a set of ℓ arcs whose deletion destroys all directed (s, t)-paths, and let T denote the set of all vertices including s joined to s by a directed path in $N - Z$. Then since $s \in T$, and $t \in \bar{T}$, $K = [T, \bar{T}]$ is a cut in N. By the definition of T, $N - Z$ can contain no arc of $[T, \bar{T}]$, and hence $K \subseteq Z$. Since K^* is a minimum cut, we conclude that cap $K^* \leq$ cap $K = |K| \leq |Z| = \ell$. Thus, cap $K^* = \ell$. □

We now state and prove the *edge version of Menger's theorem for directed graphs*.

Theorem 3.6.11. *Let x and y be two vertices of a digraph D. Then the maximum number of arc-disjoint directed (x, y)-paths in D is equal to the minimum number of arcs whose deletion destroys all directed (x, y)-paths in D.*

Proof. Apply Theorem 3.6.9 to the two results of Theorem 3.6.10. □

Theorem 3.6.12 is the *edge version of Menger's theorem for undirected graphs*.

Theorem 3.6.12. *Let x and y be two vertices of a graph G. Then the maximum number of edge-disjoint (x, y)-paths in G is equal to the minimum number of edges of G whose deletion destroys all (x, y)-paths in G.*

Proof. Construct a digraph $D(G)$ from G as follows: $V(G)$ is also the vertex set of $D(G)$ and if $u, v \in V(G)$, then $(u, v) \in A(D(G))$ if and only if u and v are adjacent in G; that is, $D(G)$ is obtained from G by replacing each edge uv of G by a symmetric pair of arcs (u, v) and (v, u). By Theorem 3.6.11, the maximum number of arc-disjoint directed (x, y)-paths in $D(G)$ is equal to the minimum number of arcs whose deletion destroys all directed (x, y)-paths in $D(G)$. But each directed (x, y)-path in $D(G)$ gives rise to a unique (x, y)-path in G, and conversely an (x, y)-path in G yields a unique directed (x, y)-path in $D(G)$. Hence, the deletion of a set of λ edges in G destroys all (x, y)-paths in G if and only if the deletion of the corresponding set of λ arcs in $D(G)$ destroys all directed (x, y)-paths in $D(G)$. □

Theorem 3.6.13 is the *vertex version of Menger's theorem for digraphs*.

Theorem 3.6.13. *Let x and y be two vertices of a digraph D such that $(x, y) \notin A(D)$. Then the maximum number of internally disjoint directed (x, y)-paths in D is equal to the minimum number of vertices whose deletion destroys all directed (x, y)-paths in D.*

Proof. Construct a new digraph D' from D as follows:

(a) Split each vertex $v \in V \setminus \{x, y\}$ into two new vertices, v' and v'', and join them by an arc (v', v''), and
(b) Replace

   (i) Each arc (u, v) of D where $u \notin \{x, y\}$ and $v \notin \{x, y\}$ by the arc (u'', v'),

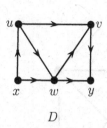
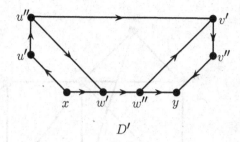

$$D \qquad\qquad\qquad\qquad D'$$

Fig. 3.16 Digraphs D and D' for proof of Theorem 3.6.13

   (ii) Each arc (x, v) of D by (x, v') and (v, x) by (v'', x), and
   (iii) Each arc (v, y) of D by (v'', y) and (y, v) by (y, v').
       (See Fig. 3.16.)

Now, to each directed (x, y)-path in D', there corresponds a directed (x, y)-path in D obtained by contracting all arcs of the type (v', v'') [that is, delete the arc (v', v'') and identify the vertices v' and v''], and, conversely, to each directed (x, y)-path in D, there corresponds a directed (x, y)-path in D' obtained by splitting each intermediate vertex of the path. Furthermore, two directed (x, y)-paths in D' are arc-disjoint if and only if the corresponding directed paths in D are internally disjoint. Hence, the maximum number of arc-disjoint directed (x, y)-paths in D' is equal to the maximum number of internally disjoint directed (x, y)-paths in D.

Similarly, the minimum number of arcs in D' whose deletion destroys all directed (x, y)-paths in D' is equal to the minimum number of vertices in D whose deletion destroys all directed (x, y)-paths in D. To see this, let A' be a minimum set of p arcs of D' whose deletion destroys all directed (x, y)-paths in D', and let B' be a minimum set of q vertices of D whose deletion destroys all directed (x, y)-paths in D. We have to show that $p = q$. Any arc of A' must contain either v' or v'' corresponding to the vertex v of D. Then the deletion of all vertices v corresponding to such arcs of D' separates x and y in D and hence $q \leq p$. Conversely, if $v \in B'$, delete the corresponding arc (v', v'') in D'. Then the deletion of the q arcs (v', v'') (which correspond to the q vertices of B') from D' destroys all directed (x, y)-paths in D', and therefore $p \leq q$. Thus, $p = q$. The result now follows from Theorem 3.6.11.   □

Theorem 3.6.14 is the *vertex version of Menger's theorem for undirected graphs.*

Theorem 3.6.14. *Let x and y be two nonadjacent vertices of a graph G. Then the maximum number of internally disjoint (x, y)-paths in G is equal to the minimum number of vertices whose deletion destroys all (x, y)-paths.*

Proof. Define $D(G)$ as in Theorem 3.6.12 and apply Theorem 3.6.13.   □

Let G be an undirected graph with $n \geq k + 1$ vertices. Suppose G satisfies the condition $(*)$:

$(*)$ Any two distinct vertices of G are connected by k internally disjoint paths in G.

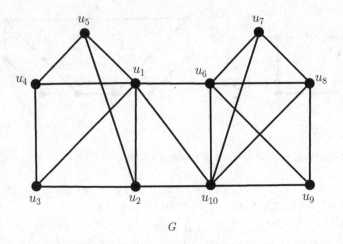

Fig. 3.17 A 2-connected and 3-edge connected graph

Then Theorem 3.6.14 implies that to separate two nonadjacent vertices x and y of G, at least k vertices are to be removed. Hence if (∗) holds, G is k-connected.

Conversely, if G is k-connected, to separate any pair of nonadjacent vertices x and y of G, at least k vertices are to be removed, and by Theorem 3.6.14, there are at least k internally disjoint (x, y)-paths in G. However, if x and y are adjacent, then since $G - xy$ is $(k-1)$-connected, there are at least k internally disjoint (x, y)-paths, including the edge xy. Thus, we have the following result of Whitney's generalizing Theorem 3.3.7.

Theorem 3.6.15 (Whitney). *A graph G with $n \geq k + 1$ vertices is k-connected if and only if any two vertices of G are connected by at least k internally disjoint paths.*

Example 3.6.16. The graph G of Fig. 3.17 is 2-connected and 3-edge connected. The pair of vertices u_5 and u_{10} are connected by the following two internally disjoint paths:

$$u_5 u_1 u_{10} \text{ and } u_5 u_4 u_3 u_2 u_{10}$$

Moreover, they are connected by the following 3-edge-disjoint paths:

$$u_5 u_1 u_{10}; \ u_5 u_2 u_{10}; \text{ and } u_5 u_4 u_1 u_6 u_{10}.$$

Exercise 6.1. If u and v are vertices of a graph G such that any two u-v paths in G have an internal common vertex, show that all the u-v paths in G have an internal common vertex.

Exercise 6.2. Show that if G is k-connected, then $G \vee K_1$ is $(k + 1)$-connected.

Exercise 6.3. Let S be a subset of the vertex set of a k-connected graph G with $|S| = k$. If $v \in V \setminus S$, show that there exist k internally disjoint paths from v to

Fig. 3.18 A θ-graph

the k vertices of S. [Remark: In particular, if C is a cycle of length at least k in a k-connected graph G, and $v \in V(G) \setminus V(C)$, then there are k internally disjoint paths from v to C.]

The remark in Exercise 6.3 yields the following theorem of Dirac [55], which generalizes Theorem 3.3.8.

Exercise 6.4. *Dirac's theorem* [55]: If a graph is k-connected ($k \geq 2$), then any set of k vertices of G lie on a cycle of G. (Note: The cycle may contain additional vertices besides these k vertices.) Hint: Use induction on k. If G is $(k + 1)$-connected, and $\{v_1, v_2, \ldots, v_k, v_{k+1}\}$ is any set of $k + 1$ vertices of G, by the induction assumption, $v_1, v_2, \ldots, v_k$ all lie on a cycle C of G. If $V(C) = \{v_1, v_2, \ldots, v_k\}$, then the k disjoint paths from v_{k+1} to C must end in $v_1, v_2, \ldots, v_k$. Otherwise, $V(C) \supsetneq \{v_1, v_2, \ldots, v_k\}$. Since G is $(k + 1)$-connected, by the pigeonhole principle, the end vertices of two of the $(k + 1)$ disjoint paths from v_{k+1} to C must belong to one of the k closed paths $[v_i, v_{i+1}]$, $1 \leq i \leq k - 1$ and $[v_k, v_1]$ on C. (Here $[v_i, v_{i+1}]$ and $[v_k, v_1]$ are those paths on C that contain no other v_j.)

Exercise 6.5. Show by means of an example that the converse of Dirac's theorem (Exercise 6.4) is false.

Exercise 6.6. Show that a k-connected simple graph on $(k + 1)$ vertices is K_{k+1}.

Exercise 6.7. *Dirac's theorem* [55]; *see also* [93]: Show that a graph G with at least $2k$ vertices is k-connected if and only if for any two disjoint sets V_1 and V_2 of k vertices each, there exist k disjoint paths from V_1 to V_2 in G.

Exercise 6.8. Show that a 2-connected non-Hamiltonian graph contains a θ-subgraph. (A θ-graph is a graph of the form $C \cup P$, where C is a cycle of length at least 4 and P is a path of length at least 2 that joins two nonadjacent vertices of C and is internally disjoint from C.) (See Fig. 3.18.)

3.7 Exercises

7.1. Prove that there exists no simple connected cubic graph with fewer than 10 vertices containing a cut edge. (For a simple connected cubic graph having exactly 10 vertices and having a cut edge, see Exercise 7.13)

7.2. Show that no vertex v of a simple graph can be a cut vertex of both G and G^c.

7.3. Show that a simple connected graph that is not a block contains at least two end blocks.

7.4. Show that a connected k-regular bipartite graph is 2-connected.

7.5. Let $b(v)$ denote the number of blocks of a simple connected graph G to which a vertex v belongs. Then prove that the number of blocks $b(G)$ of G is given by $b(G) = 1 + \sum_{v \in V(G)} (b(v) - 1)$.

   (Hint: Use induction on the number of blocks of G.)

7.6. If $c(B)$ denotes the number of cut vertices of a simple connected graph G belonging to the block B, prove that the number of cut vertices $c(G)$ of G is given by $c(G) = 1 + \sum(c(B) - 1)$, the summation being over the blocks of G.

7.7. Show that a simple connected graph with at least three vertices is a path if and only if it has exactly two vertices that are not cut vertices.

7.8. Prove that if a graph G is k-connected or k-edge connected, then $m \geq \frac{nk}{2}$.

7.9. Construct a graph with $\kappa = 3$, $\lambda = 4$, and $\delta = 5$.

7.10. For any three positive integers a, b, c, with $a \leq b \leq c$, construct a simple graph with $\kappa = a$, $\lambda = b$, and $\delta = c$.

7.11. Let G be a cubic graph with a 1-*factor* (i.e., a 1-regular spanning subgraph) F of G. Prove that any cut edge of G belongs to F.

7.12. Let G be a k-connected graph and let S be a separating set of G^2 such that $G^2 - S$ has q components. Show that $|S| \geq qk$.

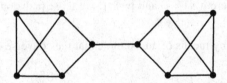

7.13. Find all the edge cuts of the above graph.

7.14. Let G be a 2-connected graph and let $v_1, v_2 \in V(G)$. Let n_1 and n_2 be positive integers with $n = n_1 + n_2$. Show that there exists a partition of V into $V_1 \cup V_2$ with $|V_i| = n_i$, $G[V_i]$ connected, and $v_i \in V_i$ for each $i = 1, 2$. (Remark: The generalization of this result to k-connected graphs is also true [55].)

Notes

Chronologically, Menger's theorem appeared first [140]. Then followed Whitney's generalizations [193] of Menger's theorem. Our proof of Menger's theorem is based on the max-flow min-cut theorem of Ford and Fulkerson [65, 66].

Chapter 4
Trees

4.1 Introduction

"Trees" form an important class of graphs. Of late, their importance has grown considerably in view of their wide applicability in theoretical computer science.

In this chapter, we present the basic structural properties of trees, their centers and centroids. In addition, we present two interesting consequences of the Tutte–Nash–Williams theorem on the existence of k pairwise edge-disjoint spanning trees in a simple connected graph. We also present Cayley's formula for the number of spanning trees in the labeled complete graph K_n. As applications, we present Kruskal's algorithm and Prim's algorithm, which determine a minimum-weight spanning tree in a connected weighted graph and discuss Dijkstra's algorithm, which determines a minimum-weight shortest path between two specified vertices of a connected weighted graph.

4.2 Definition, Characterization, and Simple Properties

Certain graphs derive their names from their diagrams. A "tree" is one such graph. Formally, a connected graph without cycles is defined as a *tree*. A graph without cycles is called an *acyclic graph* or a *forest*. So each component of a forest is a tree. A forest may consist of just a single tree! Figure 4.1 displays two pairs of isomorphic trees.

Remarks 4.2.1. 1. It follows from the definition that a forest (and hence a tree) is a simple graph.
2. A subgraph of a tree is a forest and a connected subgraph of a tree T is a *subtree* of T.

R. Balakrishnan and K. Ranganathan, *A Textbook of Graph Theory*,
Universitext, DOI 10.1007/978-1-4614-4529-6_4,
© Springer Science+Business Media New York 2012

Fig. 4.1 Examples of
isomorphic trees

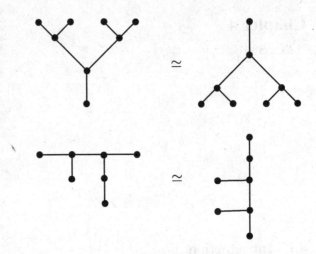

In a connected graph, any two distinct vertices are connected by at least one path. Trees are precisely those simple connected graphs in which every pair of distinct vertices is joined by a unique path.

Theorem 4.2.2. *A simple graph is a tree if and only if any two distinct vertices are connected by a unique path.*

Proof. Let T be a tree. Suppose that two distinct vertices u and v are connected by two distinct u-v paths. Then their union contains a cycle (cf. Exercise 5.9, Chap. 1) in T, contradicting that T is a tree.

Conversely, suppose that any two vertices of a graph G are connected by a unique path. Then G is obviously connected. Also, G cannot contain a cycle, since any two distinct vertices of a cycle are connected by two distinct paths. Hence G is a tree. □

A spanning subgraph of a graph G, which is also a tree, is called a *spanning tree* of G. A connected graph G and two of its spanning trees T_1 and T_2 are shown in Fig. 4.2.

The graph G of Fig. 4.2 shows that a graph may contain more than one spanning tree; each of the trees T_1 and T_2 is a spanning tree of G.

A loop cannot be an edge of any spanning tree, since such a loop constitutes a cycle (of length 1). On the other hand, a cut edge of G must be an edge of every spanning tree of G. Theorem 4.2.3 shows that every connected graph contains a spanning tree.

Theorem 4.2.3. *Every connected graph contains a spanning tree.*

Proof. Let G be a connected graph. Let $\mathscr{C}$ be the collection of all connected spanning subgraphs of G. $\mathscr{C}$ is nonempty as $G \in \mathscr{C}$. Let $T \in \mathscr{C}$ have the fewest number of edges. Then T must be a spanning tree of G. If not, T would contain a cycle of G, and the deletion of any edge of this cycle would give a (spanning)

Fig. 4.2 Graph G and two of
its spanning trees T_1 and T_2

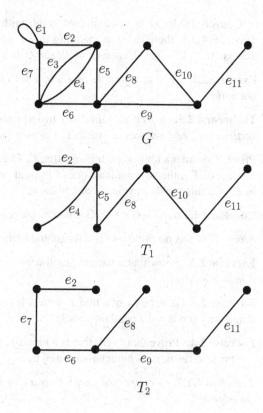

subgraph in $\mathscr{C}$ having one edge less than that of T. This contradicts the choice of T.
Hence, T has no cycles and is therefore a spanning tree of G.   □

There is a nice relation between the number of vertices and the number of edges
of any tree.

Theorem 4.2.4. *The number of edges in a tree on n vertices is $n - 1$. Conversely, a
connected graph on n vertices and $n - 1$ edges is a tree.*

Proof. Let T be a tree. We use induction on n to prove that $m = n - 1$. When $n = 1$
or $n = 2$, the result is straightforward.

Now assume that the result is true for all trees on $(n - 1)$ or fewer vertices,
$n \geq 3$. Let T be a tree with n vertices. Let $e = uv$ be an edge of T. Then uv is
the unique path in T joining u and v. Hence the deletion of e from T results in a
disconnected graph having two components T_1 and T_2. Being connected subgraphs
of a tree, T_1 and T_2 are themselves trees. As $n(T_1)$ and $n(T_2)$ are less than $n(T)$, by
an induction hypothesis, $m(T_1) = n(T_1) - 1$ and $m(T_2) = n(T_2) - 1$. Therefore,
$m(T) = m(T_1) + m(T_2) + 1 = n(T_1) - 1 + n(T_2) - 1 + 1 = n(T_1) + n(T_2) - 1 =
n(T) - 1$. Hence, the result is true for T. By induction, the result follows in one
direction.

Conversely, let G be a connected graph with n vertices and $n - 1$ edges. By Theorem 4.2.3, there exists a spanning tree T of G. T has n vertices and being a tree has $(n - 1)$ edges. Hence $G = T$, and G is a tree.   □

Exercise 2.1. Give an example of a graph with n vertices and $n - 1$ edges that is not a tree.

Theorem 4.2.5. *A tree with at least two vertices contains at least two pendant vertices (i.e., end vertices or vertices of degree 1).*

Proof. Consider a longest path P of a tree T. The end vertices of P must be pendant vertices of T; otherwise, at least one of the end vertices of P has a second neighbor in P, and this yields a cycle, a contradiction.   □

Corollary 4.2.6. *If $\delta(G) \geq 2$, G contains a cycle.*

Proof. If G has no cycles, G is a forest and hence $\delta(G) \leq 1$ by Theorem 4.2.5.   □

Exercise 2.2. Show that a simple graph with ω components is a forest if and only if $m = n - \omega$.

Exercise 2.3. A vertex v of a tree T with at least three vertices is a cut vertex of T if and only if v is not a pendant vertex.

Exercise 2.4. Prove that every tree is a bipartite graph.

Our next result is a characterization of trees.

Theorem 4.2.7. *A connected graph G is a tree if and only if every edge of G is a cut edge of G.*

Proof. If G is a tree, there are no cycles in G. Hence, no edge of G can belong to a cycle. By Theorem 3.2.7, each edge of G is a cut edge of G. Conversely, if every edge of a connected graph G is a cut edge of G, then G cannot contain a cycle, since no edge of a cycle is a cut edge of G. Hence, G is a tree.   □

Theorem 4.2.8. *A connected graph G with at least two vertices is a tree if and only if its degree sequence $(d_1, d_2, \ldots, d_n)$ satisfies the condition: $\sum_{i=1}^{n} d_i = 2(n - 1)$ with $d_i > 0$ for each i.*

Proof. Let G be a tree. As G is connected and nontrivial, it can have no isolated vertex. Hence every term of the degree sequence of G is positive. Further, by Theorem 1.4.4, $\sum_{i=1}^{n} d_i = 2m = 2(n - 1)$.

Conversely, assume that the condition $\sum_{i=1}^{n} d_i = 2(n - 1)$ holds. This implies that $m = n - 1$ as $\sum_{i=1}^{n} d_i = 2m$. Now apply Theorem 4.2.4.   □

Lemma 4.2.9. *If u and v are nonadjacent vertices of a tree T, then $T + uv$ contains a unique cycle.*

Proof. If P is the unique u-v path in T, then $P + uv$ is a cycle in $T + uv$. It is unique, as the path P is unique in T.   □

Example 4.2.10. Prove that if $m(G) = n(G)$ for a simple connected graph G, then G is unicyclic, that is, a graph containing exactly one cycle.

Proof. By Theorem 4.2.3, G contains a spanning tree T. As T has $n(G) - 1$ edges, $E(G) \backslash E(T)$ consists of a single edge e. Then $G = T \cup e$ is unicyclic.   □

Exercise 2.5. If for a simple graph G, $m(G) \geq n(G)$, prove that G contains a cycle.

Exercise 2.6. Prove that every edge of a connected graph G that is not a loop is in some spanning tree of G.

Exercise 2.7. Prove that the following statements are equivalent:

  (i) G is connected and unicyclic (i.e., G has exactly one cycle).
 (ii) G is connected and $n = m$.
(iii) For some edge e of G, $G - e$ is a tree.
 (iv) G is connected and the set of edges of G that are not cut edges forms a cycle.

Example 4.2.11. Prove that for a simple connected graph G, $L(G)$ is isomorphic to G if and only if G is a cycle.

Proof. If G is a cycle, then clearly $L(G)$ is isomorphic to G. Conversely, let $G \simeq L(G)$. Then $n(G) = n(L(G))$, and $m(G) = m(L(G))$. But since $n(L(G)) = m(G)$, we have $m(G) = n(G)$. By Example 4.2.10, G is unicyclic. Let $C = v_1 v_2 \ldots v_k v_1$ be the unique cycle in G. If $G \neq C$, there must be an edge $e \notin E(C)$ incident with some vertex v_i of C (as G is connected). Thus, there is a star with at least three edges at v_i. This star induces a clique of size at least 3 in $L(G)$ ($\simeq G$). This shows that there exists at least one more cycle in $L(G)$ distinct from the cycle corresponding to C in G. This contradicts the fact that $L(G) \simeq G$ (as G is unicyclic).   □

4.3 Centers and Centroids

There are certain parameters attached to any connected graph. These are defined below.

Definitions 4.3.1. Let G be a connected graph.

1. The *diameter* of G is defined as $\max\{d(u, v) : u, v \in V(G)\}$ and is denoted by diam(G).
2. If v is a vertex of G, its *eccentricity* $e(v)$ is defined by $e(v) = \max\{d(v, u) : u \in V(G)\}$.
3. The *radius* $r(G)$ of G is the minimum eccentricity of G; that is, $r(G) = \min\{e(v) : v \in V(G)\}$. Note that diam$(G) = \max\{e(v) : v \in V(G)\}$.
4. A vertex v of G is called a *central vertex* if $e(v) = r(G)$. The set of central vertices of G is called the *center* of G.

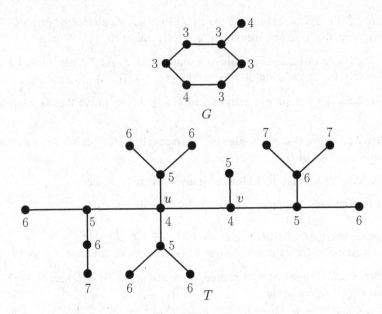

Fig. 4.3 Eccentricities of vertices for graphs G and T

Example 4.3.2. Figure 4.3 displays two graphs T and G with the eccentricities of their vertices. We find that $r(T) = 4$ and $\text{diam}(T) = 7$. Each of u and v is a central vertex of T. Also, $r(G) = 3$ and $\text{diam}(G) = 4$. Further, G has five central vertices.

Remark 4.3.3. It is obvious that $r(G) \leq \text{diam}(G)$. For a complete graph, $r(G) = \text{diam}(G) = 1$. For a complete bipartite graph $G(X, Y)$ with $|X| \geq 2$ and $|Y| \geq 2$, $r(G) = \text{diam}(G) = 2$. For the graphs of Fig. 4.3, $r(G) < \text{diam}(G)$. The terms "radius" and "diameter" tempt one to expect that $\text{diam}(G) = 2r(G)$. But this need not be the case as the complete graphs and the graphs of Fig. 4.3 show. In a tree, for any vertex u, $d(u, v)$ is maximum only when v is a pendant vertex. We use this observation in the proof of Theorem 4.3.4.

Theorem 4.3.4 (Jordan [117]). *Every tree has a center consisting of either a single vertex or two adjacent vertices.*

Proof. The result is obvious for the trees K_1 and K_2. The vertices of K_1 and K_2 are central vertices. Now let T be a tree with $n(T) \geq 3$. Then T has at least two pendant vertices (cf. Theorem 4.2.5). Clearly, the pendant vertices of T cannot be central vertices. Delete all the pendant vertices from T. This results in a subtree T' of T. As any maximum-distance path in T from any vertex of T' ends at a pendant vertex of T, the eccentricity of each vertex of T' is one less than the eccentricity of the same vertex in T. Hence the vertices of minimum eccentricity of T' are the same as those of T. In other words, T and T' have the same center. Now, if T'' is the tree obtained from T' by deleting all the pendant vertices of T', then T'' and T' have the same center. Hence the centers of T'' and T are the same. Repeat the

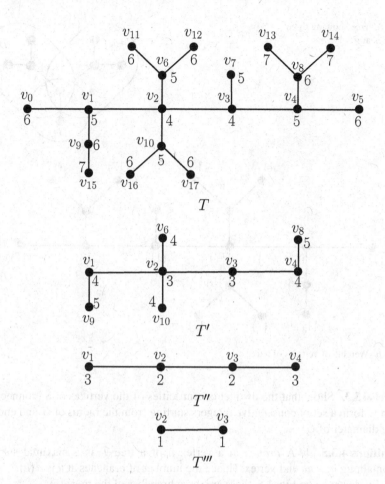

Fig. 4.4 Determining the center of tree T

process of deleting the pendant vertices in the successive subtrees of T until there results a K_1 or K_2. This will always be the case as T is finite. Hence the center of T is either a single vertex or a pair of adjacent vertices. □

The process of determining the center described above is illustrated in Fig. 4.4 for the tree T of Fig. 4.3. We observe that the center of T consists of the pair of adjacent vertices v_2 and v_3.

Exercise 3.1. Construct a tree with 85 vertices that has $\Delta = 5$ and the center consisting of a single vertex.

Exercise 3.2. Show that an automorphism of a tree on an odd number (≥ 3) of vertices has a fixed vertex; that is, for any automorphism f of a tree T with $n = 2k + 1$ ($k \geq 1$) vertices, there exists a vertex v of T with $f(v) = v$. (Hint: Use the fact that f permutes the end vertices of T.)

Fig. 4.5 Tree showing three
branches at u

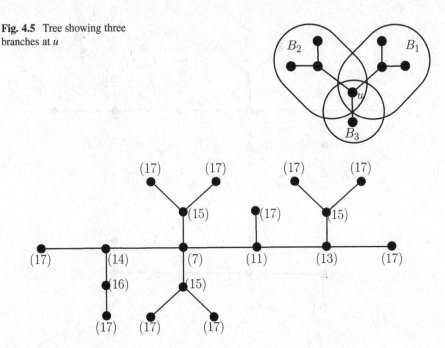

Fig. 4.6 Weights of vertices of a tree

Exercise 3.3. Show that the distinct eccentricities of the vertices of a (connected) graph G form a set of consecutive integers starting from the radius of G and ending in the diameter of G.

Definitions 4.3.5. 1. A *branch* at a vertex u of a tree T is a maximal subtree containing u as an end vertex. Hence the number of branches at u is $d(u)$.

For instance, in Fig. 4.5, there are three branches of the tree at u.

2. The *weight* of a vertex u of T is the maximum number of edges in any branch at u.

3. A vertex v is a *centroid vertex* of T if v has minimum weight. The set of all centroid vertices is called the *centroid* of T.

In Fig. 4.6 the numbers in the parentheses indicate the weights of the corresponding vertices. It is clear that all the end vertices of T have the same weight, namely, $m(T)$.

As in the case of centers, any tree has a centroid consisting of either two adjacent vertices or a single vertex. But there is no relation between the center and centroid of a tree either with regard to the number of vertices or with regard to their location.

Exercise 3.4. Give an example of

(i) A tree with just one central vertex that is also a centroidal vertex;
(ii) A tree with two central vertices, one of which is also a centroidal vertex;
(iii) A tree with two centroidal vertices, one of which is also a central vertex;

(iv) A tree with two central vertices, both of which are also centroidal vertices; and

 (v) A tree with a disjoint center and centroid.

Exercise 3.5. Show that the radius of a tree T is equal to $\left\lceil \dfrac{\mathrm{diam}(T)}{2} \right\rceil$.

Exercise 3.6. Show that in a tree, any path of maximum length contains the center of the tree.

Exercise 3.7. Show that the center of a tree consists of two adjacent vertices if and only if its diameter is even.

4.4 Counting the Number of Spanning Trees

Counting the number of spanning trees in a graph occurs as a natural problem in many branches of science. Spanning trees were used by Kirchoff to generate a "cycle basis" for the cycles in the graphs of electrical networks. In this section, we consider the enumeration of spanning trees in graphs.

The number of spanning trees of a connected labeled graph G will be denoted by $\tau(G)$. If G is disconnected, we take $\tau(G) = 0$. There is a recursive formula for $\tau(G)$. Before we establish this formula, we shall define the concept of *edge contraction* in graphs.

Definition 4.4.1. An edge e of a graph G is said to be *contracted* if it is deleted from G and its ends are identified. The resulting graph is denoted by $G \circ e$.

Edge contraction is illustrated in Fig. 4.7.

If e is not a loop of G, then $n(G \circ e) = n(G) - 1$, $m(G \circ e) = m(G) - 1$, and $\omega(G \circ e) = \omega(G)$. For a loop e, $n(G \circ e) = n(G)$, $m(G \circ e) = m(G) - 1$, and $\omega(G \circ e) = \omega(G)$. Theorem 4.4.2 gives a recursive formula for $\tau(G)$.

Theorem 4.4.2. *If e is not a loop of a connected graph G, $\tau(G) = \tau(G - e) + \tau(G \circ e)$.*

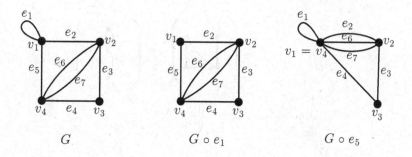

Fig. 4.7 Edge contraction

Proof. $\tau(G)$ is the sum of the number of spanning trees of G containing e and the number of spanning trees of G not containing e.

Since $V(G - e) = V(G)$, every spanning tree of $G - e$ is a spanning tree of G not containing e, and conversely, any spanning tree of G for which e is not an edge is also a spanning tree of $G - e$. Hence the number of spanning trees of G not containing e is precisely the number of spanning trees of $G - e$, that is, $\tau(G - e)$. If T is a spanning tree of G containing e, the contraction of e in both T and G results in a spanning tree $T \circ e$ of $G \circ e$.

Conversely, if T_0 is a spanning tree of $G \circ e$, there exists a unique spanning tree T of G containing e such that $T \circ e = T_0$. Thus, the number of spanning trees of G containing e is $\tau(G \circ e)$. Hence $\tau(G) = \tau(G - e) + \tau(G \circ e)$. $\qquad\square$

We illustrate below the use of Theorem 4.4.2 in calculating the number of spanning trees. In this illustration, each graph within parentheses stands for the number of its spanning trees. For example, $\boxed{}$ stands for the number of spanning trees of C_4.

Example 4.4.3. Find $\tau(G)$ for the following graph G:

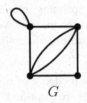

$$G$$

Proof.

$$\left(\boxed{}_e \right) = \left(\boxed{} \right) + \left(\begin{array}{c} \text{\Large 8} \end{array} \right)$$

$$= \left(\boxed{}_{e'} \right) + \left(\begin{array}{c} \text{\Large 8} \end{array} \right)$$

$$= \left\{ \left(\underset{e''}{\square} \right) + \left(\begin{matrix} 8 \end{matrix} \right) \right\} + \left(\begin{matrix} 8 \end{matrix} \right)$$

$$= \left(\sqcup \right) + \left(\triangle \right) + 2 \left(\begin{matrix} 8 \end{matrix} \right)$$

$$= 1 + 3 + 2(4)$$

$$= 12.$$

[By enumeration,

$$\left(\sqcup \right) = 1, \quad \left(\triangle \right) = 3, \quad \text{and} \quad \left(\begin{matrix} 8 \end{matrix} \right) = 4.$$

]

Hence $\tau(G) = 12$.   □

We have seen in Sect. 3.2 that every connected graph has a spanning tree. When will it have k edge-disjoint spanning trees? An answer to this interesting question was given by both Tutte [181] and Nash-Williams [145] at just about the same time.

Theorem 4.4.4 (Tutte [181]; Nash-Williams [145]). *A simple connected graph G contains k pairwise edge-disjoint spanning trees if and only if for each partition $\mathscr{P}$ of $V(G)$ into p parts, the number $m(\mathscr{P})$ of edges of G joining distinct parts is at least $k(p-1)$, $2 \le p \le |V(G)|$.*

Proof. We prove only the easier part of the theorem (necessity of the condition). Suppose G has k pairwise edge-disjoint spanning trees. If T is one of them and if $\mathscr{P} = \{V_1, \ldots, V_p\}$ is a partition of $V(G)$ into p parts, then G must have at least $|\mathscr{P}| - 1$ edges of T. As this is true for each of the k pairwise edge-disjoint trees of G, the number of edges joining distinct parts of $\mathscr{P}$ is at least $k(p-1)$.   □

For the proof of the converse part of the theorem, we refer the reader to the references cited.

As a consequence of Theorem 4.4.4, we obtain immediately at least one family of graphs that possesses the property stated in the theorem.

Corollary 4.4.5. *Every 2k-edge-connected (k ≥ 1) graph contains k pairwise edge-disjoint spanning trees.*

Proof. Let G be $2k$-edge connected, and let $\mathscr{P} = \{V_1, \ldots, V_p\}$ be a partition of V into p subsets. By hypothesis on G, there are at least $2k$ edges from each part V_i to $V \setminus V_i = \bigcup_{\substack{j=1 \\ j \neq i}}^{p} V_j$. The total number of such edges is at least kp (as each such edge is counted twice). Hence, $m(\mathscr{P}) \geq kp > k(p-1)$. Theorem 4.4.4 now ensures that there are at least k pairwise edge-disjoint spanning trees in G.   □

Setting $k = 2$ in the above corollary, we get the result of Kundu.

Corollary 4.4.6 (Kundu [128]). *Every 4-edge-connected graph contains two edge-disjoint spanning trees.*

Corollary 4.4.7. *Every 3-edge-connected graph G has three spanning trees whose intersection is a spanning totally disconnected subgraph of G.*

Proof. Let G be a 3-edge-connected graph. Duplicate each edge of G by a parallel edge. The resulting graph, say, G', is 6-edge connected, and hence by Corollary 4.4.5, G' has three pairwise edge-disjoint spanning trees, say, T_1', T_2', and T_3'. Hence $E(T_1' \cap T_2' \cap T_3') = \phi$. Let T_i, $1 \leq i \leq 3$, be the tree obtained from T_i' by replacing any parallel edge of G' by its original edge in G. Then, clearly, T_1, T_2, and T_3 are three spanning trees of G with $E(T_1 \cap T_2 \cap T_3) = \phi$ because neither an edge of G nor its parallel edge can belong to all of T_1', T_2', and T_3'.   □

4.5   Cayley's Formula

Cayley was the first mathematician to obtain a formula for the number of spanning trees of a labeled complete graph.

Theorem 4.5.1 (Cayley [33]). $\tau(K_n) = n^{n-2}$, *where K_n is a labeled complete graph on n vertices, $n \geq 2$.*

Before we prove Theorem 4.5.1, we establish two lemmas.

Lemma 4.5.2. *Let $(d_1, \ldots, d_n)$ be a sequence of positive integers with $\sum_{i=1}^{n} d_i = 2(n-1)$. Then there exists a tree T with vertex set $\{v_1, \ldots, v_n\}$ and $d(v_i) = d_i$, $1 \leq i \leq n$.*

Proof. It is easy to prove the result by induction on n.   □

Lemma 4.5.3. *Let $\{v_1, \ldots, v_n\}$, $n \geq 2$ be given and let $\{d_1, \ldots, d_n\}$ be a sequence of positive integers such that $\sum_{i=1}^{n} d_i = 2(n-1)$. Then the number of trees with $\{v_1, \ldots, v_n\}$ as the vertex set in which v_i has degree d_i, $1 \leq i \leq n$, is $\frac{(n-2)!}{(d_1-1)! \ldots (d_n-1)!}$.*

Proof. We prove the result by induction on n. For $n = 2$, $2(n - 1) = 2$, so that $d_1 + d_2 = 2$. Since $d_1 \geq 1$ and $d_2 \geq 1$, $d_1 = d_2 = 1$. Hence K_2 is the only tree in which v_i has degree d_i, $i = 1, 2$. So the result is true for $n = 2$. Now assume that the result is true for all positive integers up to $n - 1$, $n \geq 3$. Let $\{d_1, \ldots, d_n\}$ be a sequence of positive integers such that $\sum_{i=1}^{n} d_i = 2(n - 1)$, and let $\{v_1, \ldots, v_n\}$ be any set. If $d_i \geq 2$ for every i, $1 \leq i \leq n$, then $\sum_{i=1}^{n} d_i \geq 2n$. Hence, there exists an i, $1 \leq i \leq n$, for which $d_i = 1$. For the sake of definiteness, assume that $d_n = 1$. By Lemma 4.5.2, there exists a tree T with $V(T) = \{v_1, \ldots, v_n\}$ and degree of $v_i = d_i$. Let v_j be the unique vertex of T adjacent to v_n. Delete v_n from T. The resulting graph is a tree T' with $\{v_1, \ldots, v_{n-1}\}$ as its vertex set and $(d_1, \ldots, d_{j-1}, d_j - 1, d_{j+1}, \ldots, d_{n-1})$ as its degree sequence.

In the opposite direction, given a tree T' with $\{v_1, \ldots, v_{n-1}\}$ as its vertex set and $(d_1, \ldots, d_{j-1}, d_j - 1, d_{j+1}, \ldots, d_{n-1})$ as its degree sequence, a tree T with vertex set $\{v_1, \ldots, v_n\}$ and degree sequence $(d_1, \ldots, d_n)$, $d_n = 1$, can be obtained by introducing a new vertex v_n and taking $T = T' + v_j v_n$. Hence the number of trees with vertex set $\{v_1, \ldots, v_n\}$ and degree sequence $(d_1, \ldots, d_n)$ with $d_n = $ degree of $v_n = 1$ and v_n adjacent to v_j is the same as the number of trees with vertex set $\{v_1, \ldots, v_{n-1}\}$ and degree sequence $(d_1, \ldots, d_{j-1}, d_j - 1, d_{j+1}, \ldots, d_{n-1})$. By the induction hypothesis, the latter number is equal to

$$\frac{(n - 3)!}{(d_1 - 1)! \ldots (d_{j-1} - 1)! (d_j - 2)! (d_{j+1} - 1)! \ldots (d_{n-1} - 1)!}$$

$$= \frac{(n - 3)!(d_j - 1)}{(d_1 - 1)! \ldots (d_{j-1} - 1)! (d_j - 1)! (d_{j+1} - 1)! \ldots (d_{n-1} - 1)!}.$$

Summing over j, the number of trees with $\{v_1, \ldots, v_n\}$ as its vertex set and $(d_1, \ldots, d_n)$ as its degree sequence is

$$\sum_{j=1}^{n-1} \frac{(n - 3)! (d_j - 1)}{(d_1 - 1)! \ldots (d_{n-1} - 1)!}$$

$$= \frac{(n - 3)!}{(d_1 - 1)! \ldots (d_{n-1} - 1)!} \sum_{j=1}^{n-1} (d_j - 1)$$

$$= \frac{(n - 3)!}{(d_1 - 1)! \ldots (d_{n-1} - 1)!} \left[\left(\sum_{j=1}^{n-1} d_j \right) - (n - 1) \right]$$

$$= \frac{(n - 3)!}{(d_1 - 1)! \ldots (d_{n-1} - 1)!} [(2n - 3) - (n - 1)]$$

$$= \frac{(n - 3)!}{(d_1 - 1)! \ldots (d_{n-1} - 1)!} (n - 2)$$

$$= \frac{(n-2)!}{(d_1-1)! \ldots (d_{n-1}-1)!}$$

$$= \frac{(n-2)!}{(d_1-1)! \ldots (d_n-1)!} \text{ (recall that } d_n = 1).$$

This completes the proof of Lemma 4.5.2. $\qquad \square$

Proof of theorem 4.5.1. The total number of trees T_n with vertex set $\{v_1, \ldots, v_n\}$ is obtained by summing over all possible sequences $(d_1, \ldots, d_n)$ with $\sum_{i=1}^{n} d_i = 2n - 2$. Hence,

$$\tau(K_n) = \sum_{d_i \geq 1} \frac{(n-2)!}{(d_1-1)! \ldots (d_n-1)!} \text{ with } \sum_{i=1}^{n} d_i = 2n - 2$$

$$= \sum_{k_i \geq 0} \frac{(n-2)!}{k_1! \ldots k_n!} \text{ with } \sum_{i=1}^{n} k_i = n - 2, \text{ where } k_i = d_i - 1, 1 \leq i \leq n.$$

Putting $x_1 = x_2 = \cdots = x_n = 1$ and $m = n - 2$ in the multinomial expansion

$$(x_1 + x_2 + \cdots + x_n)^m = \sum_{k_i \geq 0} \frac{x_1^{k_1} x_2^{k_2} \ldots x_n^{k_n}}{k_1! k_2! \ldots k_n!} m! \text{ with } (k_1 + k_2 + \cdots + k_n) = m,$$

we get $n^{n-2} = \sum_{k_i \geq 0} \frac{(n-2)!}{k_1! k_2! \ldots, k_n!}$ with $(k_1 + k_2 + \cdots + k_n) = n - 2$. Thus,

$\tau(K_n) = n^{n-2}$. $\qquad \square$

4.6 Helly Property

Definitions 4.6.1. A family $\{A_i : i \in I\}$ of subsets of a set A is said to satisfy the *Helly property* if, whenever $J \subseteq I$ and $A_i \cap A_j \neq \phi$ for every $i, j \in J$, then $\bigcap_{j \in J} A_j \neq \phi$.

Theorem 4.6.2. *Any family of subtrees of a tree satisfies the Helly property.*

Proof. Let $\mathscr{F} = \{T_i : i \in I\}$ be a family of subtrees of a tree T. Suppose that for all $i, j \in J \subseteq I, T_i \cap T_j \neq \phi$. We have to prove that $\bigcap_{j \in J} T_j \neq \phi$. If for some $i \in J$, tree T_i is a single-vertex tree $\{v\}$ (i.e., K_1), then, clearly, $\bigcap_{j \in J} T_j = \{v\}$. We therefore suppose that each tree $T_i \in \mathscr{F}$ with $i \in J$ has at least two vertices.

We now apply induction on the number of vertices of T. Let the result be true for all trees with at most n vertices, and let T be a tree with $(n + 1)$ vertices. Let v_0 be an end vertex of T, and u_0 its unique neighbor in T. Let $T_i' = T_i - v_0, i \in J$, and $T' = T - v_0$. (If $v_0 \notin T_i$, we take $T_i' = T_i$.) By the induction assumption, the result is true for the tree T'. Moreover, $T_i' \cap T_j' \neq \phi$ for any $i, j \in J$. In fact, if T_i and T_j have a vertex u ($\neq v_0$) in common, then T_i' and T_j' also have u in common, whereas if T_i and T_j have v_0 in common, then T_i and T_j have u_0 also in common and so do T_i' and T_j'. Hence, by the induction assumption, $\bigcap_{j \in J} T_j' \neq \phi$, and therefore $\bigcap_{j \in J} T_j \neq \phi$. □

Exercise 6.1. In the cycle C_5, give a family of five paths such that the intersection of the vertex sets of any two of them is nonempty while the intersection of the vertex sets of all of them is empty.

Exercise 6.2. Prove that a connected graph G is a tree if and only if every family of paths in G satisfies the Helly property.

4.7 Applications

We conclude this chapter by presenting some immediate applications of trees in everyday life problems.

4.7.1 The Connector Problem

Problems 1. Various cities in a country are to be linked via roads. Given the various possibilities of connecting the cities and the costs involved, what is the most economical way of laying roads so that in the resulting road network, any two cities are connected by a chain of roads? Similar problems involve designing railroad networks and water-line transports.

Problems 2. A layout for a housing settlement in a city is to be prepared. Various locations of the settlement are to be linked by roads. Given the various possibilities of linking the locations and their costs, what is the minimum-cost layout so that any two locations are connected by a chain of roads?

Problems 3. A layout for the electrical wiring of a building is to be prepared. Given the costs of the various possibilities, what is the minimum-cost layout?

These three problems are particular cases of a graph-theoretical problem known as the *connector problem*.

Definition 4.7.1. Let G be a graph. To each edge e of G, we associate a nonnegative number $w(e)$ called its *weight*. The resulting graph is a *weighted graph*. If H is a subgraph of G, the sum of the weights of the edges of H is called the *weight* of H. In particular, the sum of the weights of the edges of a path is called the *weight of the path*.

We shall now concentrate on Problem 1. Problems 2 and 3 can be dealt with similarly. Let G be a graph constructed with the set of cities as its vertex set. An edge of G corresponds to a road link between two cities. The cost of constructing a road link is the weight of its corresponding edge. Then a minimum-weight spanning tree of G provides the most economical layout for the road network.

We present two algorithms, Kruskal's algorithm and Prim's algorithm, for determining a minimum-weight spanning tree in a connected weighted graph. We can assume, without loss of generality, that the graph is simple because, since no loop can be an edge of a spanning tree, we can discard all loops. Also, since we are interested in determining a minimum-weight spanning tree, we can retain, from a set of multiple edges having the same ends, an edge with the minimum weight, and we can discard all the others.

First, we describe Kruskal's algorithm [127].

4.7.2   Kruskal's Algorithm

Let G be a simple connected weighted graph with edge set $E = \{e_1, \ldots, e_m\}$. The three steps of the algorithm are as follows:

Step 1 :   Choose an edge e_1 with its weight $w(e_1)$ as small as possible.

Step 2 :   If the edges $e_1, e_2, \ldots, e_i, i \geq 1$, have already been chosen, choose e_{i+1} from the set $E \backslash \{e_1, e_2, \ldots, e_i\}$ such that

   (i) The subgraph induced by the edge set $\{e_1, e_2, \ldots, e_{i+1}\}$ is acyclic, and

   (ii) $w(e_{i+1})$ is as small as possible subject to (i).

Step 3 :   Stop when step 2 cannot be implemented further.

We now show that Kruskal's algorithm does indeed produce a minimum-weight spanning tree.

Theorem 4.7.2. *Any spanning tree produced by Kruskal's algorithm is a minimum-weight spanning tree.*

Proof. Let G be a simple connected graph of order n with edge set $E(G) = \{e_1, \ldots, e_m\}$. Let T^* be a spanning tree produced by Kruskal's algorithm and let $E(T^*) = \{e_1, \ldots, e_{n-1}\}$. For any spanning tree T of G, let $f(T)$ be the least value of i such that $e_i \notin E(T)$. Suppose T^* is not of minimum weight. Let T_0 be any minimum-weight spanning tree with $f(T_0)$ as large as possible.

Suppose $f(T_0) = k$. This means that $e_1, \ldots, e_{k-1}$ are in both T_0 and T^*, but $e_k \notin T_0$. Then $T_0 + e_k$ contains a unique cycle C. Since not every edge of C can be in T^*, C must contain an edge e_k' not belonging to T^*. Let $T_0' = T_0 + e_k - e_k'$. Then T_0' is another spanning tree of G. Moreover,

$$w(T_0') = w(T_0) + w(e_k) - w(e_k'). \tag{4.1}$$

Now, in Kruskal's algorithm, e_k was chosen as an edge with the smaller weight such that $G[\{e_1, \ldots, e_{k-1}, e_k\}]$ was acyclic. Since $G[\{e_1, \ldots, e_{k-1}, e_k'\}]$ is a subgraph of the tree T_0, it is also acyclic. Hence,

$$w(e_k) \leq w(e_k'), \tag{4.2}$$

Table 4.1 Mileage between Indian cities

	Mumbai	Hyderabad	Nagpur	Calcutta	New Delhi	Chennai
Mumbai (M)						
Hyderabad (H)	385					
Nagpur (N)	425	255				
Calcutta (Ca)	1035	740	679			
New Delhi (D)	708	773	531	816		
Chennai (Ch)	644	329		860	1095	

and therefore from (4.1) and (4.2),

$$w(T_0') = w(T_0) + w(e_k) - w(e_k')$$
$$\leq w(T_0).$$

But T_0 is of minimum weight. Hence, $w(T_0') = w(T_0)$, and so T_0' is also of minimum weight. However, as $\{e_1, \ldots, e_k\} \subset E(T_0')$,

$$f(T_0') > k = f(T_0),$$

contradicting the choice of T_0. Thus, T^* is a minimum-weight spanning tree of G. $\qquad \square$

When the graph is not weighted, we can give the weight 1 to each of its edges and then apply the algorithm. The algorithm then gives an acyclic subgraph with as many edges as possible, that is, a spanning tree of G.

Illustration The distances in miles between some of the Indian cities connected by air are given in Table 4.1.

Determine a minimum-cost operational system so that every city is connected to every other city. Assume that the cost of operation is directly proportional to the distance.

Let G be a graph with the set of cities as its vertex set. An edge corresponds to a pair of cities for which the ticketed mileage is indicated. The ticketed mileage is the weight of the corresponding edge (see Fig. 4.8).

The required operation system demands a minimum-cost spanning tree of G. We shall apply Kruskal's algorithm and determine such a system. The following is a sequence of edges selected according to the algorithm.

$$HN, HCh, HM, ND, NCa.$$

The corresponding spanning tree is shown in bold lines, and its weight is $255 + 329 + 385 + 531 + 679 = 2,179$.

We next describe Prim's algorithm [159].

HN, HCh, HM, ND, NCa.

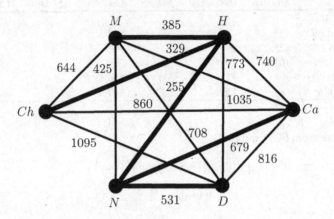

Fig. 4.8 Graph of mileage between cities

4.7.3 Prim's Algorithm

Let G be a simple connected weighted graph having n vertices. Let the vertices of G be labeled as $v_1, v_2, \ldots, v_n$. Let $W = W(G) = (w_{ij})$ be the weight matrix of G. That is, W is the $n \times n$ matrix with

(i) $w_{ii} = \infty$, for $1 \le i \le n$,
(ii) $w_{ij} = w_{ji} =$ the weight of the edge (v_i, v_j) if v_i and v_j are adjacent,
(iii) $w_{ij} = w_{ji} = \infty$ if v_i and v_j are nonadjacent.

The algorithm constructs a minimum-cost spanning tree.

Step 1: Start with v_1. Connect v_1 to v_k, where v_k is a nearest vertex to v_1 (v_k is nearest to v_1 if $v_1 v_k$ is an edge with minimum possible weight). The vertex v_k could be easily determined by observing the matrix W. Actually, v_k is a vertex corresponding to which the entry in row 1 of W is minimum.

Step 2: Having chosen v_k, let $v_i \ne v_1$ or v_k be a vertex corresponding to the smallest entry in rows 1 and k put together. Then v_i is the vertex "nearest" the edge subgraph defined by the edge $v_1 v_k$. Connect v_i to v_1 or v_k, according to whether the entry is in the first row or kth row. Suppose it is, say, in the kth row; then it is the (k, i)th entry of W.

Step 3: Consider the edge subgraph defined by the edge set $\{v_1 v_k, v_k v_i\}$. Determine the nearest neighbor to the set of vertices $\{v_1, v_k, v_i\}$.

Step 4: Continue the process until all the n vertices have been connected by $(n-1)$ edges. This results in a minimum-cost spanning tree.

Proof of correctness: Let T be a tree obtained by applying Prim's algorithm. We want to show that T is a minimum-weight spanning tree (that is, an optimal tree) of G. We prove by induction on $n = |V(G)|$. Suppose $e = v_1 v_2$ is an edge of least weight incident at v_1.

Claim. There exists a minimum-weight spanning tree of G that contains e. To see this, consider an optimal tree T' of G. Suppose T' does not contain e. As T' contains the vertex v_1, T' must contain some edge f of G incident at v_1. By Prim's algorithm, $w(e) \leq w(e')$ for every edge e' of G incident at v_1 and consequently, $w(e) \leq w(f)$, where w denotes the weight function. Hence, the spanning tree $T'' = T' + e - f$ of G has the property that $w(T'') = w(T') + w(e) - w(f) \leq w(T')$.

As T' is optimal, T'' is also optimal. But T'' contains e. This establishes our claim.

Let $G' = G \circ e$, the contraction of G obtained by contracting the edge e. Every spanning tree of G that contains e gives rise to a unique spanning tree of G'. Conversely, every spanning tree of G' gives rise to a unique spanning tree of G containing e.

Let S denote the set of vertices of the tree T_p (with $e \in E(T_p)$) grown by Prim's algorithm at the end of p steps, $p \geq 2$, and S' denote the set of vertices of $T'_p = T_p \circ e$. Then $[S, V(G) \backslash S] = [S', V(G') \backslash S']$. Therefore, an edge of minimum weight in $[S, V(G) \backslash S]$ is also an edge of minimum weight in $[S', V(G') \backslash S']$. As the final tree T is a Prim tree of G, the final tree $T \circ e$ of G' is a Prim tree of G'. But then G' has one vertex less than that of G and so $T \circ e$ is an optimal tree of $G \circ e$. Consequently, T is an optimal tree of G. $\qquad\square$

Illustration Consider the weighted graph G shown in Fig. 4.8. The weight matrix W of G is

$$
W : \quad
\begin{array}{c}
\\ M \\ H \\ N \\ Ca \\ D \\ Ch
\end{array}
\begin{array}{c}
\begin{array}{cccccc}
M & H & N & Ca & D & Ch
\end{array} \\
\left[
\begin{array}{cccccc}
\infty & 385 & 425 & 1035 & 708 & 644 \\
385 & \infty & 255 & 740 & 773 & 329 \\
425 & 255 & \infty & 679 & 531 & \infty \\
1035 & 740 & 679 & \infty & 816 & 860 \\
708 & 773 & 531 & 816 & \infty & 1095 \\
644 & 329 & \infty & 860 & 1095 & \infty
\end{array}
\right]
\end{array}
$$

In row M (i.e., in the row corresponding to the city M, namely, Mumbai), the smallest weight is 385, which occurs in column H. Hence join M and H. Now, after omitting columns M and H, 255 is the minimum weight in the rows M and H put together. It occurs in row H and column N. Hence, join H and N. Now, omitting columns M, H, and N, the smallest number in the rows M, H, and N put together is 329, and it occurs in row H and column Ch, so join H and Ch. Again, the smallest entry in rows M, H, N and Ch not belonging to the corresponding columns is 531, and it occurs in row N and column D. So join N and D. Now, the lowest entry in rows M, H, N, D and Ch not belonging to the corresponding columns is 679, and it occurs in row N and column Ca. So join N and Ca. This construction gives the same minimum-weight spanning tree of Fig. 4.8.

Remark. In each iteration of Prim's algorithm, a subtree of a minimum-weight spanning tree is obtained, whereas in any step of Kruskal's algorithm, just a subgraph of a minimum-weight spanning tree is constructed.

4.7.4   Shortest-Path Problems

A manufacturing concern has a warehouse at location X and the market for the product at another location Y. Given the various routes of transporting the product from X to Y and the cost of operating them, what is the most economical way of transporting the materials? This problem can be tackled using graph theory. All such optimization problems come under a type of graph-theoretic problem known as "shortest-path problems." Three types of shortest-path problems are well known:
    Let G be a connected weighted graph.

1. Determine a shortest path, that is, a minimum-weight path between two specified vertices of G.
2. Determine a set of shortest paths between all pairs of vertices of G.
3. Determine a set of shortest paths from a specified vertex to all other vertices of G.

We consider only the first problem. The other two problems are similar. We describe Dijkstra's algorithm [52] for determining the shortest path between two specified vertices. Once again, it is clear that in shortest-path problems, we could restrict ourselves to simple connected weighted graphs.

4.7.5   Dijkstra's Algorithm

Let G be a simple connected weighted graph having vertices $v_1, v_2, \ldots, v_n$. Let s and t be two specified vertices of G. We want to determine a shortest path from s to t. Let W be the weight matrix of G. Dijkstra's algorithm allots weights to the vertices of G. At each stage of the algorithm, some vertices have permanent weights and others have temporary weights.

To start with, the vertex s is allotted the permanent weight 0 and all other vertices the temporary weight ∞. In each iteration of the algorithm, one new vertex is allotted a permanent weight by the following rules:

Rule 1: If v_j is a vertex that has not yet been allotted a permanent weight, determine for each vertex v_i that had already been allotted a permanent weight,

$$\alpha_{ij} = \min\{\text{old weight of } v_j, (\text{old weight of } v_i) + w_{ij}\}.$$

Let $w_j = \text{Min}_i \, \alpha_{ij}$. Then w_j is a new temporary weight of v_j not exceeding the previous temporary weight.

Fig. 4.9 Graph of mileage
between cities

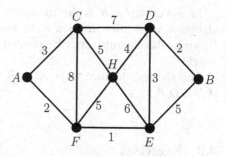

Table 4.2 Steps of algorithm for shortest path from A to B

	A	B	C	D	E	F	H
Iteration 0	⟦0⟧	∞	∞	∞	∞	∞	∞
Iteration 1	0	∞	[3]	∞	∞	[2]	∞
Iteration 2	0	∞	⟦3⟧	∞	[3]	⟦2⟧	[7]
Iteration 3	0	∞	⟦3⟧	[10]	⟦3⟧	⟦2⟧	[7]
Iteration 4	0	[8]	⟦3⟧	⟦6⟧	⟦3⟧	⟦2⟧	[7]
Iteration 5	0	[8]	⟦3⟧	⟦6⟧	⟦3⟧	⟦2⟧	⟦7⟧
Iteration 6	0	⟦8⟧	⟦3⟧	⟦6⟧	⟦3⟧	⟦2⟧	⟦7⟧

Rule 2: Determine the smallest among the w_j's. If this smallest weight is at v_k, w_k becomes the permanent weight of v_k. In case there is a tie, any one vertex is taken for allotting a permanent weight.

The algorithm stops when the vertex t gets a permanent weight.

It is clear from the algorithm that the permanent weight of each vertex is the shortest weighted distance from s to that vertex. The shortest path from s to t is constructed by working backward from the terminal vertex t. Let P be a shortest path and p_i a vertex of P. The weight of p_{i-1}, the vertex immediately preceding p_i on P, is such that the weight of the edge $p_{i-1}p_i$ equals the difference in permanent weights of p_i and p_{i-1}.

We present below an illustrative example. Let us find the shortest path from A to B in the graph of Fig. 4.9.

We present the various iterations of the algorithm by arrays of weights of the vertices, one consisting of weights before iteration and another after it. Temporary weights will be enclosed in squares and the permanent weights enclosed in double squares. The steps of the algorithm for determining the shortest path from vertex A to vertex B in graph G of Fig. 4.9 are given in Table 4.2.

In our example, B is the last vertex to get a permanent weight. Hence the algorithm stops after Iteration 6, in which B is allotted the permanent weight. However, the algorithm may be stopped as soon as vertex B gets the permanent weight.

The shortest distance from A to B is 8. A shortest path with weight 8 is $A F E D B$.

4.8 Exercises

8.1. Show that any tree of order n contains a subtree of order k for every $k \le n$.

8.2. Let u, v, w be any three vertices of a tree T. Show that either u, v, w all lie in a path of T or else there exists a unique vertex z of T which is common to the u-v, v-w, w-u paths of T.

8.3. Show that in a tree, the number of vertices of degree at least 3 is at most the number of end vertices minus 2.

8.4. Show that if G is a connected graph with at least three vertices, then G contains two vertices u and v such that $G - \{u, v\}$ is also connected.

8.5. * If H is a graph of minimum degree at least $k-1$, then prove that H contains every tree on k vertices. (Hint: Prove by induction on k.) (See [88].)

8.6. Prove that a nontrivial simple graph G is a tree if and only if for any set of r distinct vertices in G, $r \ge 2$, the minimum number of edges required to separate them is $r - 1$. (See E. Sampathkumar [168].)

8.7. Show that a simple connected graph contains at least $m - n + 1$ distinct cycles.

8.8. Prove that for a connected graph G, $r(G) \le diam(G) \le 2r(G)$. (The graphs of Fig. 4.3 show that the inequalities can be strict.)

8.9. Prove that a tree with at least three vertices has diameter 2 if and only if it is a star.

8.10. Determine the number of spanning trees of the two graphs in Fig. 4.10:

8.11. If T is a tree with at least two vertices, show that there exists a set of edge-disjoint paths covering all the vertices of T such that each of these paths has at least one end vertex that is an end vertex of T.

8.12. Let T be a tree of order n with $V(T) = \{1, 2, \ldots, n\}$, and let A be a set of transpositions defined by $A = \{(i, j) : ij \in E(T)\}$. Show that A is a minimal set of transpositions that generates the symmetric group S_n.

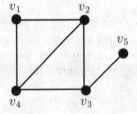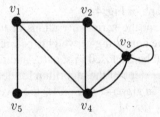

Fig. 4.10

Fig. 4.11

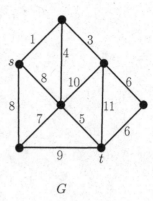

G

8.13. For the graph G of Fig. 4.11, determine two distinct minimum-weight spanning trees using

(i) Kruskal's algorithm,
(ii) Prim's algorithm.

What is the weight of such a tree? Also, determine a minimum-weight s-t path using Dijkstra's algorithm.

8.14. Apply Prim's algorithm to the illustrative example given in Sect. 4.7.3 by starting from the third row of the weight matrix.

8.15. If G is a connected weighted graph in which no two edges have the same weight, show that G has a unique minimum-weight spanning tree.

8.16. Establish the correctness of Dijkstra's algorithm.

Notes

In 1847, G. R. Kirchoff (1824–1887) developed the theory of trees for their applications in electrical networks. Ten years later, in 1857, the English mathematician A. Cayley (1821–1895) rediscovered trees while he was trying to enumerate the isomers of the saturated hydrocarbons $C_n H_{2n+2}$ (see also Chap. 1). Since then "trees" have grown both vertically and horizontally. They are widely used today in computer science.

There is also a simpler (?) proof of the converse part of Theorem 4.4.4 using matroid theory (see pp. 126–127 of [191]).

Corollary 4.4.6 is due to Kilpatrick [122], but the elegant proof given here is due to Jaeger [114]. The proof of Cayley's theorem presented here (Theorem 4.5.1) is based on Moon [142], which also contains nine other proofs. The book by Serre [170] entitled *Trees* is mainly concerned with the connection between trees and the group $SL_2(Q_p)$.

For general algorithmic results related to graphs, see [3, 72, 153].

Chapter 5
Independent Sets and Matchings

5.1 Introduction

Vertex-independent sets and vertex coverings as also edge-independent sets and edge coverings of graphs occur very naturally in many practical situations and hence have several potential applications. In this chapter, we study the properties of these sets. In addition, we discuss matchings in graphs and, in particular, in bipartite graphs. Matchings in bipartite graphs have varied applications in operations research. We also present two celebrated theorems of graph theory, namely, Tutte's 1-factor theorem and Hall's matching theorem. All graphs considered in this chapter are loopless.

5.2 Vertex-Independent Sets and Vertex Coverings

Definition 5.2.1. A subset S of the vertex set V of a graph G is called *independent* if no two vertices of S are adjacent in G. $S \subseteq V$ is a *maximum independent set* of G if G has no independent set S' with $|S'| > |S|$. A *maximal independent set* of G is an independent set that is not a proper subset of another independent set of G.

For example, in the graph of Fig. 5.1, $\{u, v, w\}$ is a maximum independent set and $\{x, y\}$ is a maximal independent set that is not maximum.

Definition 5.2.2. A subset K of V is called a *covering* of G if every edge of G is incident with at least one vertex of K. A covering K is *minimum* if there is no covering K' of G such that $|K'| < |K|$; it is *minimal* if there is no covering K_1 of G such that K_1 is a proper subset of K.

In the graph W_5 of Fig. 5.2, $\{v_1, v_2, v_3, v_4, v_5\}$ is a covering of W_5 and $\{v_1, v_3, v_4, v_6\}$ is a minimal covering. Also, the set $\{x, y\}$ is a minimum covering of the graph of Fig. 5.1.

R. Balakrishnan and K. Ranganathan, *A Textbook of Graph Theory*,
Universitext, DOI 10.1007/978-1-4614-4529-6_5,
© Springer Science+Business Media New York 2012

Fig. 5.1 Graph with
maximum independent set
$\{u, v, w\}$ and maximal
independent set $\{x, y\}$

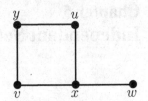

Fig. 5.2 Wheel W_5

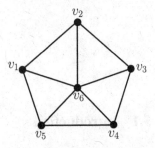

The concepts of covering and independent sets of a graph arise very naturally in practical problems. Suppose we want to store a set of chemicals in different rooms. Naturally, we would like to store incompatible chemicals, that is, chemicals that are likely to react violently when brought together, in distinct rooms. Let G be a graph whose vertex set represents the set of chemicals and let two vertices be made adjacent in G if and only if the corresponding chemicals are incompatible. Then any set of vertices representing compatible chemicals forms an independent set of G.

Now consider the graph G whose vertices represent the various locations in a factory and whose edges represent the pathways between pairs of such locations. A light source placed at a location supplies light to all the pathways incident to that location. A set of light sources that supplies light to all the pathways in the factory forms a covering of G.

Theorem 5.2.3. *A subset S of V is independent if and only if $V \setminus S$ is a covering of G.*

Proof. S is independent if and only if no two vertices in S are adjacent in G. Hence, every edge of G must be incident to a vertex of $V \setminus S$. This is the case if and only if $V \setminus S$ is a covering of G. □

Definition 5.2.4. The number of vertices in a maximum independent set of G is called the *independence number* (or the *stability number*) of G and is denoted by $\alpha(G)$. The number of vertices in a minimum covering of G is the *covering number* of G and is denoted by $\beta(G)$. We denote these numbers simply by α and β when there is no confusion.

Corollary 5.2.5. *For any graph G, $\alpha + \beta = n$.*

Proof. Let S be a maximum independent set of G. By Theorem 5.2.3, $V \backslash S$ is a covering of G and therefore $|V \backslash S| = n - \alpha \geq \beta$. Similarly, let K be a minimum covering of G. Then $V \backslash K$ is independent and so $|V \backslash K| = n - \beta \leq \alpha$. These two inequalities together imply that $n = \alpha + \beta$. $\square$

5.3 Edge-Independent Sets

Definitions 5.3.1. 1. A subset M of the edge set E of a loopless graph G is called
   independent if no two edges of M are adjacent in G.
2. A *matching* in G is a set of independent edges.
3. An *edge covering* of G is a subset L of E such that every vertex of G is incident
   to some edge of L. Hence, an edge covering of G exists if and only if $\delta > 0$.
4. A matching M of G is *maximum* if G has no matching M' with $|M'| > |M|$.
   M is *maximal* if G has no matching M' strictly containing M. $\alpha'(G)$ is the
   cardinality of a maximum matching and $\beta'(G)$ is the size of a minimum edge
   covering of G.
5. A set S of vertices of G is said to be *saturated* by a matching M of G or M-
   saturated if every vertex of S is incident to some edge of M. A vertex v of G is
   M-*saturated* if $\{v\}$ is M-saturated. v is M-*unsaturated* if it is not M-saturated.
   For example, in the wheel W_5 (Fig. 5.2), $M = \{v_1 v_2, v_4 v_6\}$ is a maximal
   matching; $\{v_1 v_5, v_2 v_3, v_4 v_6\}$ is a maximum matching and a minimum edge covering;
   the vertices v_1, v_2, v_4, and v_6 are M-saturated, whereas v_3 and v_5 are M-unsaturated.

Remark 5.3.2. The edge analog of Theorem 5.2.3 is not true, however. For instance, in the graph G of Fig. 5.3, the set $E' = \{e_3, e_4\}$ is independent, but $E \backslash E' = \{e_1, e_2, e_5\}$ is not an edge covering of G. Also, $E'' = \{e_1, e_3, e_4\}$ is an edge covering of G, but $E \backslash E''$ is not independent in G. Again, E' is a matching in G that saturates v_2, v_3, v_4 and v_5 but does not saturate v_1.

Theorem 5.3.3. *For any graph G for which $\delta > 0$, $\alpha' + \beta' = n$.*

Proof. Let M be a maximum matching in G so that $|M| = \alpha'$. Let U be the set of M-unsaturated vertices in G. Since M is maximum, U is an independent set of vertices with $|U| = n - 2\alpha'$. Since $\delta > 0$, we can pick one edge for each vertex in

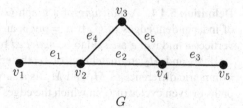

Fig. 5.3 Graph illustrating
edge relationships

G

Fig. 5.4 Herschel graph

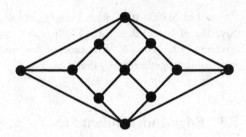

U incident with it. Let F be the set of edges thus chosen. Then $M \cup F$ is an edge covering of G. Hence, $|M \cup F| = |M| + |F| = \alpha' + n - 2\alpha' \geq \beta'$, and therefore

$$n \geq \alpha' + \beta'. \tag{5.1}$$

Now let L be a minimum edge covering of G so that $|L| = \beta'$. Let $H = G[L]$ be the edge subgraph of G defined by L, and let M_H be a maximum matching in H. Denote the set of M_H-unsaturated vertices in H by U. As L is an edge covering of G, H is a spanning subgraph of G. Consequently, $|L| - |M_H| = |L \backslash M_H| \geq |U| = n - 2|M_H|$ and so $|L| + |M_H| \geq n$. But since M_H is a matching in G, $|M_H| \leq \alpha'$. Thus,

$$n \leq |L| + |M_H| \leq \beta' + \alpha'. \tag{5.2}$$

Inequalities (5.1) and (5.2) imply that $\alpha' + \beta' = n$. □

Exercise 3.1. Determine the values of the parameters α, α', β, and β' for

1. K_n,
2. The Petersen graph P,
3. The Herschel graph (see Fig. 5.4).

Exercise 3.2. For any graph G with $\delta > 0$, prove that $\alpha \leq \beta'$ and $\alpha' \geq \beta$.

Exercise 3.3. Show that for a bipartite graph G, $\alpha \beta \geq m$ and that equality holds if and only if G is complete.

5.4 Matchings and Factors

Definition 5.4.1. A *matching* of a graph G is (as given in Definition 5.3.1) a set of independent edges of G. If $e = uv$ is an edge of a matching M of G, the end vertices u and v of e are said to be *matched* by M.

If M_1 and M_2 are matchings of G, the edge subgraph defined by $M_1 \triangle M_2$, the symmetric difference of M_1 and M_2, is a subgraph H of G whose components are paths or even cycles of G in which the edges alternate between M_1 and M_2.

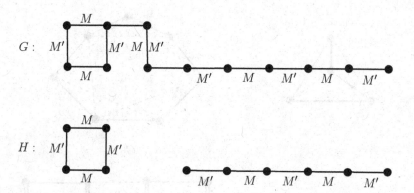

Fig. 5.5 Graphs for proof of Theorem 5.4.4

Definition 5.4.2. An *M-augmenting path* in *G* is a path in which the edges alternate between $E \setminus M$ and M and its end vertices are M-unsaturated. An *M-alternating path* in *G* is a path whose edges alternate between $E \setminus M$ and M.

Example 5.4.3. In the graph *G* of Fig. 5.2, $M_1 = \{v_1v_2, v_3v_4, v_5v_6\}$, $M_2 = \{v_1v_2, v_3v_6, v_4v_5\}$, and $M_3 = \{v_3v_4, v_5v_6\}$ are matchings of *G*. Moreover, $G[M_1 \triangle M_2]$ is the even cycle $(v_3v_4v_5v_6v_3)$. The path $v_2v_3v_4v_6v_5v_1$ is an M_3-augmenting path in *G*.

Maximum matchings have been characterized by Berge [19].

Theorem 5.4.4. *A matching M of a graph G is maximum if and only if G has no M-augmenting path.*

Proof. Assume first that *M* is maximum. If *G* has an *M*-augmenting path *P* : $v_0v_1v_2 \ldots v_{2t+1}$ in which the edges alternate between $E \setminus M$ and M, then *P* has one edge of $E \setminus M$ more than that of *M*. Define

$$M' = (M \cup \{v_0v_1, v_2v_3, \ldots, v_{2t}v_{2t+1}\}) \setminus \{v_1v_2, v_3v_4, \ldots, v_{2t-1}v_{2t}\}.$$

Clearly, M' is a matching of *G* with $|M'| = |M| + 1$, which is a contradiction since *M* is a maximum matching of *G*.

Conversely, assume that *G* has no *M*-augmenting path. Then *M* must be maximum. If not, there exists a matching M' of *G* with $|M'| > |M|$. Let *H* be the edge subgraph $G[M \triangle M']$ defined by the symmetric difference of *M* and M'. Then the components of *H* are paths or even cycles in which the edges alternate between *M* and M'. Since $|M'| > |M|$, at least one of the components of *H* must be a path starting and ending with edges of M'. But then such a path is an *M*-augmenting path of *G*, contradicting the assumption (see Fig. 5.5).   □

Definition 5.4.5. A *factor* of a graph *G* is a spanning subgraph of *G*. A *k-factor* of *G* is a factor of *G* that is *k*-regular. Thus, a 1-*factor* of *G* is a matching that saturates

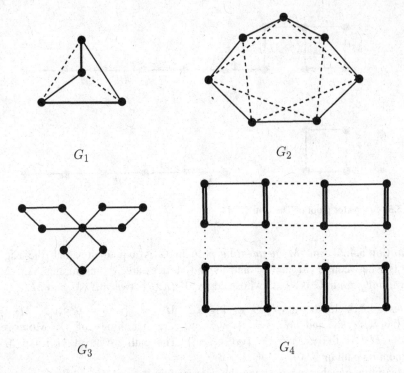

Fig. 5.6 Graphs illustrating factorability

all the vertices of G. For this reason, a 1-factor of G is called a *perfect matching* of G. A 2-factor of G is a factor of G that is a disjoint union of cycles of G. A graph G is *k-factorable* if G is an edge-disjoint union of k-factors of G.

Example 5.4.6. In Fig. 5.6, G_1 is 1-factorable and G_2 is 2-factorable, whereas G_3 has neither a 1-factor nor a 2-factor. The dotted, solid, and ordinary lines of G_1 give the three distinct 1-factors, and the dotted and ordinary lines of G_2 give its two distinct 2-factors.

Exercise 4.1. Give an example of a cubic graph having no 1-factor.

Exercise 4.2. Show that $K_{n,n}$ and K_{2n} are 1-factorable.

Exercise 4.3. Show that the number of 1-factors of

(i) $K_{n,n}$ is $n!$,
(ii) K_{2n} is $\frac{(2n)!}{2^n n!}$.

Exercise 4.4. The *n-cube* Q_n is the graph whose vertices are binary n-tuples. Two vertices of Q_n are adjacent if and only if they differ in exactly one place. Show that Q_n $(n \geq 2)$ has a perfect matching. (The 3-cube Q_3 and the 4-cube Q_4 are displayed in Fig. 5.7.) It is easy to see that $Q_n \simeq K_2 \square K_2 \square \ldots \square K_2$ (n times).

Fig. 5.7 (a) 3-cube Q_3 and
(b) 4-cube Q_4

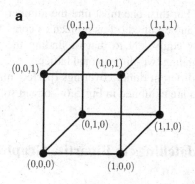

The 3-cube Q_3

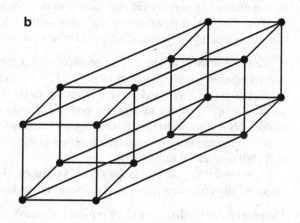

The 4-cube Q_4

Exercise 4.5. Show that the Petersen graph P is not 1-factorable. (Hint: Look at the possible types of 1-factors of P.)

Exercise 4.6. Show that every tree has at most one perfect matching.

Exercise 4.7*. Show that if a 2-edge-connected graph has a 1-factor, then it has at least two distinct 1-factors.

Exercise 4.8. Show that the graph G_4 of Fig. 5.6 is not 1-factorable.

An Application to Physics 5.4.7. In crystal physics, a crystal is represented by a three-dimensional lattice in which each face corresponds to a two-dimensional lattice. Each vertex of the lattice represents an atom of the crystal, and an edge between two vertices represents the bond between the two corresponding atoms.

In crystallography, one is interested in obtaining an analytical expression for certain surface properties of crystals consisting of diatomic molecules (also called dimers). For this, one must find the number of ways in which all the atoms of the crystal can be paired off as molecules consisting of two atoms each. The problem is clearly equivalent to that of finding the number of perfect matchings of the corresponding two-dimensional lattice.

Two different dimer coverings (perfect matchings) of the lattice defined by the graph G_4 are exhibited in Fig. 5.6—one in solid lines and the other in parallel lines.

5.5   Matchings in Bipartite Graphs

Assignment Problem 5.5.1. Suppose in a factory there are n jobs $j_1, j_2, \ldots, j_n$ and s workers $w_1, w_2, \ldots, w_s$. Also suppose that each job j_i can be performed by a certain number of workers and that each worker w_j has been trained to do a certain number of jobs. Is it possible to assign each of the n jobs to a worker who can do that job so that no two jobs are assigned to the same worker?

We convert this job assignment problem into a problem in graphs as follows: Form a bipartite graph G with bipartition (J, W), where $J = \{j_1, j_2, \ldots, j_n\}$ and $W = \{w_1, w_2, \ldots, w_s\}$, and make j_i adjacent to w_j if and only if worker w_j can do the job j_i. Then our assignment problem translates into the following graph problem: Is it possible to find a matching in G that saturates all the vertices of J?

A solution to the above matching problem in bipartite graphs has been given by Hall [90] (see also Hall, Jr. [91]).

For a subset $S \subseteq V$ in a graph G, $N(S)$ denotes the neighbor set of S, that is, the set of all vertices each of which is adjacent to at least one vertex in S.

Theorem 5.5.2 (Hall). *Let G be a bipartite graph with bipartition (X, Y). Then G has a matching that saturates all the vertices of X if and only if*

$$|N(S)| \geq |S| \tag{5.3}$$

for every subset S of X.

Proof. If G has a matching that saturates all the vertices of X, then distinct vertices of X are matched to distinct vertices of Y. Hence, trivially, $|N(S)| \geq |S|$ for every subset $S \subseteq X$.

Conversely, assume that the condition (5.3) above holds but that G has no matching that saturates all the vertices of X. Let M be a maximum matching of G. As M does not saturate all the vertices of X, there exists a vertex $x_0 \in X$ that is M-unsaturated. Let Z denote the set of all vertices of G connected to x_0 by M-alternating paths. Since M is a maximum matching, by Theorem 5.4.4, G has no M-augmenting path. As x_0 is M-unsaturated, x_0 is the only vertex of Z that is M-unsaturated. Let $A = Z \cap X$ and $B = Z \cap Y$. Then the vertices of $A \backslash \{x_0\}$ get matched under M to the vertices of B, and $N(A) = B$. Thus, since $|B| = |A| - 1$, $|N(A)| = |B| = |A| - 1 < |A|$, and this contradicts the assumption (5.3) (see Fig. 5.8).   □

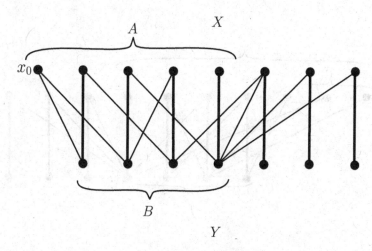

Fig. 5.8 Figure for proof of Theorem 5.5.2 (matching edges are *boldfaced*)

We now give some important consequences of Hall's theorem.

Theorem 5.5.3. *A k ($\geq$ 1)-regular bipartite graph is 1-factorable.*

Proof. Let G be k-regular with bipartition (X, Y). Then $E(G)$ = the set of edges incident to the vertices of X = the set of edges incident to the vertices of Y. Hence, $k|X| = |E(G)| = k|Y|$, and therefore $|X| = |Y|$. If $S \subseteq X$, then $N(S) \subseteq Y$, and $N(N(S))$ contains S. Let E_1 and E_2 be the sets of edges of G incident to S and $N(S)$, respectively. Then $E_1 \subseteq E_2$, $|E_1| = k|S|$, and $|E_2| = k|N(S)|$. Hence, as $|E_2| \geq |E_1|$, $|N(S)| \geq |S|$. So by Hall's theorem (Theorem 5.5.2), G has a matching that saturates all the vertices of X; that is, G has a perfect matching M. Deletion of the edges of M from G results in a $(k - 1)$-regular bipartite graph. Repeated application of the above argument shows that G is 1-factorable. $\qquad\square$

König's theorem: Consider any matching M of a graph G. If K is any (vertex) covering for the graph, then it is clear that to cover each edge of M, we have to choose at least one vertex of K. Thus, $|M| \leq |K|$. In particular, if M^* is a maximum matching and K^* is a minimum covering of G, then

$$|M^*| \leq |K^*|. \qquad (5.4)$$

König's theorem asserts that for bipartite graphs, equality holds in relation (5.4). Before we establish this theorem, we present a lemma that is interesting in its own right and is similar to Lemma 3.6.8.

Lemma 5.5.4. *Let K be any covering and M any matching of a graph G with $|K| = |M|$. Then K is a minimum covering and M is a maximum matching.*

Proof. Let M^* be a maximum matching and K^* a minimum covering of G. Then $|M| \leq |M^*|$ and $|K| \geq |K^*|$. Hence, by (5.4) we have $|M| \leq |M^*| \leq |K^*| \leq |K|$. Since $|M| = |K|$, we must have $|M| = |M^*| = |K^*| = |K|$, proving the lemma. $\qquad\square$

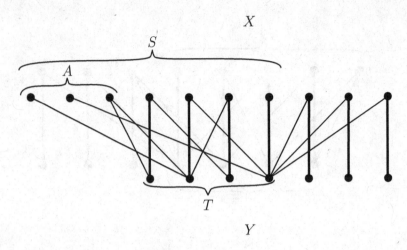

Fig. 5.9 Graph for proof of Theorem 5.5.5

Theorem 5.5.5 (König). *In a bipartite graph the minimum number of vertices that cover all the edges of G is equal to the maximum number of independent edges; that is, $\alpha'(G) = \beta(G)$.*

Proof. Let G be a bipartite graph with bipartition (X, Y). Let M be a maximum matching in G. Denote by A the set of vertices of X unsaturated by M (see Fig. 5.9). As in the proof of Theorem 5.5.2, let Z stand for the set of vertices connected to A by M-alternating paths starting in A. Let $S = X \cap Z$ and $T = Y \cap Z$. Then clearly, $T = N(S)$ and $K = T \cup (X \backslash S)$ is a covering of G, because if there is an edge e not incident to any vertex in K, then one of the end vertices of e must be in S and the other in $Y \backslash T$, contradicting the fact that $N(S) = T$. Clearly, $|K| = |M|$, and so by Lemma 5.5.4, M is a maximum matching and K a minimum covering of G. □

Let A be a binary matrix (so that each entry of A is 0 or 1). A *line* of A is a row or column of A. A line covers all of its entries. Two 1's of A are called *independent* if they do not lie in the same line of A. The matrix version of König's theorem is given in Theorem 5.5.6.

Theorem 5.5.6 (Matrix version of König's theorem). *In a binary matrix, the minimum number of lines that cover all the 1's is equal to the maximum number of independent 1's.*

Proof. Let $A = (a_{ij})$ be a binary matrix of size p by q. Form a bipartite graph G with bipartition (X, Y), where X and Y are sets of cardinality p and q, respectively, say, $X = \{v_1, v_2, \ldots, v_p\}$ and $Y = \{w_1, w_2, \ldots, w_q\}$. Make v_i adjacent to w_j in G if and only if $a_{ij} = 1$. Then an entry 1 in A corresponds to an edge of G, and two independent 1's in A correspond to two independent edges of G. Further, each vertex of G corresponds to a line of A. Thus, the matrix version of König's theorem is actually a restatement of König's theorem. □

A consequence of Theorem 5.5.2 is the theorem on the existence of a *system of distinct representatives* (SDR) for a family of subsets of a given finite set.

Definition 5.5.7. Let $\mathscr{F} = \{A_\alpha : \alpha \in J\}$ be a family of sets. An *SDR* for the family $\mathscr{F}$ is a family of elements $\{x_\alpha : \alpha \in J\}$ such that $x_\alpha \in A_\alpha$ for every $\alpha \in J$ and $x_\alpha \neq x_\beta$ whenever $\alpha \neq \beta$.

Example 5.5.8. For instance, if $A_1 = \{1\}$, $A_2 = \{2,3\}$, $A_3 = \{3,4\}$, $A_4 = \{1,2,3,4\}$, and $A_5 = \{2,3,4\}$, then the family $\{A_1, A_2, A_3, A_4\}$ has $\{1,2,3,4\}$ as an SDR, whereas the family $\{A_1, A_2, A_3, A_4, A_5\}$ has no SDR. It is clear that for $\mathscr{F}$ to have an SDR, it is necessary that for any positive integer k, the union of any k sets of $\mathscr{F}$ must contain at least k elements. That this condition is also sufficient when $\mathscr{F}$ is a finite family of finite sets is the assertion of Hall's theorem on the existence of an SDR.

Theorem 5.5.9 (Hall's theorem on the existence of an SDR [90]). *Let $\mathscr{F} = \{A_i : 1 \leq i \leq r\}$ be a family of finite sets. Then $\mathscr{F}$ has an SDR if and only if the union of any k members of $\mathscr{F}$, $1 \leq k \leq r$, contains at least k elements.*

Proof. We need only prove the sufficiency part. Let $\bigcup_{i=1}^{r} A_i = \{y_1, y_2, \ldots, y_n\}$. Form a bipartite graph $G = G(X, Y)$ with $X = \{x_1, x_2, \ldots, x_r\}$, where x_i corresponds to the set A_i, $1 \leq i \leq r$, and $Y = \{y_1, y_2, \ldots, y_n\}$. Make x_i adjacent to y_j in G if and only if $y_j \in A_i$. Then it is clear that $\mathscr{F}$ has an SDR if and only if G has a matching that saturates all the vertices of X. But this is the case, by Theorem 5.5.2, if for each $S \subseteq X$, $|N(S)| \geq |S|$, that is, if and only if $|\bigcup_{x_i \in S} A_i| \geq |S|$, which is precisely the condition stated in the theorem. $\qquad\square$

Exercise 5.1. Prove Theorem 5.5.5 (König's theorem) assuming Theorem 5.5.9.

Exercise 5.2. Show that a bipartite graph has a 1-factor if and only if $|N(S)| \geq |S|$ for every subset S of V. Does this hold for any graph G?

When does a graph have a 1-factor? Tutte's celebrated 1-*factor theorem* answers this question. The proof given here is due to Lovász [135]. A component of a graph is *odd* or *even* according to whether it has an odd or even number of vertices. Let $O(G)$ denote the number of odd components of G.

Theorem 5.5.10 (Tutte's 1-factor theorem [179]). *A graph G has a 1-factor if and only if*

$$O(G - S) \leq |S|, \tag{5.5}$$

for all $S \subseteq V$.

Proof. While considering matchings in graphs, we are interested only in the adjacency of pairs of vertices. Hence, we may assume without loss of generality that G is simple. If G has a 1-factor M, each of the odd components of $G - S$ must have at least one vertex, which is to be matched only to a vertex of S under M. Hence, for each odd component of $G - S$, there exists an edge of the matching with one end in S. Hence, the number of vertices in S should be at least as large as the number of odd components in $G - S$; that is, $O(G - S) \leq |S|$.

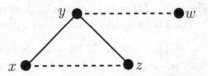

Fig. 5.10 Supergraph G^* for proof of Theorem 5.5.10. *Unbroken lines* correspond to edges of G^* and *broken lines* correspond to edges not belonging to G^*

Conversely, assume that condition (5.5) holds. If G has no 1-factor, we join pairs of nonadjacent vertices of G until we get a maximal supergraph G^* of G with G^* having no 1-factor. Condition (5.5) holds clearly for G^* as

$$O(G^* - S) \leq O(G - S). \tag{5.6}$$

(When two odd components are joined by an edge, the result is an even component.)

Taking $S = \phi$ in (5.5), we see that $O(G) = 0$, and so $n(G^*)\,(= n(G)) = n$ is even. Further, for every pair of nonadjacent vertices u and v of G^*, $G^* + uv$ has a 1-factor, and any such 1-factor must necessarily contain the edge uv.

Let K be the set of vertices of G^* of degree $(n - 1)$. $K \neq V$, since otherwise $G^* = K_n$ has a perfect matching. We claim that each component of $G^* - K$ is complete. Suppose to the contrary that some component G_1 of $G^* - K$ is not complete. Then in G_1 there are vertices x, y and z such that $xy \in E(G^*)$, $yz \in E(G^*)$, but xz does not belong to $E(G^*)$ (Exercise 5.11 of Chap. 1). Moreover, since $y \in V(G_1)$, $d_{G^*}(y) < n - 1$ and hence there exists a vertex w of G^* with $yw \notin E(G^*)$. Necessarily, w does not belong to K. (See Fig. 5.10.)

By the choice of G^*, each of $G^* + xz$ and $G^* + yw$ has a 1-factor, say M_1 and $M2$, respectively. Necessarily, $xz \in M_1$ and $yw \in M_2$. Let H be the subgraph of $G^* + \{xz, yw\}$ induced by the edges in the symmetric difference $M_1 \triangle M_2$ of M_1 and M_2. Since M_1 and M_2 are 1-factors, each vertex of G^* is saturated by both M_1 and M_2, and H is a disjoint union of even cycles in which the edges alternate between M_1 and M_2. There arise two cases:

Case 1. xz and yw belong to different components of H (Fig. 5.11a). If yw belongs to the even cycle C, then the edges of M_1 in C together with the edges of M_2 not belonging to C form a 1-factor in G^*, contradicting the choice of G^*.

Case 2. xz and yw belong to the same component C of H. Since each component of H is a cycle, C is a cycle (Fig. 5.11b). By the symmetry of x and z, we may suppose that the vertices x, y, w, and z occur in that order on C. Then the edges of M_1 belonging to the $yw \ldots z$ section of C together with the edge yz and the edges of M_2 not in the $yw \ldots z$ section of C form a 1-factor of G^*, again contradicting the choice of G^*. Thus, each component of $G^* - K$ is complete.

By condition (5.6), $O(G^* - K) \leq |K|$. Hence, a vertex of each of the odd components of $G^* - K$ is matched to a vertex of K. (This is possible since each

Fig. 5.11 1-factors M_1 and M_2 for (**a**) case 1 and (**b**) case 2 in proof of Theorem 5.5.10. *Ordinary lines* correspond to edges of M_1 and *bold lines* correspond to edges of M_2

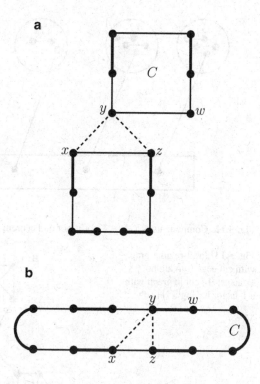

vertex of K is adjacent to every other vertex of G^*.) Also, the remaining vertices in each of the odd and even components of $G^* - K$ can be matched among themselves (see Fig. 5.12). The total number of vertices thus matched is even. Since $|V(G^*)|$ is even, the remaining vertices, if any, of K can be matched among themselves. This gives a 1-factor of G^*. Note that if $K = \emptyset$, $O(G^*) = 0$, and the existence of a 1-factor in G^* is trivially true. But by choice, G^* has no 1-factor. This contradiction proves that G has a 1-factor.   □

Corollary 5.5.11 (Petersen [158]).   *Every connected 3-regular graph having no cut edges has a 1-factor.*

Proof. Let G be a connected 3-regular graph without cut edges. Let $S \subseteq V$. Denote by $G_1, G_2, \ldots, G_k$ the odd components of $G - S$. Let m_i be the number of edges of G having one end in $V(G_i)$ and the other end in S. Since G is a cubic graph,

$$\sum_{v \in V(G_i)} d(v) = 3n(G_i), \tag{5.7}$$

and

$$\sum_{v \in S} d(v) = 3|S|. \tag{5.8}$$

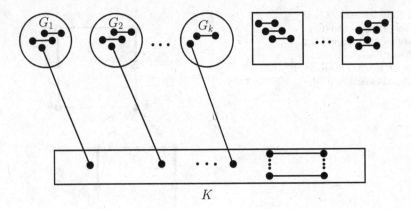

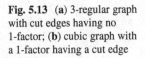

K

Fig. 5.12 Components of $G^* - K$ for proof of Theorem 5.5.10

Fig. 5.13 (a) 3-regular graph
with cut edges having no
1-factor; (b) cubic graph with
a 1-factor having a cut edge

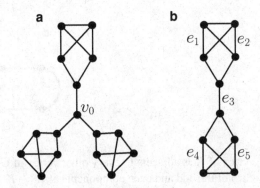

Now $E(G_i) = [V(G_i), V(G_i) \cup S] \setminus [V(G_i), S]$, where $[A, B]$ denotes the set of
edges having one end in A and the other end in B, $A \subseteq V$, $B \subseteq V$. Hence, $m_i =$
$|[V(G_i), S]| = \sum_{v \in V(G_i)} d(v) - 2m(G_i)$, and since $d(v)$ is 3 for each v and $V(G_i)$
is an odd component, m_i is odd for each i. Further, as G has no cut edges, $m_i \geq 3$.
Thus, $O(G - S) = k \leq \frac{1}{3} \sum_{i=1}^{k} m_i \leq \frac{1}{3} \sum_{v \in S} d(v) = \frac{1}{3} 3 |S| = |S|$. Therefore,
by Tutte's theorem (Theorem 5.5.10), G has a 1-factor.   □

Example 5.5.12. A 3-regular graph with cut edges may not have a 1-factor (see
Fig. 5.13a). Again, a cubic graph with a 1-factor may have cut edges (see Fig. 5.13b).

   In Fig. 5.13a, if $S = \{v_0\}$, $O(G - S) = 3 > 1 = |S|$, and so G has no 1-factor.
In Fig. 5.13b, $\{e_1, e_2, e_3, e_4, e_5\}$ is a 1-factor, and e_3 is a cut edge of G.

   If G has no 1-factor, by Theorem 5.5.10 there exists $S \subset V(G)$ with $O(G - S) >$
$|S|$. Such a set S is called an *antifactor* set of G; clearly, S is a proper subset of
$V(G)$.

   Let G be a graph of even order n and let S be an antifactor set of G. Then $|S|$
and $O(G - S)$ have the same parity, and therefore $O(G - S) \equiv |S| \pmod 2$. Thus,
we make the following observation.

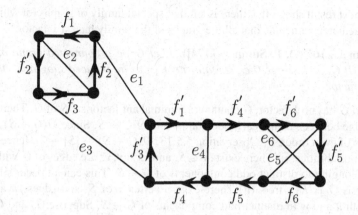

Fig. 5.14 Figure for the proof of Corollary 5.5.14

Observation 5.5.13. If S is an antifactor set of a graph G of even order, then $O(G - S) \geq |S| + 2$.

Corollary 5.5.14 (W. H. Cunnigham; see [119]). *The edge set of a simple 2-edge-connected cubic graph G can be partitioned into paths of length 3.*

Proof. By Corollary 5.5.11, G is a union of a 1-factor and a 2-factor. Orient the edges of each cycle of the above 2-factor in any manner so that each cycle becomes a directed cycle. Then if e_1 is any edge of the 1-factor, and f_1, f_1' are the two arcs of G having their tails at the end vertices of e_1, then $\{e_1, f_1, f_1'\}$ forms a typical 3-path of the edge partition of G (see Fig. 5.14).   □

Corollary 5.5.15. *A $(p - 1)$-regular connected simple graph on $2p$ vertices has a 1-factor.*

Proof. Proof is by contradiction. Let G be a $(p - 1)$-regular connected simple graph on $2p$ vertices having no 1-factor. Then G has an antifactor set S. By Observation 5.5.13, $O(G - S) \geq |S| + 2$. Hence, $|S| + (|S| + 2) \leq 2p$, and therefore $|S| \leq p - 1$. Let $|S| = p - r$. Then $r \neq 1$ since if $r = 1$, $|S| = p - 1$, and therefore $O(G - S) = p + 1$. (Recall that G has $2p$ vertices.) Hence, each odd component of $G - S$ is a singleton, and therefore each such vertex must be adjacent to all the $p-1$ vertices of S [as G is $(p-1)$-regular]. But this means that every vertex of S is of degree at least $p + 1$, a contradiction. Thus, $|S| = p - r$, $2 \leq r \leq p - 1$. If G' is any component of $G - S$ and $v \in V(G')$, then v can be adjacent to at most $|S|$ vertices of S. Therefore, as G is $(p - 1)$-regular, v must be adjacent to at least $(p-1)-(p-r) = r-1$ vertices of G'. Thus, $|V(G')| \geq r$. Counting the vertices of all the odd components of $G-S$ and the vertices of S, we get $(|S|+2)r+|S| \leq 2p$, or $(p - r + 2)r + (p - r) \leq 2p$. This gives $(r - 1)(r - p) \geq 0$, violating the condition on r.   □

Our next result shows that there is another special family of graphs for which we can immediately conclude that all the graphs of the family have a 1-factor.

Theorem 5.5.16* (D. P. Sumner [174]). *Let G be a connected graph of even order n. If G is claw-free (i.e., contains no $K_{1,3}$ as an induced subgraph), then G has a 1-factor.*

Proof. If G has no 1-factor, G contains a minimal antifactor set S of G. There must be an edge between S and each odd component of $G - S$. Since $O(G - S) > |S|$ and G is of even order, by Observation 5.5.13, $O(G - S) \geq |S| + 2$. Hence, there are two possibilities: (i) There exists $v \in S$, and vx, vy, vz are edges of G with x, y and z belonging to distinct odd components of $G - S$. This cannot occur since by hypothesis G is $K_{1,3}$-free. (ii) There exist a vertex v of S, and edges vu and vw of G with u and w in distinct odd components of $G - S$. Suppose G_u and G_w are the odd components containing u and w, respectively. Then $< G_u \cup G_w \cup \{v\} >$ is an odd component of $G - S_1$, where $S_1 = S - \{v\}$. Further, $O(G - S_1) = O(G - S) - 1 > |S| - 1 = |S_1|$, and hence S_1 is an antifactor set of G with $|S_1| = |S| - 1$, a contradiction to the choice of S. Thus, G must have a 1-factor. [Note that by Observation 5.5.13, the case $|S| = 1$ and $O(G-S) = 2$ cannot arise.] □

Exercise 5.3. Find a 1-factorization of (i) Q_3, (ii) Q_4.

Exercise 5.4. Prove that Q_n, $n \geq 2$, is 1-factorable.

Exercise 5.5. Display a 2-factorization of K_9.

Exercise 5.6. Show that a k-regular $(k-1)$-edge-connected graph of even order has a 1-factor. (This result of F. Babler generalizes Petersen's result (Corollary 5.5.11) and can be shown by imitating the proof of Corollary 5.5.11).

Exercise 5.7. If G is a k-connected graph of even order having no $K_{1,k+1}$ as an induced subgraph, show that G has a 1-factor.

Exercise 5.8. Show that if G is a connected graph of even order, then G^2 has a 1-factor.

Exercise 5.9. (A square matrix $A = (a_{ij})$ is called *doubly stochastic* if $a_{ij} \geq 0$ for each i and j, and the sum of the entries in each row and column of A is 1.) Let $A = (a_{ij})$ be a doubly stochastic matrix of order n. Let $G = G(X, Y)$ be the bipartite graph with $|X| = |Y| = n$ obtained by setting $x_i x_j \in E(G)$ if and only if $a_{ij} = 0$. Prove that G has a perfect matching. (Hint: Apply Hall's theorem.)

5.6* Perfect Matchings and the Tutte Matrix

It has been established by Tutte that the existence of a perfect matching in a simple graph is related to the nonsingularity of a certain square matrix. This matrix is called the "Tutte matrix" of the graph. We now define the Tutte matrix.

Definition 5.6.1. Let $G = (V, E)$ be a simple graph of order n and let $V = \{v_1, v_2, \ldots, v_n\}$. Let $\{x_{ij} : 1 \le i < j \le n\}$ be a set of indeterminates. Then the *Tutte matrix* of G is defined to be the n by n matrix $T = (t_{ij})$, where

$$
t_{ij} = \begin{cases} x_{ij} & \text{if } v_i v_j \in E(G) \text{ and } i < j \\ -x_{ji} & \text{if } v_i v_j \in E(G) \text{ and } i > j \\ 0 & \text{otherwise} \end{cases}
$$

Thus, T is a skew-symmetric matrix of order n.

Example 5.6.2. For example, if G is the graph

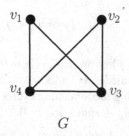

$$G$$

then

$$
T = \begin{bmatrix} t_{11} & t_{12} & t_{13} & t_{14} \\ t_{21} & t_{22} & t_{23} & t_{24} \\ t_{31} & t_{32} & t_{33} & t_{34} \\ t_{41} & t_{42} & t_{43} & t_{44} \end{bmatrix} = \begin{bmatrix} 0 & 0 & x_{13} & x_{14} \\ 0 & 0 & x_{23} & x_{24} \\ -x_{13} & -x_{23} & 0 & x_{34} \\ -x_{14} & -x_{24} & -x_{34} & 0 \end{bmatrix}. \tag{5.9}
$$

Now, by the definition of a determinant of a square matrix, the determinant of $T(= \det T)$ is given by $\det T = \sum_{\pi \in S_n} \operatorname{sgn}(\pi) \, t_{1\pi(1)} \, t_{2\pi(2)} \cdots t_{n\pi(n)}$, where $\pi \in S_n$ (i.e., π is a permutation on $\{1, 2, \ldots, n\}$), and $\operatorname{sgn}(\pi) = 1$ or -1, according to whether π is an even or odd permutation. We denote the expression $t_{1\pi(1)} t_{2\pi(2)} \cdots t_{n\pi(n)}$ by t_π. Hence, $\det T = \sum_{\pi \in S_n} \operatorname{sgn}(\pi) t_\pi$. Further, if n is odd, say, $\pi = (123)$, then $t_\pi = t_{12} t_{23} t_{31} = x_{12} x_{23} (-x_{13})$ [Note: We take $x_{ij} = 0$ if $v_i v_j \notin E(G)$). Also, $\pi^{-1} = (321)$ and $t_{\pi^{-1}} = t_{13} t_{21} t_{32} = x_{13}(-x_{12})(-x_{23})$, so that $t_\pi + t_{\pi^{-1}} = 0$. It is clear that the same relation is true for any odd $n \ge 3$.]

Now, for the Tutte matrix of relation (5.9), we have

$$
\det T = x_{13}^2 x_{24}^2 + x_{14}^2 x_{23}^2 - 2 x_{13} x_{24} x_{14} x_{23}.
$$

In this expression, the term $x_{13}^2 x_{24}^2$ is obtained by choosing the entries $x_{13}, x_{24}, -x_{13} = x_{31}$, and $-x_{24} = x_{42}$ of T, and hence it corresponds to the 1-factor $\{v_1 v_3, v_2 v_4\}$. Similarly, the term $x_{14}^2 x_{23}^2$ corresponds to the 1-factor $\{v_1 v_4, v_2 v_3\}$, and the term $x_{13} x_{24} x_{14} x_{23}$ corresponds to the cycle $(v_1 v_3 v_2 v_4)$ consisting of the edges $v_1 v_3, v_3 v_2, v_2 v_4$, and $v_4 v_1$.

We are now ready to prove Tutte's theorem, but before doing so, we make two useful observations.

Observation 5.6.3. If $\pi \in S_n$ is a product of disjoint even cycles, then $\text{sgn}(\pi)t_\pi =$ is a product of squares of the form x_{ij}^2.

Indeed, in this case, n is even, and the edges of G corresponding to the alternate transpositions in all of the even cycles of π form a 1-factor of G. [For example, for the even cycle (1234), we take the alternate transpositions (12) and (34).] Further, if v_i, v_j $(i < j)$ is an edge of this 1-factor, the partial product $t_{ij}t_{ji} = -x_{ij}^2$ occurs in t_π. The number of such products is $\frac{n}{2}$, and therefore

$$\text{sgn}(\pi)t_\pi = (-1)^{\frac{n}{2}}(-1)^{\frac{n}{2}} \prod x_{ij}^2 = \prod x_{ij}^2,$$

where the product runs over all pairs (i, j) with $i < j$ such that v_i, v_j is an edge of the 1-factor corresponding to π.

Observation 5.6.4. If $\pi \in S_n$ has an odd cycle α in its decomposition into the product of disjoint cycles, consider $\pi_1 \in S_n$, where π_1 is obtained from π by replacing α by α^{-1} and retaining the remaining cycles in π. Then, from our earlier remarks, it is clear that $\text{sgn}(\pi)t_\pi + \text{sgn}(\pi_1)t_{\pi_1} = 0$.

Theorem 5.6.5 (W. T. Tutte). *A simple graph G has a 1-factor if and only if its Tutte matrix is invertible.*

Proof. Let G be a simple graph having T as its Tutte matrix. Suppose that $\det T \neq 0$. Then by Observation 5.6.4 and the fact that $\det T = \sum_{\pi \in S_n} \text{sgn}(\pi) t_{1\pi(1)} t_{2\pi(2)} \cdots t_{n\pi(n)}$, there exists a $\pi \in S_n$ containing no odd cycle in its cycle decomposition. Then π is a product of even cycles and, by Observation 5.6.3, $\text{sgn}(\pi)t_\pi = \prod x_{ij}^2$. The alternate transpositions of the even cycles of π then yield a 1-factor of G.

Conversely, assume that G has a 1-factor. Let $\pi \in S_n$ be the product of those transpositions corresponding to the 1-factor of G. [If $v_i v_j$ is an edge of the 1-factor, the corresponding transposition is (ij).] Then by Observation 5.6.3, $\text{sgn}(\pi)t_\pi = \prod x_{ij}^2$. Now set

$$x_{ij} = \begin{cases} 1 \text{ if } x_{ij}^2 \text{ appears in the product for } \text{sgn}(\pi)t_\pi \\ 0 \text{ otherwise.} \end{cases}$$

Then $\text{sgn}(\pi)t_\pi = 1$, and for these values of x_{ij}, $\text{sgn}(\sigma)t_\sigma = 0$ for any $\sigma \in S_n$, $\sigma \neq \pi$. This means that the polynomial $\det T$ is not the zero polynomial. $\square$

Remark 5.6.6. Actually, our definition of the Tutte matrix of G depends on the order of the vertices of G. That is to say, the definition of T is based on regarding G as a labeled graph. However, if T is nonsingular with regard to one labeling of G, then the Tutte matrix of G will remain nonsingular with regard to any other labeling of G. This is because if T and T' are the Tutte matrices of G with regard to two labelings of G, $T' = PTP^{-1}$, where P is a permutation matrix of order n. Hence, T is nonsingular if and only if T' is nonsingular.

Exercise 6.1. By evaluating the Tutte matrix of the following graph G, show that G has a 1-factor.

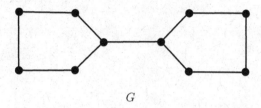

G

Notes

Readers who are more interested in matching theory can consult [91, 136], and [155]. Our proof of Tutte's 1-factor theorem is due to Lovász [135] (see also [27]).

Chapter 6
Eulerian and Hamiltonian Graphs

6.1 Introduction

The study of Eulerian graphs was initiated in the 18th century and that of Hamiltonian graphs in the 19th century. These graphs possess rich structures; hence, their study is a very fertile field of research for graph theorists. In this chapter, we present several structure theorems for these graphs.

6.2 Eulerian Graphs

Definition 6.2.1. An *Euler trail* in a graph G is a spanning trail in G that contains all the edges of G. An *Euler tour* of G is a closed Euler trail of G. G is called *Eulerian* (Fig. 6.1a) if G has an *Euler tour*. It was Euler who first considered these graphs, and hence their name.

It is clear that an Euler tour of G, if it exists, can be described from any vertex of G. Clearly, every Eulerian graph is connected.

Euler showed in 1736 that the celebrated *Königsberg bridge problem* has no solution. The city of Königsberg (now called Kaliningrad) has seven bridges linking two islands A and B and the banks C and D of the Pregel (now called Pregalya) River, as shown in Fig. 6.2.

The problem was to start from any one of the four land areas, take a stroll across the seven bridges, and get back to the starting point without crossing any bridge a second time. This problem can be converted into one concerning the graph obtained by representing each land area by a vertex and each bridge by an edge. The resulting graph H is the graph of Fig. 6.1b. The Königsberg bridge problem will have a solution provided that this graph H is Eulerian. But this is not the case since it has vertices of odd degrees (see Theorem 6.2.2).

Eulerian graphs admit, among others, the following two elegant characterizations, Theorems 6.2.2 and 6.2.3*.

R. Balakrishnan and K. Ranganathan, *A Textbook of Graph Theory*,
Universitext, DOI 10.1007/978-1-4614-4529-6_6,
© Springer Science+Business Media New York 2012

Fig. 6.1 (a) Eulerian graph
G; (b) non-Eulerian graph H

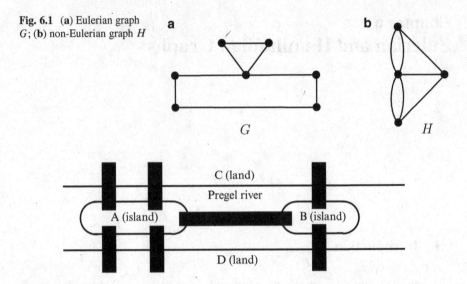

Fig. 6.2 Königsberg bridge problem

Theorem 6.2.2. *For a nontrivial connected graph* G, *the following statements are equivalent:*

(i) G *is Eulerian.*

(ii) *The degree of each vertex of* G *is an even positive integer.*

(iii) G *is an edge-disjoint union of cycles.*

Proof. (i) $\Rightarrow$ (ii): Let T be an Euler tour of G described from some vertex $v_0 \in V(G)$. If $v \in V(G)$, and $v \neq v_0$, then every time T enters v, it must move out of v to get back to v_0. Hence two edges incident with v are used during a visit to v, and therefore, $d(v)$ is even. At v_0, every time T moves out of v_0, it must get back to v_0. Consequently, $d(v_0)$ is also even. Thus, the degree of each vertex of G is even.

(ii) $\Rightarrow$ (iii): As $\delta(G) \geq 2$, G contains a cycle C_1 (Exercise 11.11 of Chap. 1). In $G \backslash E(C_1)$, remove the isolated vertices if there are any. Let the resulting subgraph of G be G_1. If G_1 is nonempty, each vertex of G_1 is again of even positive degree. Hence $\delta(G_1) \geq 2$, and so G_1 contains a cycle C_2. It follows that after a finite number, say r, of steps, $G \backslash E(C_1 \cup \ldots \cup C_r)$ is totally disconnected. Then G is the edge-disjoint union of the cycles $C_1, C_2, \ldots, C_r$.

(iii) $\Rightarrow$ (i): Assume that G is an edge-disjoint union of cycles. Since any cycle is Eulerian, G certainly contains an Eulerian subgraph. Let G_1 be a longest closed trail in G. Then G_1 must be G. If not, let $G_2 = G \backslash E(G_1)$. Since G is an edge-disjoint union of cycles, every vertex of G is of even degree ≥ 2. Further, since G_1 is Eulerian, each vertex of G_1 is of even degree ≥ 2. Hence each vertex of G_2 is of even degree. Since G_2 is not totally disconnected and G is connected, G_2 contains a cycle C having a vertex v in common with G_1. Describe the Euler tour of G_1

Fig. 6.3 Eulerian graph with
edge e belonging to three
cycles

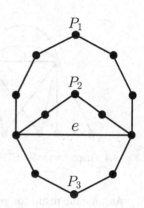

starting and ending at v and follow it by C. Then $G_1 \cup C$ is a closed trail in G
longer than G_1. This contradicts the choice of G_1, and so G_1 must be G. Hence G
is Eulerian.   □

If $G_1, \ldots, G_r$ are subgraphs of a graph G that are pairwise edge-disjoint and
their union is G, then this fact is denoted by writing $G = G_1 \oplus \ldots \oplus G_r$. In the
above equation, if $G_i = C_i$, a cycle of G for each i, then $G = C_1 \oplus \ldots C_r$. The
set of cycles $S = \{C_1, \ldots, C_r\}$ is then called a *cycle decomposition* of G. Thus,
Theorem 6.2.2 implies that *a connected graph is Eulerian if and only if it admits a
cycle decomposition.*

There is yet another characterization of Eulerian graphs due to McKee [138] and
Toida [175]. Our proof is based on Fleischner [63, 64].

Theorem 6.2.3*. *A graph G is Eulerian if and only if each edge e of G belongs to
an odd number of cycles of G.*

For instance, in Fig. 6.3, e belongs to the three cycles $P_1 \cup e$, $P_2 \cup e$, and $P_3 \cup e$.

Proof. Denote by γ_e the number of cycles of G containing e. Assume that γ_e is odd
for each edge e of G. Since a loop at any vertex v of G is in exactly one cycle of G
and contributes 2 to the degree of v in G, we may suppose that G is loopless.

Let $S = \{C_1, \ldots, C_p\}$ be the set of cycles of G. Replace each edge e of G
by γ_e parallel edges and replace e in each of the γ_e cycles containing e by one of
these parallel edges, making sure that none of the parallel edges is repeated. Let the
resulting graph be G_0 and let the new set of cycles be $S_0 = \{C_1^0, \ldots, C_p^0\}$. Clearly,
S_0 is a cycle decomposition of G_0. Hence, by Theorem 6.2.2, G_0 is Eulerian. But
then $d_{G_0}(v) \equiv 0 \pmod 2$ for each $v \in V(G_0) = V(G)$. Moreover, $d_G(v) =
d_{G_0}(v) - \sum_e (\gamma_e - 1)$, where e is incident at v in G and hence $d_G(v) \equiv 0 \pmod 2$,
γ_e being odd for each $e \in E(G)$. Thus, G is Eulerian.

Conversely, assume that G is Eulerian. We proceed by induction on $n = |V(G)|$.
If $n = 1$, each edge is a loop and hence belongs to exactly one cycle of G.

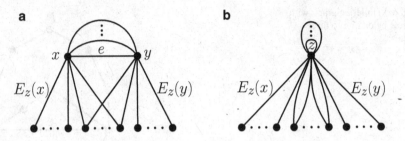

Fig. 6.4 Graph for proof of Theorem 6.2.3

Assume the result for graphs with fewer than n (≥ 2) vertices. Let G be a graph with n vertices. Let $e = xy$ be an edge of G and let $\lambda(e)$ be the multiplicity of e in G.

The graph $G \circ e$ obtained from G by contracting the edge e (cf. Sect. 4.4 of Chap. 4) is also Eulerian. Denote by z the new vertex of $G \circ e$ obtained by identifying the vertices x and y of G. The set of edges incident with z in $G \circ e$ is partitioned into three subsets (see Fig. 6.4):

1. $E_z(x)$ = set of edges arising out of edges of G incident with x but not with y
2. $E_z(y)$ = set of edges arising out of edges of G incident with y but not with x
3. $E_z(xy)$ = set of $\lambda(e) - 1$ loops of $G \circ e$ corresponding to the edges parallel to e in G

Let $k = |E_z(x)|$. Since G is Eulerian,

$$k + \lambda(e) = d_G(x) \equiv 0 (\mathrm{mod}\, 2). \qquad (6.1)$$

Let Γ_f and $\Gamma(e_i, e_j)$ denote, respectively, the number of cycles in $G \circ e$ containing the edge f and the pair (e_i, e_j) of edges. Since $|V(G \circ e)| = n - 1$, and since $G \circ e$ is Eulerian by the induction assumption, Γ_f is odd for each edge f of $G \circ e$. Now, any cycle of G containing e either consists of e and an edge parallel to e in G (and there are $\lambda(e) - 1$ of them) or contains e, an edge e_i of $E_z(x)$, and an edge e'_j of $E_z(y)$. These correspond in $G \circ e$, respectively, to a loop at z and to a cycle containing the edges of $G \circ e$ that correspond to the edges e_i and e'_j of G. By abuse of notation, we denote these corresponding edges of $G \circ e$ also by e_i and e'_j, respectively. Moreover, any cycle of $G \circ e$ containing an edge e_i of $E_z(x)$ will also contain either an edge e_j of $E_z(x)$ or an edge e'_j of $E_z(y)$, but not both. A cycle of the former type is counted once in Γ_{e_i} and once in Γ_{e_j}, and these will not give rise to cycles in G containing e. Thus,

$$\gamma_e = (\lambda(e) - 1) + \sum_{e_i \in E_z(x)} \Gamma_{e_i} - \sum_{\substack{\{i,j\} \\ i \neq j \\ e_i, e_j \in E_z(x)}} \Gamma(e_i, e_j).$$

Now, by the induction hypothesis, $\Gamma_{e_i} \equiv 1 \pmod 2$ for each e_i, and $\Gamma(e_i, e_j) = \Gamma(e_j, e_i)$ in the last sum on the right, and hence this latter sum is even. Thus, $\gamma_e \equiv (\lambda(e) - 1) + k \pmod 2 \equiv 1 \pmod 2$ by relation (6.1). ☐

A consequence of Theorem 6.2.3 is a result of Bondy and Halberstam [26], which gives yet another characterization of Eulerian graphs.

Corollary 6.2.4*. *A graph is Eulerian if and only if it has an odd number of cycle decompositions.*

Proof. In one direction, the proof is trivial. If G has an odd number of cycle decompositions, then it has at least one, and hence G is Eulerian.

Conversely, assume that G is Eulerian. Let $e \in E(G)$ and let $C_1, \ldots, C_r$ be the cycles containing e. By Theorem 6.2.3, r is odd. We proceed by induction on $m = |E(G)|$ with G Eulerian.

If G is just a cycle, then the result is true. Assume then that G is not a cycle. This means that for each i, $1 \le i \le r$, by the induction assumption, $G_i = G - E(C_i)$ has an odd number, say s_i, of cycle decompositions. (If G_i is disconnected, apply the induction assumption to each of the nontrivial components of G_i.) The union of each of these cycle decompositions of G_i and C_i yields a cycle decomposition of G. Hence the number of cycle decompositions of G containing C_i is s_i, $1 \le i \le r$. Let $s(G)$ denote the number of cycle decompositions of G. Then

$$s(G) = \sum_{i=1}^{r} s_i \equiv r \pmod 2 \text{ (since } s_i \equiv 1 \pmod 2)$$

$$\equiv 1 \pmod 2.$$

☐

Exercise 2.1. Find an Euler tour in the graph G below.

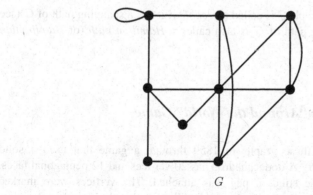

G

Fig. 6.5 (a) Hamiltonian graph; (b) non-Hamiltonian but traceable graph

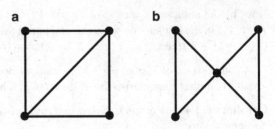

Exercise 2.2. Does there exist an Eulerian graph with

(i) An even number of vertices and an odd number of edges?
(ii) An odd number of vertices and an even number of edges? Draw such a graph if it exists.

Exercise 2.3. Prove that a connected graph is Eulerian if and only if each of its blocks is Eulerian.

Exercise 2.4. If G is a connected graph with $2k(k > 0)$ vertices of odd degree, show that $E(G)$ can be partitioned into k open (i.e., not closed) trails.

Exercise 2.5. Prove that a connected graph is Eulerian if and only if each of its edge cuts has an even number of edges.

6.3   Hamiltonian Graphs

Definition 6.3.1. A graph is called *Hamiltonian* if it has a spanning cycle (see Fig. 6.5a). These graphs were first studied by Sir William Hamilton, a mathematician. A spanning cycle of a graph G, when it exists, is often called a *Hamilton cycle* (or *Hamiltonian cycle*) of G.

Definition 6.3.2. A graph G is called *traceable* if it has a spanning path of G (see Fig. 6.5b). A spanning path of G is also called a *Hamilton path* (or *Hamiltonian path*) of G.

6.3.1   Hamilton's "Around the World" Game

Hamilton introduced these graphs in 1859 through a game that used a solid dodecahedron (Fig. 6.6). A dodecahedron has 20 vertices and 12 pentagonal faces. At each vertex of the solid, a peg was attached. The vertices were marked Amsterdam, Ann Arbor, Berlin, Budapest, Dublin, Edinburgh, Jerusalem, London, Melbourne, Moscow, Novosibirsk, New York, Paris, Peking, Prague, Rio di Janeiro,

Fig. 6.6 Solid dodecahedron for Hamilton's "Around the World" problem

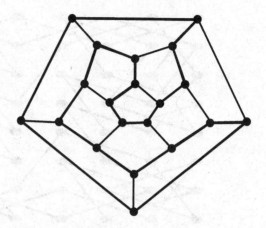

Rome, San Francisco, Tokyo, and Warsaw. Further, a string was also provided. The object of the game was to start from any one of the vertices and keep on attaching the string to the pegs as we move from one vertex to another along a particular edge with the condition that we have to get back to the starting city without visiting any intermediate city more than once. In other words, the problem asks one to find a Hamilton cycle in the graph of the dodecahedron (see Fig. 6.6). Hamilton solved this problem as follows: When a traveler arrives at a city, he has the choice of taking the edge to his right or left. Denote the choice of taking the edge to the right by R and that of taking the edge to the left by L. Let 1 denote the operation of staying where he is.

Define the product $O_1 O_2$ of two operations O_1 and O_2 as O_1 followed by O_2. For example, LR denotes going left first and then going right. Two sequences of operations are *equal* if, after starting at a vertex, the two sequences lead to the same vertex. The product defined above is associative but not commutative. Further, it is clear (see Fig. 6.6) that

$$R^5 = L^5 = 1$$

$$RL^2R = LRL,$$

$$LR^2L = RLR,$$

$$RL^3R = L^2, \text{ and}$$

$$LR^3L = R^2.$$

These relations give

$$1 = R^5 = R^2R^3 = (LR^3L)R^3 - (LR^3)(LR^3) = (LR^3)^2 = (LR^2R)^2$$
$$= (L(LR^3L)R)^2 = (L^2R^3LR)^2 = (L^2((LR^3L)R)LR)^2 = (L^3R^3LRLR)^2$$
$$= LLLRRRLRLRLLLRRRLRLR. \tag{6.2}$$

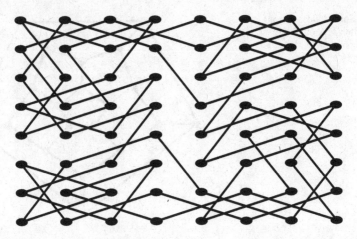

Fig. 6.7 A knight's tour in a chessboard

The last sequence of operations contains 20 operations and contains no partial sequence equal to 1. Hence, this sequence must represent a Hamilton cycle. Thus, starting from any vertex and following the sequence of operations (6.2), we do indeed get a Hamilton cycle of the graph of Fig. 6.6.

Knight's Tour in a Chessboard 6.3.3. The knight's tour problem is the problem of determining a closed tour through all 64 squares of an 8×8 chessboard by a knight with the condition that the knight does not visit any intermediate square more than once. This is equivalent to finding a Hamilton cycle in the corresponding graph of $64 \, (= \, 8 \times 8)$ vertices in which two vertices are adjacent if and only if the knight can move from one vertex to the other following the rules of the chess game. Figure 6.7 displays a knight's tour.

Even though Eulerian graphs admit an elegant characterization, no decent characterization of Hamiltonian graphs is known as yet. In fact, it is one of the most difficult unsolved problems in graph theory. (Actually, it is an NP-complete problem; see reference [71].) Many sufficient conditions for a graph to be Hamiltonian are known; however, none of them happens to be an elegant necessary condition.

We begin with a necessary condition. Recall that $\omega(H)$ stands for the number of components of the graph H.

Theorem 6.3.4. *If G is Hamiltonian, then for every nonempty proper subset S of V, $\omega(G - S) \leq |S|$.*

Proof. Let C be a Hamilton cycle in G. Then, since C is a spanning subgraph of G, $\omega(G - S) \leq \omega(C - S)$. If $|S| = 1$, $C - S$ is a path, and therefore $\omega(C - S) = 1 = |S|$. The removal of a vertex from a path P results in one or two components, according to whether the removed vertex is an end vertex or an internal vertex of P.

Fig. 6.8 Theta graph

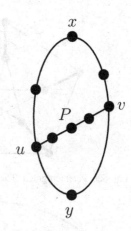

Hence, by induction, the number of components in $C - S$ cannot exceed $|S|$. This proves that $\omega(G - S) \leq \omega(C - S) \leq |S|$.   □

It follows directly from the definition of a Hamiltonian graph or from Theorem 6.3.4 that any Hamiltonian graph must be 2-connected. [If G has a cut vertex v, then taking $S = \{v\}$, we see that $\omega(G - S) > |S|$.] The converse, however, is not true. For example, the theta graph of Fig. 6.8 is 2-connected but not Hamiltonian. Here, P stands for a u-v path of any length ≥ 2 containing neither x nor y.

Exercise 3.1. Show by means of an example that the condition in Theorem 6.3.4 is not sufficient for G to be Hamiltonian.

Exercise 3.2. Use Theorem 6.3.4 to show that the Herschel graph (shown in Fig. 5.4) is non-Hamiltonian.

Exercise 3.3. Do Exercise 3.2 by using Theorem 1.5.10 (characterization theorem for bipartite graphs).

If a cubic graph G has a Hamilton cycle C, then $G \backslash E(C)$ is a 1-factor of G. Hence, for a cubic graph G to be Hamiltonian, G must have a 1-factor F such that $G \backslash E(F)$ is a Hamilton cycle of G. Now, the Petersen graph P (shown in Fig. 1.7) has two different types of 1-factors (see Fig. 6.9), and for any such 1-factor F of P, $P \backslash E(F)$ consists of two disjoint 5-cycles. Hence P is non-Hamiltonian.

Theorem 6.3.5 is a basic result due to Ore [150] which gives a sufficient condition for a graph to be Hamiltonian.

Theorem 6.3.5 (Ore [150]). *Let G be a simple graph with $n \geq 3$ vertices. If, for every pair of nonadjacent vertices u, v of G, $d(u) + d(v) \geq n$, then G is Hamiltonian.*

Proof. Suppose that G satisfies the condition of the theorem, but G is not Hamiltonian. Add edges to G (without adding vertices) and get a supergraph G^* of G such that G^* is a maximal simple graph that satisfies the condition of the

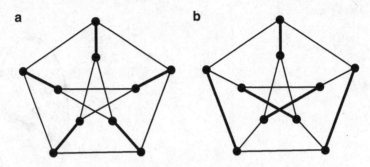

Fig. 6.9 Petersen graph. The solid edges form a 1-factor of P

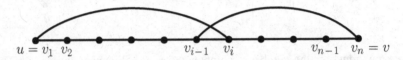

Fig. 6.10 Hamilton path for proof of Theorem 6.3.5

theorem, but G^* is non-Hamiltonian. Such a graph G^* must exist since G is non-Hamiltonian while the complete graph on $V(G)$ is Hamiltonian. Hence, for any pair u and v of nonadjacent vertices of G^*, $G^* + uv$ must contain a Hamilton cycle C. This cycle C would certainly contain the edge $e = uv$. Then $C - e$ is a Hamilton path $u = v_1 v_2 v_3 \ldots v_n = v$ of G^* (see Fig. 6.10).

Now, if $v_i \in N(u)$, $v_{i-1} \notin N(v)$; otherwise, $v_1 v_2 \ldots v_{i-1} v_n v_{n-1} v_{n-2} \ldots v_{i+1}$ $v_i v_1$ would be a Hamilton cycle in G^*. Hence, for each vertex v_i adjacent to u, the vertex v_{i-1} of $V - \{v\}$ is nonadjacent to v. But then

$$d_{G^*}(v) \le (n-1) - d_{G^*}(u).$$

This gives that $d_{G^*}(u) + d_{G^*}(v) \le n - 1$, and therefore $d_G(u) + d_G(v) \le n - 1$, a contradiction.   □

Corollary 6.3.6 (Dirac [54]). *If G is a simple graph with $n \ge 3$ and $\delta \ge \frac{n}{2}$, then G is Hamiltonian.*   □

Corollary 6.3.7. *Let G be a simple graph with $n \ge 3$ vertices. If $d(u) + d(v) \ge n - 1$ for every pair of nonadjacent vertices u and v of G, then G is traceable.*

Proof. Choose a new vertex w and let G' be the graph $G \vee \{w\}$. Then each vertex of G has its degree increased by one, and therefore in G', $d(u) + d(v) \ge n + 1$ for every pair of nonadjacent vertices. Since $|V(G')| = n + 1$, by Theorem 6.3.5, G' is Hamiltonian. If C' is a Hamilton cycle of G', then $C' - w$ is a Hamilton path of G. Thus, G is traceable.   □

Exercise 3.4. Show by means of an example that the conditions of Theorem 6.3.5 and its Corollary 6.3.6 are not necessary for a simple connected graph to be Hamiltonian.

Exercise 3.5. Show that if a cubic graph G has a spanning closed trail, then G is Hamiltonian.

Exercise 3.6. Prove that the n-cube Q_n is Hamiltonian for every $n \geq 2$.

Exercise 3.7. Prove that the wheel W_n is Hamiltonian for every $n \geq 4$.

Exercise 3.8. Prove that a simple k-regular graph on $2k-1$ vertices is Hamiltonian.

Exercise 3.9. For any vertex v of the Petersen graph P, show that $P - v$ is Hamiltonian. (A non-Hamiltonian graph G with this property, namely, for any vertex v of G the subgraph $G - v$ of G is Hamiltonian, is called a hypo-Hamiltonian graph. In fact, P is the lowest-order graph with this property.)

Exercise 3.10. For any vertex v of the Petersen graph P, show that a Hamilton path exists starting at v.

Exercise 3.11. If $G = G(X, Y)$ is a bipartite Hamiltonian graph, show that $|X| = |Y|$.

Exercise 3.12. Let G be a simple graph on $2k$ vertices with $\delta(G) \geq k$. Show that G has a perfect matching.

Exercise 3.13. Prove that a simple graph of order n with n even and $\delta \geq \frac{(n+2)}{2}$ has a 3-factor.

Bondy and Chvátal [25] observed that the proof of Theorem 6.3.5 is essentially based on the following result.

Theorem 6.3.8. *Let G be a simple graph of order $n \geq 3$ vertices. Then G is Hamiltonian if and only if $G + uv$ is Hamiltonian for every pair of nonadjacent vertices u and v with $d(u) + d(v) \geq n$.*

The last result has been instrumental for Bondy and Chvátal to define the closure of a graph G.

Definition 6.3.9. The *closure* of a graph G, denoted $\mathrm{cl}(G)$, is defined to be that supergraph of G obtained from G by recursively joining pairs of nonadjacent vertices whose degree sum is at least n until no such pair exists.

This recursive definition does not stipulate the order in which the new edges are added. Hence, we must first show that the definition does not depend upon the order of the newly added edges. Figure 6.11 explains the construction of $\mathrm{cl}(G)$.

Theorem 6.3.10. *The closure $\mathrm{cl}(G)$ of a graph G is well defined.*

Proof. Let G_1 and G_2 be two graphs obtained from G by recursively joining pairs of nonadjacent vertices whose degree sum is at least n until no such pair exists. We have to prove that $G_1 = G_2$.

Fig. 6.11 Closure of a graph

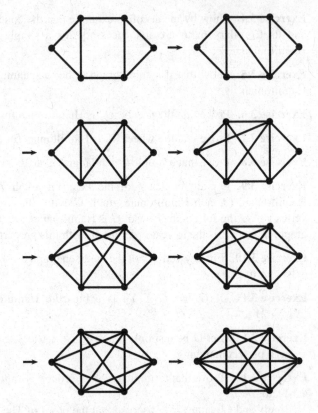

Let $\{e_1, \ldots, e_p\}$ and $\{f_1, \ldots, f_q\}$ be the sets of new edges added to G in these sequential orderings to get G_1 and G_2, respectively. We want to show that each e_i is some f_j (and therefore belongs to G_2) and that each f_k is some e_l (and therefore belongs to G_1). Let e_i be the first edge in $\{e_1, \ldots, e_p\}$ not belonging to G_2. Then $\{e_1, \ldots, e_{i-1}\}$ are all in both G_1 and G_2, and $uv = e_i \notin E(G_2)$. Let $H = G + \{e_1, \ldots, e_{i-1}\}$. Then H is a subgraph of both G_1 and G_2. By the way cl(G) is defined,

$$d_H(u) + d_H(v) \geq n,$$

and hence,

$$d_{G_2}(u) + d_{G_2}(v) \geq n.$$

But this is a contradiction since u and v are nonadjacent vertices of G_2, and G_2 is a closure of G. Thus $e_i \in E(G_2)$ for each i and similarly, $f_k \in E(G_1)$ for each k. $\qquad\qquad\square$

An immediate consequence of Theorem 6.3.8 is the following.

Theorem 6.3.11. *If cl(G) is Hamiltonian, then G is Hamiltonian.*

Corollary 6.3.12. *If cl(G) is complete, then G is Hamiltonian.*

Exercise 3.14. Determine the closure of the following graph.

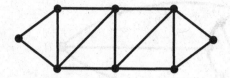

We conclude this section with a result of Chvátal and Erdös [39].

Theorem 6.3.13 (Chvátal and Erdös).   *If, for a simple 2-connected graph G, $\alpha \leq \kappa$, then G is Hamiltonian. (α is the independence number of G and κ is the connectivity of G.)*

Proof. Suppose $\alpha \leq \kappa$ but G is not Hamiltonian. Let C : $v_0 v_1 \ldots v_{p-1}$ be a longest cycle of G. We fix this orientation on C. By Dirac's theorem (Exercise 6.4 of Chap. 3), $p \geq \kappa$. Let $v \in V(G) \setminus V(C)$. Then by Menger's theorem (see also Exercise 6.3 of Chap. 3), there exist κ internally disjoint paths $P_1, \ldots, P_\kappa$ from v to C. Let $v_{i_1}, v_{i_2}, \ldots, v_{i_\kappa}$ be the end vertices (with suffixes in the increasing order) of these paths on C. No two of the consecutive vertices $v_{i_1}, v_{i_2}, \ldots, v_{i_\kappa}, v_{i_1}$ can be adjacent vertices of C, since otherwise we get a cycle of G longer than C. Hence, between any two consecutive vertices of $\{v_{i_1}, v_{i_2}, \ldots, v_{i_\kappa}, v_{i_1}\}$, there exists at least one vertex of G. Let u_{i_j} be the vertex next to v_{i_j} in the v_{i_j}-$v_{i_{j+1}}$ path along C (see Fig. 6.12a).

We claim that $\{u_{i_1}, \ldots, u_{i_\kappa}\}$ is an independent set of G. Suppose u_{i_j} is adjacent to u_{i_m}, $m > j$ (suffixes taken modulo κ); then

$$u_{i_j} \ldots v_{i_{j+1}} \ldots v_{i_m} P_m^{-1} v P_j v_{i_j} \ldots v_{i_{j-1}} \ldots u_{i_m} u_{i_j}$$

is a cycle of G longer than C, a contradiction.

Further, $\{v, u_{i_1} \ldots, u_{i_\kappa}\}$ is also an independent set of G. [Otherwise, $v u_{i_m} \in E(G)$ for some m. See Fig. 6.12b. Then

$$v u_{i_m} \ldots v_{i_{m+1}} \ldots v_{i_\kappa} \ldots v_{i_1} \ldots v_{i_m} P_m^{-1} v$$

is a cycle longer than C, a contradiction.] But this implies that $\alpha > \kappa$, a contradiction to our hypothesis. Thus G is Hamiltonian. $\square$

This theorem, although interesting, is not powerful in that for the cycle $C_n, \kappa = 2$ while $\alpha = \lfloor \frac{n}{2} \rfloor$ and hence increases with n.

A graph G with at least three vertices is *Hamiltonian-connected* if any two vertices of G are connected by a Hamilton path in G. For example, for $n \geq 3$, K_n is Hamiltonian-connected, whereas for $n \geq 4$, C_n is not Hamiltonian-connected.

Theorem 6.3.14.   *If G is a simple graph with $n \geq 3$ vertices such that $d(u) + d(v) \geq n + 1$ for every pair of nonadjacent vertices of G, then G is Hamiltonian-connected.*

Fig. 6.12 Graphs for proof
of Theorem 6.3.13

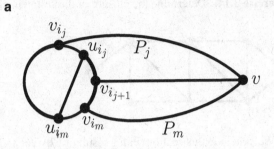

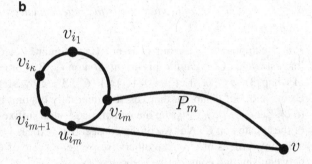

Proof. Let u and v be any two vertices of G. Our aim is to show that a Hamilton path exists from u to v in G.

Choose a new vertex w, and let $G^* = G \cup \{wu, wv\}$. We claim that $\mathrm{cl}(G^*) = K_{n+1}$. First, the recursive addition of the pairs of nonadjacent vertices u and v of G with $d(u) + d(v) \geq n + 1$ gives K_n. Further, each vertex of K_n is of degree $n - 1$ in K_n and $d_{G^*}(w) = 2$. Hence, $\mathrm{cl}(G^*) = K_{n+1}$. So by Corollary 6.3.12, G^* is Hamiltonian. Let C be a Hamilton cycle in G^*. Then $C - w$ is a Hamilton path in G from u to v. $\qquad\square$

6.4* Pancyclic Graphs

Definition 6.4.1. A graph G of order n (≥ 3) is *pancyclic* if G contains cycles of all lengths from 3 to n. G is called *vertex-pancyclic* if each vertex v of G belongs to a cycle of every length l, $3 \leq l \leq n$.

Example 6.4.2. Clearly, a *vertex-pancyclic graph is pancyclic*. However, the converse is not true. Figure 6.13 displays a pancyclic graph that is not vertex-pancyclic.

The study of pancyclic graphs was initiated by Bondy [24], who showed that Ore's sufficient condition for a graph G to be Hamiltonian (Theorem 6.3.5) actually implies much more. Note that if $\delta \geq \frac{n}{2}$, then $m \geq \frac{n^2}{4}$.

Fig. 6.13 Pancyclic graph
that is not vertex-pancyclic

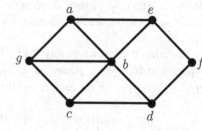

Fig. 6.14 Graph for proof
of Theorem 6.4.3

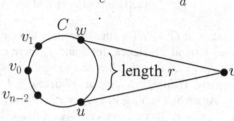

Theorem 6.4.3. *Let G be a simple Hamiltonian graph on n vertices with at least $\lceil \frac{n^2}{4} \rceil$ edges. Then G either is pancyclic or else is the complete bipartite graph $K_{\frac{n}{2},\frac{n}{2}}$. In particular, if G is Hamiltonian and $m > \frac{n^2}{4}$, then G is pancyclic.*

Proof. The result can directly be verified for $n = 3$. We may therefore assume that $n \geq 4$. We apply induction on n. Suppose the result is true for all graphs of order at most $n - 1 (n \geq 4)$, and let G be a graph of order n.

First, assume that G has a cycle $C = v_0 v_1 \ldots v_{n-2} v_0$ of length $n - 1$. Let v be the (unique) vertex of G not belonging to C. If $d(v) \geq \frac{n}{2}$, v is adjacent to two consecutive vertices on C, and hence G has a cycle of length 3. Suppose for some r, $2 \leq r \leq \frac{n-1}{2}$, C has no pair of vertices u and w on C adjacent to v in G with $d_C(u, w) = r$. Then, if $v_{i_1}, v_{i_2}, \ldots v_{i_{d(v)}}$ are the vertices of C that are adjacent to v in G (recall that C contains all the vertices of G except v,) then $v_{i_1+r}, v_{i_2+r}, \ldots, v_{i_{d(v)}+r}$ are nonadjacent to v in G, where the suffixes are taken modulo $(n - 1)$. Hence, $2d(v) \leq n - 1$, a contradiction. Thus, for each r, $2 \leq r \leq \frac{n-1}{2}$, C has a pair of vertices u and w on C adjacent to v in G with $d_C(u, w) = r$. So for each r, $2 \leq r \leq \frac{n-1}{2}$, G has a cycle of length $r + 2$ as well as a cycle of length $n - 1 - r + 2 = n - r + 1$ (see Fig. 6.14). Consequently, G is pancyclic. (Recall that G is already Hamiltonian.)

If $d(v) \leq \frac{n-1}{2}$, then $G[V(C)]$, the subgraph of G induced by $V(C)$ has at least $\frac{n^2}{4} - d(v) \geq \frac{n^2}{4} - \frac{n-1}{2} > \frac{(n-1)^2}{4}$ edges. So by the induction assumption, $G[V(C)]$ is pancyclic and hence G is pancyclic. (By hypothesis, G is Hamiltonian.)

Next, assume that G has no cycle of length $n - 1$. Then G is not pancyclic. In this case, we show that G is $K_{\frac{n}{2},\frac{n}{2}}$.

Let $C = v_0 v_1 v_2 \ldots v_{n-1} v_0$ be a Hamiltonian cycle of G. We claim that of the two pairs $v_i v_k$ and $v_{i+1} v_{k+2}$ (where suffixes are taken modulo n), at most one of them can be an edge of G. Otherwise, $v_k v_{k-1} v_{k-2} \ldots v_{i+1} v_{k+2} v_{k+3} v_{k+4} \ldots v_i v_k$ is an $(n - 1)$-cycle in G (as it misses only the vertex v_{k+1} of G), a contradiction.

Hence, if $d(v_i) = r$, then there are r vertices adjacent to v_i in G and hence at least r vertices that are nonadjacent to v_{i+1}. Thus, $d(v_i + 1) \leq n - r$, and $d(v_i) + d(v_{i+1}) \leq n$.

Summing the last inequality over i from 0 to $n - 1$, we get $4m \leq n^2$. But by hypothesis, $4m \geq n^2$. Hence, $m = \frac{n^2}{4}$ and so n must be even. Again, this yields $d(v_i) + d(v_{i+1}) = n$ for each i, and therefore for each i and k,

$$\text{exactly one of } v_i v_k \text{ and } v_{i+1} v_{k+2} \text{ is an edge of } G. \tag{$*$}$$

Thus, if $G \neq K_{\frac{n}{2}, \frac{n}{2}}$, then certainly there exist i and j such that $v_i v_j \in E$ and $i \equiv j \pmod 2$. Hence, for some j, there exists an *even* positive integer s such that $v_{j+1} v_{j+1+s} \in E$. Choose s to be the least *even* positive integer with the above property. Then $v_j v_{j+s-1} \notin E$. Hence, $s \geq 4$ (as $s = 2$ would mean that $v_j v_{j+1} \notin E$). Again by $(*)$, $v_{j-1} v_{j+s-3} = v_{j-1} v_{j-1+(s-2)} \in E(G)$ contradicting the choice of s. Thus, $G = K_{\frac{n}{2}, \frac{n}{2}}$. The last part follows from the fact that $|E(K_{\frac{n}{2}, \frac{n}{2}})| = \frac{n^4}{4}$. $\square$

Corollary 6.4.4. *Let $G \neq K_{\frac{n}{2}, \frac{n}{2}}$, be a simple graph with $n \geq 3$ vertices, and let $d(u) + d(v) \geq n$ for every pair of nonadjacent vertices of G. Then G is pancyclic.*

Proof. By Ore's theorem (Theorem 6.3.5), G is Hamiltonian. We show that G is pancyclic by first proving that $m \geq \frac{n^2}{2}$ and then invoking Theorem 6.4.3. This is true if $\delta \geq \frac{n}{2}$ (as $2m = \sum_{i=1}^{n} d_i \geq \delta n \geq n^2/2$). So assume that $\delta < \frac{n}{2}$.

Let S be the set of vertices of degree δ in G. For every pair (u, v) of vertices of degree δ, $d(u) + d(v) < \frac{n}{2} + \frac{n}{2} = n$. Hence, by hypothesis, S induces a clique of G and $|S| \leq \delta + 1$. If $|S| = \delta + 1$, then G is disconnected with $G[S]$ as a component, which is impossible (as G is Hamiltonian). Thus $|S| \leq \delta$. Further if $v \in S$, v is nonadjacent to $n - 1 - \delta$ vertices of G. If u is such a vertex, $d(v) + d(u) \geq n$ implies that $d(u) \geq n - \delta$. Further, v is adjacent to at least one vertex $w \notin S$ and $d(w) \geq \delta + 1$ by the choice of S. These facts give that

$$2m = \sum_{i=1}^{n} d_i \geq (n - \delta - 1)(n - \delta) + \delta^2 + (\delta + 1),$$

where the last $(\delta + 1)$ comes out of the degree of w. Thus,

$$2m \geq n^2 - n(2\delta + 1) + 2\delta^2 + 2\delta + 1$$

which implies that

$$4m \geq 2n^2 - 2n(2\delta + 1) + 4\delta^2 + 4\delta + 2$$
$$= (n - (2\delta + 1))^2 + n^2 + 1$$
$$\geq n^2 + 1, \quad \text{since } n > 2\delta.$$

Consequently, $m > \frac{n^2}{4}$, and by Theorem 6.4.3, G is pancyclic. $\square$

6.5 Hamilton Cycles in Line Graphs

We now turn our attention to the existence of Hamilton cycles in line graphs.

Theorem 6.5.1. *If G is Eulerian, then $L(G)$, the line graph of G is both Hamiltonian and Eulerian.*

Proof. As G is Eulerian, it is connected and hence $L(G)$ is also connected. If $e_1 e_2 \ldots e_m$ is the edge sequence of an Euler tour in G, and if vertex u_i in $L(G)$ represents the edge e_i, $1 \leq i \leq m$, then $u_1 u_2 \ldots u_m u_1$ is a Hamilton cycle of $L(G)$. Further, if $e = v_1 v_2 \in E(G)$ and the vertex u in $L(G)$ represents the edge e, then $d_{L(G)}(u) = d_G(v_1) + d_G(v_2) - 2$, which is even (and ≥ 2) since both $d_G(v_1)$ and $d_G(v_2)$ are even (and ≥ 2). Hence in $L(G)$ every vertex is of even degree (≥ 2). So $L(G)$ is also Eulerian. $\qquad\square$

Exercise 5.1. Disprove the converse of Theorem 6.5.1 by a counterexample.

Definition 6.5.2. A *dominating trail* of a graph G is a closed trail T in G (which may be just a single vertex) such that every edge of G not in T is incident with T.

Example 6.5.3. For instance, in the graph of Fig. 6.13, the trail *abcdbea* is a dominating trail.

Harary and Nash–Williams [94] characterized graphs that have Hamiltonian line graphs.

Theorem 6.5.4 (Harary and Nash–Williams). *The line graph of a graph G with at least three edges is Hamiltonian if and only if G has a dominating trail.*

Proof. Let T be a dominating trail of G and let $\{e_1, e_2, \ldots, e_s\}$ be the edge sequence representing T. Then every edge of G not in T is incident to some vertex of T. Assume that e_1 and e_2 are incident at v_1. Replace the subsequence $\{e_1, e_2\}$ of $\{e_1, e_2, \ldots, e_s\}$ by the sequence $\{e_1, e_{11}, e_{12}, \ldots, e_{1r_1}, e_2\}$, where $e_{11}, e_{12}, \ldots e_{1r_1}$ are the edges of $E(G)\backslash E(T)$ incident at v_1 other than e_1 and e_2. Assume that we have already replaced the subsequence $\{e_i, e_{i+1}\}$ by $\{e_i, e_{i1}, \ldots, e_{ir_i}, e_{i+1}\}$. Then replace $\{e_{i+1}, e_{i+2}\}$ by the sequence $\{e_{i+1}, e_{(i+1)1}, \ldots, e_{(i+1)r_{(i+1)}}, e_{i+2}\}$ in $E(G)\backslash E(T)$, where the new edges $e_{(i+1)1} \ldots, e_{(i+1)r_{(i+1)}}$ have not appeared in the previous i subsequences. (Here we take $e_{s+1} = e_1$.) The resulting edge sequence is $e_1 e_{11} e_{12} \ldots e_{1r_1}$ $e_2 e_{21} e_{22} \ldots e_{2r_2} e_3 \ldots e_s e_{s1} e_{s2} \ldots e_{sr_s} e_1$ and this gives the Hamilton cycle $u_1 u_{11} u_{12} \ldots u_{1r_1} u_2 u_{21} u_{22} \ldots u_{2r_2} u_3 \ldots u_s u_{s1} u_{s2} \ldots u_{sr_s} u_1$ in $L(G)$. [Here u_1 is the vertex of $L(G)$ that corresponds to the edge e_1 of G, and so on.]

Conversely, assume that $L(G)$ has a Hamilton cycle C. Let $C = u_1 u_2 \ldots u_m u_1$ and let e_i be the edge of G corresponding to the vertex u_i of $L(G)$. Let T_0 be the edge sequence $e_1 e_2 \ldots e_m e_1$. We now delete edges from T_0 one after another as follows: Let $e_i e_j e_k$ be the first three distinct consecutive edges of T_0 that have a common vertex; then delete e_j from the sequence. Let $T_0' = T_0 - e_j$ $= \{e_1 e_2 \ldots e_i e_k \ldots e_m e_1\}$.

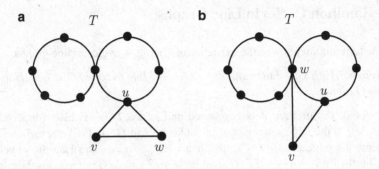

Fig. 6.15 Graphs for proof of Theorem 6.5.6. **(a)** $T \cup \{uv, vw, wu\}$ is longer than T; **(b)** $(T \backslash \{uw\}) \cup \{uv, vw\}$ is longer than T

Now proceed with T_0' as we did with T_0. Continue this process until no such triad of edges exists. Then the resulting subsequence of T_0 must be a dominating trail or a pair of adjacent edges incident at a vertex, say, v_0. In the latter case, all the edges of G are incident at v_0, and hence we take $\{v_0\}$ as the dominating trail of G. $\qquad\square$

Corollary 6.5.5. *The line graph of a Hamiltonian graph is Hamiltonian.*

Proof. Let G be a Hamiltonian graph with Hamilton cycle C. Then C is a dominating trail of G. Hence, $L(G)$ is Hamiltonian. $\qquad\square$

Exercise 5.2. Show that the line graph of a graph G has a Hamilton path if and only if G has a trail T such that every edge of G not in T is incident with T.

Exercise 5.3. Draw the line graph of the graph of Fig. 6.13 and display a Hamilton cycle in it.

Theorem 6.5.6 ([12]). *Let G be any connected graph. If each edge of G belongs to a triangle in G, then G has a spanning, Eulerian subgraph.*

Proof. Since G has a triangle, G has a closed trail. Let T be a longest closed trail in G. Then T must be a spanning Eulerian subgraph of G. If not, there exists a vertex v of G with $v \notin T$ and v is adjacent to a vertex u of T.

By hypothesis, uv belongs to a triangle, say uvw. If none of the edges of this triangle is in T, then $T \cup \{uv, vw, wu\}$ yields a closed trail longer than T (see Fig. 6.15). If $uw \in T$, then $(T - uw) \cup \{uv, vw\}$ would be a closed trail longer than T. These contradictions prove that T is a spanning closed trail of G. $\qquad\square$

Corollary 6.5.7. *Let G be any connected graph. If each edge of G belongs to a triangle, then $L(G)$ is Hamiltonian.*

Proof. The proof is an immediate consequence of Theorems 6.5.4 and 6.5.6. $\qquad\square$

Corollary 6.5.8 (Chartrand and Wall [35]). *If G is connected and $\delta(G) \geq 3$, then $L^2(G)$ is Hamiltonian.*

(Note: For $n > 1$, $L^n(G) = L(L^{n-1}(G))$, and $L^0(G) = G$.)

Proof. Since $\delta(G) \geq 3$, each vertex of $L(G)$ belongs to a clique of size at least three, and hence each edge of $L(G)$ belongs to a triangle. Now apply Corollary 6.5.7.   □

Corollary 6.5.9 (Nebesky [146]).   *If G is a connected graph with at least three vertices, then $L(G^2)$ is Hamiltonian.*

Proof. Since G is a connected graph with at least three vertices, every edge of G^2 belongs to a triangle. Hence, $L(G^2)$ is Hamiltonian by Corollary 6.5.7.   □

Theorem 6.5.10.   *Let G be a connected graph in which every edge belongs to a triangle. If e_1 and e_2 are edges of G such that $G \setminus \{e_1, e_2\}$ is connected, then there exists a spanning trail of G with e_1 and e_2 as its initial and terminal edges.*

Proof. The proof is essentially the same as for Theorem 6.5.6 and is based on considering the longest trail T in G with e_1 and e_2 as its initial and terminal edges, respectively.   □

Corollary 6.5.11 ([12]).   *Let G be any connected graph with $\delta(G) \geq 4$. Then $L^2(G)$ is Hamiltonian-connected.*

Proof. The edges incident to a vertex v of G will yield a clique of size $d(v)$ in $L(G)$. Since $\delta(G) \geq 4$, each vertex of $L(G)$ belongs to a clique of order at least 4, and hence $L(G)$ is 3-edge connected. Therefore, for any pair of distinct edges e_1 and e_2 of $L(G)$, $L(G) \setminus \{e_1, e_2\}$ is connected. Further, each edge of $L(G)$ belongs to a triangle. Hence, by Theorem 6.5.10, a spanning trail T in $L(G)$ exists having e_1 and e_2 as the initial and terminal edges, respectively. Thus, if there are any edges of $L(G)$ not belonging to T, they can only be "chords" of T. It follows (see Exercise 5.2) that in $L^2(G)$ there exists a Hamilton path starting and ending at the vertices corresponding to e_1 and e_2, respectively. Since e_1 and e_2 are arbitrary, $L^2(G)$ is Hamiltonian-connected.   □

Corollary 6.5.12 (Jaeger [113]).   *The line graph of a 4-edge-connected graph is Hamiltonian.*

To prove Corollary 6.5.12, we need the following lemma.

Lemma 6.5.13.   *Let S be a set of vertices of a nontrivial tree T, and let $|S| = 2k$, $k \geq 1$. Then there exists a set of k pairwise edge-disjoint paths in T whose end vertices are all the vertices of S.*

Proof. Certainly there exists a set of k paths in T whose end vertices are all the vertices of S. (This is because between any two vertices of T, there is a unique path in T.) Choose such a set of k paths, say $\mathscr{P} = \{P_1, P_2, \ldots, P_k\}$ with the additional condition that the sum of their lengths is minimum.

We claim that the paths of $\mathscr{P}$ are pairwise edge-disjoint. If not, there exists a pair $\{P_i, P_j\}$, $i \neq j$, with P_i and P_j having an edge in common. In this case, P_i and P_j have one or more disjoint paths of length at least 1 in common. Then $P_i \mathbin{\Delta} P_j$, the symmetric difference of P_i and P_j, properly contains a disjoint union of two paths,

Fig. 6.16 Graph for proof of
Lemma 6.5.13

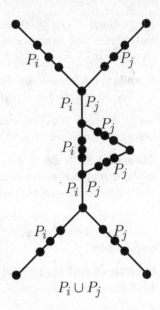

$$P_i \cup P_j$$

say Q_i and Q_j, with their end vertices being disjoint pairs of vertices belonging to
S (Fig. 6.16).

If we replace P_i and P_j by Q_i and Q_j in $\mathscr{P}$, then the resulting set of paths has
the property that their end vertices belong to S and that the sum of the lengths of
Q_i and Q_j is less than that of the sum of the lengths of the paths P_i and P_j in $\mathscr{P}$.
This contradicts the choice of $\mathscr{P}$, and hence $\mathscr{P}$ has the stated property.   □

Proof of Corollary 6.5.12. Let G be a 4-edge-connected graph. In view of
Theorem 6.5.4, it suffices to show that G contains a dominating trail.

By Corollary 4.4.6, G contains two edge-disjoint spanning trees T_1 and T_2. Let S
be the set of vertices of odd degree in T_1. Then $|S|$ is even. Let $|S| = 2k$, $k \geq 1$. By
Lemma 6.5.13, there exists a set of k pairwise edge-disjoint paths $\{P_1, P_2, \ldots, P_k\}$
in T_2 with the property stated in Lemma 6.5.13. Then $G_0 = T_1 \cup (P_1 \cup P_2 \cup \ldots P_k)$
is a connected spanning subgraph of G in which each vertex is of even degree. Thus,
G_0 is a dominating trail of G.   □

We conclude this section with a theorem on locally connected graphs (see
Definition 1.5.9 of Chap. 1).

Theorem 6.5.14* (Oberly and Sumner [149]). *A connected, locally connected,
nontrivial $K_{1,3}$-free graph is Hamiltonian.*

Proof. Let G be a connected, locally connected, nontrivial $K_{1,3}$-free graph. We may
assume that G has at least four vertices. Since G is locally connected, G certainly
has a cycle. Let C be a longest cycle of G. If C is not a Hamilton cycle, there exists
a vertex $v \in V(G)\backslash V(C)$ that is adjacent to a vertex u of C. Let u_1 and u_2 be the
neighbors of u on C. Then, as the edges uv, uu_1 and uu_2 do not induce a $K_{1,3}$ in G,

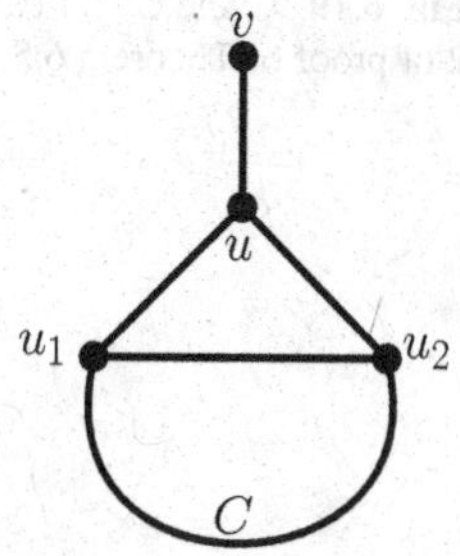

Fig. 6.17 Graph for proof
of Theorem 6.5.14

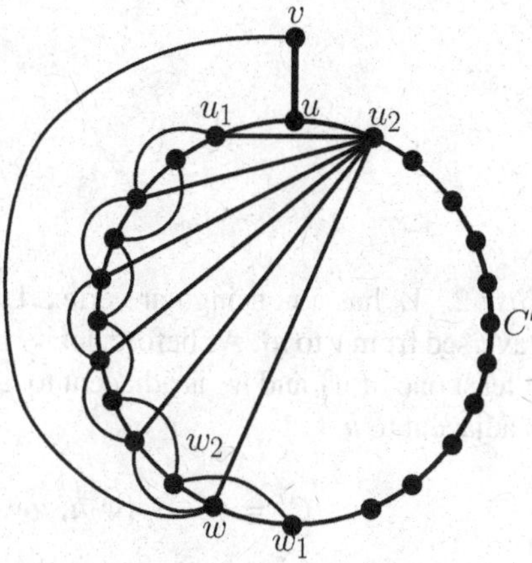

Fig. 6.18 Graph for case 1
of proof of Theorem 6.5.14

$u_1 u_2 \in E(G)$, since otherwise v is adjacent either to u_1 or u_2 and we get a cycle longer than C, a contradiction (see Fig. 6.17).

For each $x \in V(G)$, denote by $G_0(x)$ the subgraph $G[N_G(x)]$ of G. By hypothesis, $G_0(u)$ is connected, and hence there exists either a v-u_1 path P in $G_0(u)$ not containing u_2 or a v-u_2 path Q in $G_0(u)$ not containing u_1. Let us say it is the former. For the purpose of the proof of this theorem, we call a vertex $w \in V_0 = (V(C) \cap V(P)) \backslash \{u_1\}$ *singular* if neither of the two neighbors of w on C is in $N_G(u)$.

Case 1. Each vertex of V_0 is singular. Then for any $w \in V_0$, w is adjacent to u (since $w \in V(P) \subseteq V(G_0(u))$) but since w is singular, neither of the neighbors w_1 and w_2 of w on C is adjacent to u in G. Then considering the $K_{1,3}$ subgraph $\{ww_1, ww_2, wu\}$, we see that $w_1 w_2 \in E(G)$. Now, describe the cycle C' as follows: Start from u_2, move away from u along C, and whenever we encounter a singular vertex w, bypass it by going through the edge $w_1 w_2$. After reaching u_1, retrace the u_1-v path P^{-1} and follow it up by the path vuu_2. Then C' traverses through each vertex of $C \cup P$ exactly once. Thus, C' is a cycle longer than C, a contradiction to the choice of C (see Fig. 6.18).

Fig. 6.19 Cycle C' for case
2 of proof of Theorem 6.5.14

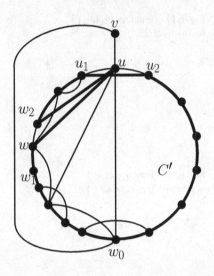

Case 2. V_0 has a nonsingular vertex. Let w be the first nonsingular vertex as P is traversed from v to u_1. As before, let w_1 and w_2 be the neighbors of w along C. Then at least one of w_1 and w_2 is adjacent to u. Without loss of generality, assume that w_2 is adjacent to u. Let

$$C' = (C \cup \{w_2u, uw, u_1u_2\}) \setminus \{w_2w, uu_1, uu_2\}.$$

(See Fig. 6.19.)

Clearly, C and C' are of the same length, and therefore C' is also a longest cycle of G. Then, by the choice w, the v-w section of P contains the only nonsingular vertex w. Let w_0 be the first singular vertex on this section. Consider the cycle C'' described as follows: Start from w_2 and move along C' away from u until we reach the vertex preceding w_0. Bypass w_0 by moving through the neighbors of w_0 along C' (as in case 1), and repeat it for each nonsingular vertex after w_0. After reaching w, move along the w-v section of P^{-1} and follow it by the path vuw_2 (see Fig. 6.20). Then C'' is a cycle longer than C' (as in case 1), a contradiction.

Hence, in any case, C cannot be a longest cycle of G. Thus, G is Hamiltonian. $\qquad\qquad\qquad\qquad\qquad\qquad\qquad\qquad\qquad\qquad\qquad\qquad\qquad\square$

6.6   2-Factorable Graphs

It is clear that if a graph G is r-factorable with k r-factors, then the degree of each vertex of G is rk. In particular, if G is 2-factorable, then G is regular of even degree, say, $2k$. That the converse is also true is a result due to Petersen [158].

Theorem 6.6.1 (Petersen). *Every $2k$-regular graph, $k \geq 1$, is 2-factorable.*

Fig. 6.20 Cycle C'' for case
2 of proof of Theorem 6.5.14

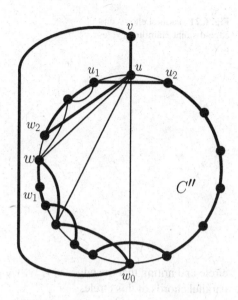

Proof. Let G be a $2k$-regular graph with $V = \{v_1, v_2, \ldots, v_n\}$. We may assume without loss of generality that G is connected. (Otherwise, we can consider the components of G separately.) Since each vertex of G is of even degree, by Theorem 6.2.2, G is Eulerian. Let T be an Euler tour of G. Form a bipartite graph H with bipartition (V, W), where $V = \{v_1, v_2, \ldots, v_n\}$ and $W = \{w_1, w_2, \ldots, w_n\}$ and in which v_i is made adjacent to w_j if and only if v_j follows v_i immediately in T. Since at every vertex of G there are k incoming edges and k outgoing edges along T, H is k-regular. Hence, by Theorem 5.5.3, H is 1-factorable. Let the k 1-factors be $M_1, \ldots, M_k$. Label the edges of M_i with the label i, $1 \le i \le k$. Then the k edges incident at each v_i of H receive the k labels $1, 2, \ldots, k$, and hence if the edges $v_i w_j$ and $v_j w_r$ are in M_p, $1 \le p \le k$, identifying the vertex w_j with the vertex v_j for each j in M_p gives an edge labeling to G in which the edges $v_i v_j$ and $v_j v_r$ receive the label p. It is then clear that the edges of M_p yield a 2-factor of G with label p. Note that v_i is nonadjacent to w_i in H, $1 \le i \le k$. Since this is true for each of the 1-factors M_p, $1 \le p \le k$, we get a 2-factorization of G into k 2-factors. □

A special case of Theorem 6.6.1 is the 2-factorization of K_{2p+1}, which is $2p$-regular. Actually, K_{2p+1} has a 2-factorization into Hamilton cycles.

Theorem 6.6.2. K_{2p+1} *is 2-factorable into p Hamilton cycles.*

Proof. Label the vertices of K_{2p+1} as $v_0, v_1, \ldots, v_{2p}$. For $i = 0, 1, \ldots, p$, let P_i be the path $v_i v_{i-1} v_{i+1} v_{i-2} v_{i+2} \cdots v_{i+p-1} v_{i-(p-1)}$ (suffixes taken modulo $2p$), and let C_i be the Hamilton cycle obtained from P_i by joining v_{2p} to the end vertices of P_i. The cycles C_i are edge-disjoint. This may be seen by placing the $2p$ vertices $v_0, v_1, \ldots, v_{2p-1}$ symmetrically on a circle and placing v_{2p} at the center of the

Fig. 6.21 Parallel chords and edge-disjoint Hamilton cycles in K_7

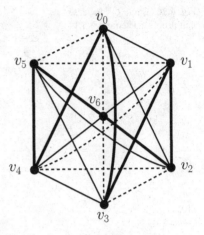

circle and noting that the edges $v_i\, v_{i-1}$, $v_{i+1}\, v_{i-2}$, $\ldots$, $v_{i+p-1}\, v_{i-p}$ form a set of p parallel chords of this circle. $\square$

Figure 6.21 displays the three sets of parallel chords and three edge-disjoint Hamilton cycles in K_7. The 2-factors are

$$F_1 : v_6\, v_0\, v_5\, v_1\, v_4\, v_2\, v_3\, v_6,$$

$$F_2 : v_6\, v_1\, v_0\, v_2\, v_5\, v_3\, v_4\, v_6,$$

$$F_3 : v_6\, v_2\, v_1\, v_3\, v_0\, v_4\, v_5\, v_6.$$

6.7 Exercises

7.1. Prove: A Hamiltonian-connected graph is Hamiltonian. (Note: The converse is not true. See the next exercise.)

7.2. Show that a Hamiltonian-connected graph is 3-connected. Display a Hamiltonian graph of connectivity 3 that is not Hamiltonian-connected.

7.3. If G is traceable, show that for every proper subset S of $V(G)$, $\omega(G - S) \leq |S| + 1$. Disprove the converse by a counterexample.

7.4. If G is simple and $\delta \geq \frac{n-1}{2}$ show that G is traceable. Disprove the converse.

7.5. If G is simple and $\delta \geq \frac{n+1}{2}$ show that G is Hamiltonian-connected. Is the converse true?

7.6. Give an example of a non-Hamiltonian simple graph G of order n ($n \geq 3$) such that for every pair of nonadjacent vertices u and v, $d(u) + d(v) \geq n - 1$. [This shows that the condition in Ore's theorem (Theorem 6.3.5) cannot be weakened further.]

7.7. Show that if a cubic graph is Hamiltonian, then it has three disjoint 1-factors.

7.8.* Show that if a cubic graph has a 1-factor, then it has at least three distinct 1-factors.

7.9. Show that a complete k-partite graph G is Hamiltonian if and only if $|V(G) \setminus N| \geq |N|$, where N is the size of a maximum part of G. (See Aravamudhan and Rajendran [9].)

7.10. A graph is called *locally Hamiltonian* if $G[N(v)]$ is Hamiltonian for each vertex v of G. Show that a locally Hamiltonian graph is 3-connected.

7.11. If $|V(G)| \geq 5$, prove that $L(G)$ is locally Hamiltonian if and only if $G \cong K_{1,n}$.

7.12. If G is a 2-connected graph that is both $K_{1,3}$-free and $(K_{1,3} + e)$-free, prove that G is Hamiltonian. (Recall that a graph G is *H-free* if G does not contain an isomorphic copy of H as an induced subgraph.)

7.13. Let G be a simple graph of order $2n$ ($n \geq 2$). If for every pair of nonadjacent vertices u and v, $d(u) + d(v) > 2n + 2$, show that G contains a spanning cubic graph.

7.14. Show by means of an example that the square of a 1-connected (i.e., connected) graph need not be Hamiltonian. (A celebrated result of H. Fleischner states that the square of any 2-connected graph is Hamiltonian—a result that was originally conjectured by M. D. Plummer.)

7.15.* Let G be a simple graph with degree sequence $(d_1, d_2, \ldots, d_n)$, where $d_1 \leq d_2 \leq \ldots \leq d_n$ and $n \geq 3$. Suppose that there is no value of r less than $\frac{n}{2}$ for which $d_r \leq r$ and $d_{n-r} < n - r$. Show that G is Hamiltonian. (See Chvátal [38] or reference [27].)

7.16. Does there exist a simple non-Hamiltonian graph with degree sequence $(2, 3, 5, 5, 5, 6, 6, 6, 6, 6)$?

7.17. Draw a non-Hamiltonian simple graph with degree sequence $(3, 3, 3, 6, 6, 6, 9, 9, 9)$.

7.18. Let G be a $(2k + 1)$-regular graph with the property that $|[S, \bar{S}]| \geq 2k$ for every proper nonempty set S of V. Prove that G has k edge-disjoint 2-factors. (Note that when $k = 1$, this is just Petersen's result: Corollary 5.5.11. Hint: Use Tutte's 1-factor Theorem 5.6.5 to show that G has a 1-factor. Then apply Petersen's Theorem 6.6.1.)

Notes

Königsberg was part of East Prussia before Germany's defeat in World War II. It has been renamed Kaliningrad, and perhaps before long it will get back its original name. It is also the birthplace of the German mathematician David Hilbert as well as the German philosopher Immanuel Kant. It is interesting to note that even though the Königsberg bridge problem did give birth to Eulerian graphs, Euler himself did not use the concept of Eulerian graphs to solve this problem; instead, he relied on an exhaustive case-by-case verification (see reference [24]).

Ore's theorem (Theorem 6.3.5) can be restated as follows: *If G is a simple graph with $n \geq 3$ vertices and $|N(u)| + |N(v)| \geq n$, for every pair of nonadjacent vertices of G, then G is Hamiltonian.* This statement replaces $d(u)$ in Theorem 6.3.5 by $|N(u)|$. There are several sufficient conditions for a graph to be Hamiltonian using the neighborhood conditions. A nice survey of these results is given in Lesniak [131]. To give a flavor of these results, we give three results here of Faudree, Gould, Jacobson, and Lesniak:

Theorem 1. *If G is a 2-connected graph of order n such that $|N(u) \cap N(u)| \geq \frac{2n-1}{3}$ for each pair u, v of nonadjacent vertices of G, then G is Hamiltonian.*

Theorem 2. *If G is a connected graph of order n such that $|N(u) \cap N(v)| \geq \frac{2n-2}{3}$ for each pair u, v of nonadjacent vertices of G, then G is traceable.*

Theorem 3. *If G is a 3-connected graph of order n such that $|N(u) \cap N(v)| > \frac{2n}{3}$ for each pair u, v of nonadjacent vertices of G, then G is Hamiltonian-connected.*

Chapter 7
Graph Colorings

7.1 Introduction .

Graph theory would not be what it is today if there had been no coloring problems. In fact, a major portion of the 20th-century research in graph theory has its origin in the four-color problem. (See Chap. 8 for details.)

In this chapter, we present the basic results concerning vertex colorings and edge colorings of graphs. We present two important theorems on graph colorings, namely, Brooks' theorem and Vizing's theorem. We also present a brief discussion on "snarks" and Kirkman's schoolgirl problem. In addition, a detailed description of the Mycielskian of a graph is also presented.

7.2 Vertex Colorings

7.2.1 Applications of Graph Coloring

We begin with a practical application of graph coloring known as the *storage problem.* Suppose a university's Department of Chemistry wants to store its chemicals. It is quite probable that some chemicals cause violent reactions when brought together. Such chemicals are *incompatible chemicals.* For safe storage, incompatible chemicals should be kept in distinct rooms. The easiest way to accomplish this is, of course, to store one chemical in each room. But this is certainly not the best way of doing it since we will be using more rooms than are really needed (unless, of course, all the chemicals are mutually incompatible!). So we ask: What is the minimum number of rooms required to store all the chemicals so that in each room only compatible chemicals are stored?

We convert the above storage problem into a problem in graphs. Form a graph $G = (V, E)$ by making V correspond bijectively to the set of available chemicals and making u adjacent to v if and only if the chemicals corresponding to u and

R. Balakrishnan and K. Ranganathan, *A Textbook of Graph Theory,*
Universitext, DOI 10.1007/978-1-4614-4529-6_7,
© Springer Science+Business Media New York 2012

v are incompatible. Then, any set of compatible chemicals correspond to a set of independent vertices of G. Thus, a safe storing of chemicals corresponds to a partition of V into independent subsets of G. The cardinality of such a minimum partition of V is then the required number of rooms. The minimum cardinality is called the *chromatic number* of the graph G.

Definition 7.2.1. The *chromatic number* $\chi(G)$ of a graph G is the minimum number of independent subsets that partition the vertex set of G. Any such minimum partition is called a *chromatic partition* of $V(G)$.

The storage problem just described is actually a vertex coloring problem of G. A *vertex coloring* of G is a map $f : V \to S$, where S is a set of distinct colors; it is *proper* if adjacent vertices of G receive distinct colors of S. This means that if $uv \in E(G)$, then $f(u) \neq f(v)$. Thus, $\chi(G)$ is the minimum cardinality of S for which there exists a proper vertex coloring of G by colors of S. Clearly, in any proper vertex coloring of G, the vertices that receive the same color are independent. The vertices that receive a particular color make up a *color class*. This allows an equivalent way of defining the chromatic number.

Definition 7.2.2. The *chromatic number* of a graph G is the minimum number of colors needed for a proper vertex coloring of G. G is *k-chromatic* if $\chi(G) = k$.

Definition 7.2.3. A *k-coloring* of a graph G is a vertex coloring of G that uses at most k colors.

Definition 7.2.4. A graph G is said to be *k-colorable* if G admits a *proper* vertex coloring using at most k colors.

In considering the chromatic number of a graph, only the adjacency of vertices is taken into account. Hence, multiple edges and loops may be discarded while considering chromatic numbers, unless needed otherwise. As a consequence, we may restrict ourselves to simple graphs when dealing with (vertex) chromatic numbers.

It is clear that $\chi(K_n) = n$. Further, $\chi(G) = 2$ if and only if G is bipartite having at least one edge. In particular, $\chi(T) = 2$ for any tree T with at least one edge (since any tree is bipartite). Further (see Fig. 7.1),

$$\chi(C_n) = \begin{cases} 2 & \text{if } n \text{ is even} \\ 3 & \text{if } n \text{ is odd.} \end{cases} \tag{7.1}$$

Exercise 2.1. Prove $\chi(G) = 2$ if and only if G is a bipartite graph with at least one edge.

Exercise 2.2. Determine the chromatic number of

(i) The Petersen graph
(ii) Wheel W_n (see Sect. 1.7, Chap. 1)
(iii) The Herschel graph (see Fig. 5.4)
(iv) The Grötzsch graph (see Fig. 7.6)

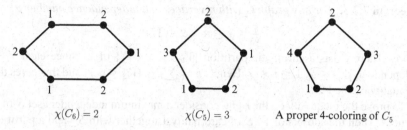

$$\chi(C_6) = 2 \qquad\qquad \chi(C_5) = 3 \qquad\qquad \text{A proper 4-coloring of } C_5$$

Fig. 7.1 Illustration of proper vertex coloring

We next consider another application of graph coloring. Let S be a set of students. Each student of S is to take a certain number of examinations for which he or she has registered. Undoubtedly, the examination schedule must be such that all students who have registered for a particular examination will take it at the same time.

Let $\mathbb{P}$ be *the set of examinations* and for $p \in \mathbb{P}$, let $S(p)$ be *the set of students* who have to take the examination p. Our aim is to draw up an examination schedule involving only the minimum number of days on the supposition that papers a and b can be given on the same day provided they have no common candidate and that no candidate shall have more than one examination on any day.

Form a graph $G = G(\mathbb{P}, E)$, where $a, b \in \mathbb{P}$ are adjacent if and only if $S(a) \cap S(b) \neq \emptyset$. Then each proper vertex coloring of G yields an examination schedule with the vertices in any color class representing the schedule on a particular day. Thus, $\chi(G)$ gives the minimum number of days required for the examination schedule.

Exercise 2.3. Draw up an examination schedule involving the minimum number of days for the following problem:

Set of students	Examination subjects
S_1	Algebra, real analysis, and topology
S_2	Algebra, operations research, and complex analysis
S_3	Real analysis, functional analysis, and complex analysis
S_4	Algebra, graph theory, and combinatorics
S_5	Combinatorics, topology, and functional analysis
S_6	Operations research, graph theory, and coding theory
S_7	Operations research, graph theory, and number theory
S_8	Algebra, number theory, and coding theory
S_9	Algebra, operations research, and real analysis

Exercise 2.4. If G is k-regular, prove that $\chi(G) \geq \frac{n}{n-k}$.

Theorem 7.2.5 gives upper and lower bounds for the chromatic number of a graph G in terms of its independence number and order.

Theorem 7.2.5. *For any graph G with n vertices and independence number* α,

$$\frac{n}{\alpha} \le \chi \le n - \alpha + 1.$$

Proof. There exists a chromatic partition $\{V_1, V_2, \ldots, V_\chi\}$ of V. Since each V_i is independent, $|V_i| \le \alpha$, $1 \le i \le \chi$. Hence, $n = \sum_{i=1}^{\chi} |V_i| \le \alpha \chi$, and this gives the inequality on the left.

To prove the inequality on the right, consider a maximum independent set S of α vertices. Then the subsets of $V \backslash S$ of cardinality 1 together with S yield a partition of V into $(n - \alpha) + 1$ independent subsets.   □

Remark 7.2.6. Unfortunately, none of the above bounds is a good one. For example, if G is the graph obtained by connecting C_{2r} with a disjoint K_{2r} $(r \ge 2)$, by an edge, we have $n = 4r$, $\alpha = r + 1$, and $\chi = 2r$, and the above inequalities become $\frac{4r}{r+1} \le 2r \le 3r$. For a simple graph G, the number $\chi^c = \chi^c(G) = \chi(G^c)$, the chromatic number of G^c is the minimum number of subsets in a partition of $V(G)$ into subsets each inducing a complete subgraph of G. Bounds on the sum and product of $\chi(G)$ and $\chi^c(G)$ were obtained by Nordhaus and Gaddum [148] (see also reference [93]), as given in Theorem 7.2.7.

Theorem 7.2.7 (Nordhaus and Gaddum [148]). *For any simple graph G,*

$$2\sqrt{n} \le \chi + \chi^c \le n + 1, \text{and} \, n \le \chi \chi^c \le \left(\frac{n+1}{2}\right)^2.$$

Proof. Let $\chi(G) = k$ and let $V_1, V_2, \ldots, V_k$ be the k color classes in a chromatic partition of G. Then $\sum_{i=1}^{k} |V_i| = n$, and so $max_{1 \le i \le k} |V_i| \ge \frac{n}{k}$. Since each V_i is an independent set of G, it induces a complete subgraph in G^c. Hence, $\chi^c \ge max_{1 \le i \le k} |V_i|$, and so $\chi \chi^c = k \chi^c \ge k \circ max_{1 \le i \le k} |V_i| \ge k \circ \frac{n}{k} = n$. Further, since the arithmetic mean of χ and χ^c is greater than or equal to their geometric mean, $\frac{\chi + \chi^c}{2} \ge \sqrt{\chi \chi^c} \ge \sqrt{n}$. Hence, $\chi + \chi^c \ge 2\sqrt{n}$. This establishes both the lower bounds.

To show that $\chi + \chi^c \le n + 1$, we use induction on n. When $n = 1$, $\chi = \chi^c = 1$, and so we have equality in this case. So assume that $\chi + \chi^c \le (n - 1) + 1 = n$ for all graphs G having $n - 1$ vertices, $n \ge 2$. Let H be any graph with n vertices, and let v be any vertex of H. Then $G = H - v$ is a graph with $n - 1$ vertices and $G^c = (H - v)^c = H^c - v$. By the induction assumption, $\chi(G) + \chi(G^c) \le n$.

Now $\chi(H) \le \chi(G) + 1$ and $\chi(H^c) \le \chi(G^c) + 1$. If either $\chi(H) \le \chi(G)$ or $\chi(H^c) \le \chi(G^c)$, then $\chi(H) + \chi(H^c) \le \chi(G) + \chi(G^c) + 1 \le n + 1$. Suppose then $\chi(H) = \chi(G) + 1$ and $\chi(H^c) = \chi(G^c) + 1$. $\chi(H) = \chi(G) + 1$ implies that removal of v from H decreases the chromatic number, and hence $d_H(v) \ge \chi(G)$. [If $d_H(v) < \chi(G)$, then in any proper coloring of G with $\chi(G)$ colors at most $\chi(G) - 1$ colors would have been used to color the neighbors of v in G, and hence v can be given one of the left-out colors, and therefore we have a coloring of H with $\chi(G)$ colors. Hence, $\chi(H) = \chi(G)$, a contradiction.] For a similar reason, $\chi(H^c) = \chi(G^c) + 1$ implies that $n - 1 - d_H(v) = d_{H^c}(v) \ge \chi(G^c)$; thus, $\chi(G) + \chi(G^c) \le d_H(v) + n - 1 - d_H(v) = n - 1$. This implies, however, that $\chi(H) + \chi(H^c) = \chi(G) + \chi(G^c) + 2 \le n + 1$.

Finally, applying the inequality $\sqrt{\chi \chi^c} \leq \frac{\chi + \chi^c}{2}$, we get $\chi \chi^c \leq (\frac{\chi + \chi^c}{2})^2 \leq (\frac{n+1}{2})^2$. $\qquad\qquad\qquad\qquad\qquad\qquad\qquad\qquad\qquad\qquad\qquad\qquad\qquad\qquad$ $\square$

Note 7.2.8. Since the publication of Theorem 7.2.7, there had been similar results for other graph parameters (see, for instance, [115] for the domination number γ). All these results have now come to be known as Nordhaus–Gaddum inequalities, with reference to the parameters in question.

Exercise 2.5. For a simple graph G, prove that $\chi(G^c) \geq \alpha(G)$.

Exercise 2.6. Prove $\chi(G) \leq \ell + 1$, where ℓ is the length of a longest path in G. For each positive integer ℓ, give a graph G with chromatic number $\ell + 1$ and in which any longest path has length ℓ.

Exercise 2.7. Which numbers can be chromatic numbers of unicyclic graphs? Draw a unicyclic graph on 15 vertices with $\Delta = 3$ and having each of these numbers as its chromatic number.

Exercise 2.8. If G is connected and $m \leq n$, show that $\chi(G) \leq 3$.

Exercise 2.9. Let G_n be the graph defined by $V(G_n) = \{(i, j) : 1 \leq i < j \leq n\}$, and $E(G_n) = \{((i, j)(k, l)) : i < j = k < l\}$. Prove

(i) $\omega(G_n) = 2$.
(ii) $\chi(G_n) = \lceil \log_2 n \rceil$. [Note that $\chi(G_n) \to \infty$ as $n \to \infty$.]

Exercise 2.10. Prove that $\chi(G \square H) = \max(\chi(G), \chi(H))$.

Exercise 2.11. Prove $\chi(G \times H) \leq \min(\chi(G), \chi(H)))$ (A celebrated conjecture of Hedetniemi [104] states that $\chi(G \times H) = \min(\chi(G), \chi(H)))$.

7.3 Critical Graphs

Definition 7.3.1. A graph G is called *critical* if for every proper subgraph H of G, $\chi(H) < \chi(G)$. Equivalently, $\chi(G - e) < \chi(G)$ for each edge e of G. Also, G is *k-critical* if it is k-chromatic and critical.

Remarks 7.3.2. If $\chi(G) = 1$, then G is either trivial or totally disconnected. Hence, G is 1-critical if and only if G is K_1. Again, $\chi(G) = 2$ implies that G is bipartite and has at least one edge. Hence, G is 2-critical if and only if G is K_2. For an odd cycle C, $\chi(C) = 3$, and if G contains an odd cycle C properly, G cannot be 3-critical.

Exercise 3.1. Prove that any critical graph is connected.

Exercise 3.2. Prove that for any graph G, $\chi(G - v) = \chi(G)$ or $\chi(G) - 1$ for any $v \in V$, and $\chi(G - e) = \chi(G)$ or $\chi(G) - 1$ for any $e \in E$.

Exercise 3.3. Show that if G is k-critical, $\chi(G - v) = \chi(G - e) = k - 1$ for any $v \in V$ and $e \in E$.

Exercise 3.4. [If $\chi(G - e) < \chi(G)$ for any e of G, G is sometimes called *edge-critical*, and if $\chi(G - v) < \chi(G)$ for any vertex v of G, G is called *vertex-critical*.] Show that a nontrivial connected graph is vertex-critical if it is edge-critical. Disprove the converse by a counterexample.

Exercise 3.5. Show that a graph is 3-critical if and only if it is an odd cycle. It is clear that any k-chromatic graph contains a k-critical subgraph. (This is seen by removing vertices and edges in succession, whenever possible, without diminishing the chromatic number.)

Theorem 7.3.3. *If G is k-critical, then $\delta(G) \geq k - 1$.*

Proof. Suppose $\delta(G) \leq k - 2$. Let v be a vertex of minimum degree in G. Since G is k-critical, $\chi(G - v) = \chi(G) - 1 = k - 1$ (see Exercise 3.3). Hence, in any proper $(k - 1)$-coloring of $G - v$, at most $(k - 2)$ colors would have been used to color the neighbors of v in G. Thus, there is at least one color, say c, that is left out of these $k - 1$ colors. If v is given the color c, a proper $(k - 1)$-coloring of G is obtained. This is impossible since G is k-chromatic. Hence, $\delta(G) \geq (k - 1)$.   □

Corollary 7.3.4. *For any graph G, $\chi(G) \leq 1 + \Delta(G)$.*

Proof. Let G be a k-chromatic graph, and let H be a k-critical subgraph of G. Then $\chi(H) = \chi(G) = k$. By Theorem 7.3.3, $\delta(H) \geq k - 1$, and hence $k \leq 1 + \delta(H) \leq 1 + \Delta(H) \leq 1 + \Delta(G)$.   □

Exercise 3.6. Give another proof of Corollary 7.3.4 by using induction on $n = |V(G)|$.

Exercise 3.7. If $\chi(G) = k$, show that G contains at least k vertices each of degree at least $k - 1$.

Exercise 3.8. Prove or disprove: If G is k-chromatic, then G contains a K_k.

Exercise 3.9. Prove: Any k ($\geq$ 2)-critical graph contains a $(k - 1)$-critical subgraph.

Exercise 3.10. For each of the graphs G of Exercise 2.2, find a critical subgraph H of G with $\chi(H) = \chi(G)$.

Exercise 3.11. Prove that the wheel $W_{2n-1} = C_{2n-1} \vee K_1$ is a 4-critical graph for each $n \geq 2$. Does a similar statement apply to W_{2n}?

Theorem 7.3.5. *In a critical graph G, no vertex cut is a clique.*

Proof. Suppose G is a k-critical graph and S is a vertex cut of G that is a clique of G (i.e., a complete subgraph of G). Let H_i, $1 \leq i \leq r$, be the components of $G \backslash S$, and let $G_i = G[V(H_i) \cup S]$. Then each G_i is a proper subgraph of G and hence admits a proper $(k - 1)$-coloring. Since S is a clique, its vertices must receive distinct colors in any proper $(k - 1)$-coloring of G_i. Hence, by fixing the colors for the vertices of S, and coloring for each i the remaining vertices of G_i so as to give a proper $(k - 1)$-coloring of G_i, we obtain a proper $(k - 1)$-coloring of G. This contradicts the fact that G is k-chromatic (see Fig. 7.2).   □

Fig. 7.2 $G[S] \simeq K_4$
$(r = 2)$

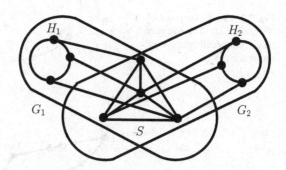

Corollary 7.3.6. *Every critical graph is a block.*

Exercise 3.12.* Prove that every k-critical graph is $(k - 1)$-edge connected (Dirac [53]).

Exercise 3.13. Show by means of an example that criticality is essential in Exercise 3.12; that is, a k-chromatic graph need not be $(k - 1)$-edge connected.

7.3.1 Brooks' Theorem

We next consider *Brooks'* [31] *theorem*. Recall Corollary 7.3.4, which states that $\chi(G) \leq 1 + \Delta(G)$. If G is an odd cycle, $\chi(G) = 3 = 1 + 2 = 1 + \Delta(G)$, and if G is a complete graph, say K_k, $\chi(G) = k = 1 + (k - 1) = 1 + \Delta(G)$. That these are the only extremal families of graphs for which $\chi(G) = 1 + \Delta(G)$ is the assertion of Brooks' theorem.

Theorem 7.3.7 (Brooks' theorem). *If a connected graph G is neither an odd cycle nor a complete graph, then $\chi(G) \leq \Delta(G)$.*

Proof. If $\Delta(G) \leq 2$, then G is either a path or a cycle. For a path G (other than K_1 and K_2), and for an even cycle G, $\chi(G) = 2 = \Delta(G)$. According to our assumption, G is not an odd cycle. So let $\Delta(G) \geq 3$.

The proof is by contradiction. Suppose the result is not true. Then there exists a minimal graph G of maximum degree $\Delta(G) = \Delta \geq 3$ such that G is not Δ-colorable, but for any vertex v of G, $G - v$ is Δ-colorable.

Claim 1. Let v be any vertex of G. Then in any proper Δ-coloring of $G - v$, all the Δ colors must be used for coloring the neighbors v in G. Otherwise, if some color i is not represented in $N_G(v)$, then v could be colored using i, and this would give a Δ-coloring of G, a contradiction to the choice of G. Thus, G is a Δ-regular graph satisfying Claim 1.

For $v \in V(G)$, let $N(v) = \{v_1, v_2, \ldots, v_\Delta\}$. In a proper Δ-coloring of $G - v = H$, let v_i receive color i, $1 \leq i \leq \Delta$. For $i \neq j$, let H_{ij} be the subgraph of H induced by the vertices receiving the ith and jth colors.

Fig. 7.3 Graphs for proof of
Theorem 7.3.7 (The numbers
inside the parentheses denote
the vertex colors)

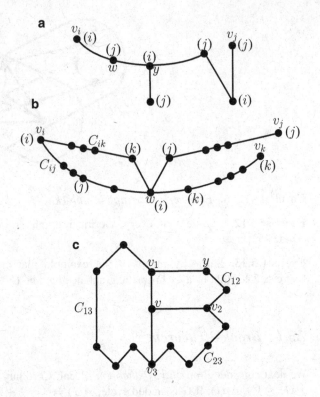

Claim 2. v_i and v_j belong to the same component of H_{ij}. Otherwise, the colors
i and j can be interchanged in the component of H_{ij} that contains the vertex v_j.
Such an interchange of colors once again yields a proper Δ-coloring of H. In this
new coloring, both v_i and v_j receive the same color, namely, i, a contradiction to
Claim 1. This proves Claim 2.

Claim 3. If C_{ij} is the component of H_{ij} containing v_i and v_j, then C_{ij} is a path in
H_{ij}. As before, $N_H(v_i)$ contains exactly one vertex of color j. Further, C_{ij} cannot
contain a vertex, say y, of degree at least 3; for, if y is the first such vertex on a
$v_i - v_j$ path in C_{ij} that has been colored, say, with i, then at least three neighbors
of y in C_{ij} have the color j. Hence, we can recolor y in H with a color different
from both i and j, and in this new coloring of H, v_i and v_j would belong to distinct
components of H_{ij} (see Fig. 7.3a). (Note that by our choice of y, any $v_i - v_j$ path
in H_{ij} must contain y.) But this contradicts Claim 3.

Claim 4. $C_{ij} \cap C_{ik} = \{v_i\}$ for $j \neq k$. Indeed, if $w \in C_{ij} \cap C_{ik}$, $w \neq v_i$, then
w is adjacent to two vertices of color j on C_{ij} and two vertices of color k on C_{ik}
(see Fig. 7.3b). Again, we can recolor w in H by giving a color different from the
colors of the neighbors of w in H. In this new coloring of H, v_i and v_j belong to
distinct components of H_{ij}, a contradiction to Claim 2. This completes the proof of
Claim 4.

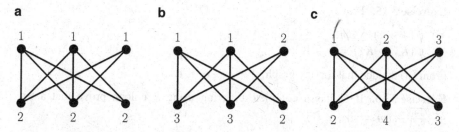

Fig. 7.4 Different colorings of $K_{3,3} - e$

We are now in a position to complete the proof of the theorem. By hypothesis, G is not complete. Hence, G has a vertex v, and a pair of nonadjacent vertices v_1 and v_2 in $N_G(v)$ (see Exercise 5.11, Chap. 1). Then the $v_1 - v_2$ path C_{12} in H_{12} of $H = G - v$ contains a vertex y ($\neq v_2$) adjacent to v_1. Naturally, y would receive color 2. Since $\Delta \geq 3$, by Claim 1, there exists a vertex $v_3 \in N_G(v)$. Now interchange colors 1 and 3 in the path C_{13} of H_{13}. This would result in a new coloring of $H = G - v$. Denote the v_i-v_j path in H under this new coloring by C'_{ij} (see Fig. 7.3c). Then $y \in C'_{23}$ since v_1 receives color 3 in the new coloring (whereas y retains color 2). Also, $y \in C_{12} - v_1 \subset C'_{12}$. Thus, $y \in C'_{23} \cap C'_{12}$. This contradicts Claim 4 (since $y \neq v_2$), and the proof is complete. □

7.3.2 Other Coloring Parameters

There are several other vertex coloring parameters of a graph G. We now mention three of them. Let f be a k-coloring (not necessarily proper) of G, and let $(V_1, V_2, \ldots, V_k)$ be the color classes of G induced by f. Coloring f is *pseudo-complete* if between any two distinct color classes, there is at least one edge of G. f is *complete* if it is pseudocomplete and each V_i, $1 \leq i \leq k$, is an independent set of G. Thus, $\chi(G)$ is the minimum k for which G has a complete k-coloring f.

Definition 7.3.8. The *achromatic number* $a(G)$ of a graph G is the maximum k for which G has a complete k-coloring.

Definition 7.3.9. The *pseudoachromatic number* $\psi(G)$ of G is the maximum k for which G has a pseudocomplete k-coloring.

Example 7.3.10. Figure 7.4 gives (a) a chromatic, (b) an achromatic, and (c) a pseudoachromatic coloring of $K_{3,3} - e$.

It is clear that for any graph G, $\chi(G) \leq a(G) \leq \psi(G)$.

Exercise 3.14. Let G be a graph and H a subgraph of G. Prove that $\chi(H) \leq \chi(G)$ and $\psi(H) \leq \psi(G)$. Show by means of an example that $a(H) \leq a(G)$ need not always be true.

Exercise 3.15. Prove

(i) $\psi(\psi - 1) \leq 2m$.
(ii) $\psi(K_a \vee K_b^c) = a + 1$.

From (ii) deduce that for any graph, $\psi \leq n - \alpha + 1$.

Exercise 3.16. If G has a complete coloring using k colors, prove that $k \leq \frac{1 + \sqrt{1+8m}}{2}$. ($m$ = size of G).

Exercise 3.17. Prove that for a complete bipartite graph G, $a(G) = 2$.

Exercise 3.18. What is the minimum number of edges that a connected graph with pseudoachromatic number ψ can have? Construct one such tree.

Exercise 3.19. If G is a subgraph of H, prove that $\psi(G) \leq \psi(H)$.

Exercise 3.20. Prove: $\psi(K_{n,n}) = n + 1$.

7.3.3   b-Colorings

Definition 7.3.11. A b-*coloring* of a graph G is a proper coloring with the additional property that each color class contains a color-dominating vertex (c.d.v.), that is, a vertex that has a neighbor in all the other color classes. The b-*chromatic number* of G is the largest k such that G has a b-coloring using k colors; it is denoted by b(G).

The concept of b-coloring was introduced by Irving and Manlove [111].

Exercise 3.21 guarantees the existence of the b-chromatic number for any graph G and shows that $\chi(G) \leq$ b(G). Note that b$(K_n) = n$ while b$(K_{m,n}) = 2$.

Exercise 3.21. Show that the chromatic coloring of a graph G is a b-coloring of G.

Exercise 3.22. Prove that $K_{n,n} - F$, $n \geq 2$, where F is a 1-factor of $K_{n,n}$, has a b-coloring using 2 colors and n colors but none with k colors for any k in $2 < k < n$.

Exercise 3.23. Prove b$(G) \leq 1 + \Delta(G)$. A better upper bound for b(G) is given in the next exercise.

Exercise 3.24. Let $d_1 \leq d_2 \leq \ldots \leq d_n$ be the degree sequence of the graph G with vertex set $V = \{v_1, \ldots, v_n\}$, and $d_i = d(v_i)$, $1 \leq i \leq n$. Let $M(G) = \max\{i : d_i \geq i - 1, 1 \leq i \leq n\}$. Prove that b$(G) \leq M(G)$. Show further that the number of vertices of degree at least $M(G)$ in G is at most $M(G)$.

Exercise 3.25. Let Q_p be the hypercube of dimension p. Prove b$(Q_1) =$ b$(Q_2)=2$, and b$(Q_3) = 4$. [A result of Kouider and Mahéo [125] states that for $p \geq 3$, b$(Q_p) = p + 1$.]

We complete this section by presenting a result of Kratochvíl, Tuza, and Voigt [126] that characterizes graphs with b-chromatic number 2. Let G be a bipartite graph with bipartition (X, Y). A vertex $x \in X$ (respectively, $y \in Y$) is called a *full vertex* (or a *charismatic vertex*) of X (respectively, Y) if it is adjacent to all the vertices of Y (respectively, X).

Theorem 7.3.12 ([126]). *Let G be a nontrivial connected graph. Then* $b(G) = 2$ *if and only if G is bipartite and has a full vertex in each part of the bipartition.*

Proof. Suppose G is bipartite and has a full vertex in each part, say $x \in X$ and $y \in Y$. Naturally, in any b-coloring, the color class containing x, say W_1, is a subset of X and that containing y, say W_2, is a subset of Y. If G has a third color class W_3 disjoint from W_1 and W_2, then W_3 must have a c.d.v. adjacent to a vertex of W_1 and a vertex of W_2. This is impossible, as G is bipartite. Therefore, $b(G) = 2$.

Conversely, let $b(G) = 2$. Then $\chi(G) = 2$ and therefore G is bipartite. Let (X, Y) be the bipartition of G. Assume that G does not have a full vertex in at least one part, say, $\dot{X}$. Let $x_1 \in X$. As x_1 is not a full vertex, there exists a vertex $y_1 \in Y$ to which it is not adjacent. Let X_1 be the maximal subset of X such that $V_1 = X_1 \cup \{y_1\}$ is independent in G. Now choose a new vertex $x_2 \in X \backslash X_1$. Again, as X has no full vertex, we can find a $y_2 \in Y \backslash \{y_1\}$ to which x_2 is not adjacent. Let X_2 be the maximal subset of $X \backslash X_1$ such that $V_2 = X_2 \cup \{y_2\}$ is independent in G. In this way, all the vertices of X would be exhausted and let $V_1, V_2, \ldots, V_k$ be the independent sets thus formed. Also, let Y_0 denote the set of uncovered vertices of Y, if any. Since G is connected, $G \neq \langle V_i \cup V_j \rangle$, and $G \neq \langle V_l \cup Y_0 \rangle$, $i, j, l \in \{1, 2, \ldots, k\}$. Hence, $k \geq 2$ when $Y_0 \neq \emptyset$ and $k \geq 3$ when $Y_0 = \emptyset$. Thus, the partition $V = V_1 \cup V_2 \cup \ldots \cup V_k \cup \{V_{k+1} = Y_0\}$ has at least 3 parts. If each of these parts has a c.d.v., we get a contradiction to the fact that $b(G) = 2$. If not, assume that the class V_l has no c.d.v. Then for each vertex x of V_l, there exists a color class V_j, $j \neq l$, having no neighbor of x. Then x could be moved to the class V_j. In this way, the vertices in V_l can be moved to the other V_i's without disturbing independence. Let us call the new classes $V_1', V_2', \ldots, V_{l-1}', V_{l+1}', \ldots, V_{k+1}'$. If each of these color classes contains a c.d.v., we get a contradiction as $k \geq 3$. Otherwise, argue as before and reduce the number of color classes. As G is connected, successive reductions should end up in at least three classes, contradicting the hypothesis that $b(G) = 2$. $\square$

A description of several other coloring parameters can be found in Jensen and Toft [116].

7.4 Homomorphisms and Colorings

Homomorphisms of graphs generalize the concept of graph colorings.

Definition 7.4.1. Let G and H be simple graphs. A *homomorphism* from G to H is a map $f : V(G) \rightarrow V(H)$ such that $f(x)f(y) \in E(H)$ whenever $xy \in E(G)$. The map f is an *isomorphism* if f is bijective and $xy \in E(G)$ if and only if $f(x)f(y) \in E(H)$.

We write $f : G \to H$ to denote the fact that f is a homomorphism from G
to H and write $G \simeq H$ to denote that G is isomorphic to H. If $f : G \to H$
is a graph homomorphism, then $\langle \{ f(x) : x \in V(G) \} \rangle$, the subgraph induced in
H by the image set $f(V(G))$ is the image of f. If $f(V(G)) = V(H)$, f is an
onto-homomorphism. If $f : G \to H$, then for any vertex v of H, $f^{-1}(v)$ is an
independent set of G. (If $f^{-1}(v) = \emptyset$, then $f^{-1}(v)$ is an independent subset of
$V(G)$, while if $f^{-1}(v)$ contains an edge uw, then $f(u) = v = f(w)$, and hence H
has a loop at v, a contradiction to the fact that H is a simple graph).

Lemma 7.4.2. *Let G_1, G_2, and G_3 be graphs and let $f_1 : G_1 \to G_2$, and $f_2 :
G_2 \to G_3$ be homomorphisms. Then $f_2 \circ f_1 : G_1 \to G_3$ is also a homomorphism.
[Here $(f_2 \circ f_1)(g) = f_2(f_1(g))$.]*

Proof. Follows by direct verification. □

A graph homomorphism is a generalization of graph coloring. Suppose G ia a
given graph and there exists a homomorphism $f : G \to K_k$, where k is the least
positive integer with this property. Then f is onto and the sets $S_i = \{ f^{-1}(v_i) :
v_i \in K_k \}$, $1 \le i \le k$, form a partition of $V(G)$. Moreover, between any two
sets S_i and S_j, $i \ne j$, there must be an edge of G. Otherwise, $A = S_i \cup S_j$ is
an independent set of G, and we can define a homomorphism from G to K_{k-1} by
mapping A to the same vertex of K_{k-1}. Thus, $\chi(G) = k$. We state this result as a
theorem.

Theorem 7.4.3. *Let G be a simple graph. Suppose there exists a homomorphism
$f : G \to K_k$, and let k be the least positive integer with this property. Then
$\chi(G) = k$.*

Corollary 7.4.4. *If there exists a homomorphism $f : G \to K_p$, then $\chi(G) \le p$.*

Corollary 7.4.5. *Let $f : G \to H$ be a graph homomorphism. Then $\chi(G) \le
\chi(H)$.*

Proof. Let $\chi(H) = k$. Then there exists a homomorphism $g : H \to K_k$. By
Lemma 7.4.2, $g \circ f : G \to K_k$ is a homomorphism. Now apply Corollary 7.4.4. □

Example 7.4.6. Let $V_1 = \{u_1, \dots, u_7\}$ and $V_2 = \{v_1, \dots, v_5\}$ be the vertex sets of
the cycles C_7 and C_5, respectively. Then the map $f(u_1) = v_1$, $f(u_2) = v_2$, $f(u_3) =
v_3$, $f(u_4) = v_2$, $f(u_5) = v_3$, $f(u_6) = v_4$, and $f(u_7) = v_5$ is a homomorphism of
C_7 to C_5.

7.4.1 Quotient Graphs

Let $f : G \to H$ be a graph homomorphism from G onto H. Let $V(H) =
\{v_1, \dots, v_k\}$, and $S_i = f^{-1}(v_i)$, $1 \le i \le k$. Then no S_i is empty. The quotient
graph G/f is defined to be the graph with the sets S_i as its vertices and in which two

vertices S_i and S_j are adjacent if $v_i v_j \in E(H)$. This defines a natural isomorphism $\tilde{f} : G/f \simeq H$.

A consequence of the above remarks is the fact that a complete k-coloring of G is just a homomorphism of G onto K_k. Recall that both the chromatic and achromatic colorings are complete colorings. We now establish the coloring interpolation theorem for the complete coloring.

Theorem 7.4.7 (Interpolation theorem for complete coloring). *If a graph G admits a complete k-coloring and a complete l-coloring, then it admits a complete i-coloring for every i between k and l.*

Proof. Let $A_1, A_2, \ldots, A_k$ and $B_1, B_2, \ldots, B_l$ be the color partitions in the two complete colorings. We assume without loss of generality that $k < l$. Clearly, it suffices to construct a complete $(k + 1)$-coloring of G. For each $i = 0, 1, 2, \ldots, l$, let $C_i = \bigcup_{1 \leq j \leq i} B_j$. Let Θ_i denote the partition of $V(G)$ by the nonempty sets of the sequence $B_1, B_2, \ldots, B_i$; $A_1 - C_i, A_2 - C_i, \ldots, A_k - C_i$. The partition Θ_0 has parts $A_1, A_2, \ldots, A_k$; the partition Θ_l has parts $B_1, B_2, \ldots, B_l$ (since $C_l = V(G)$, $A_i - C_l = \emptyset$ for each j). Hence, $G/\Theta_0 \simeq K_k$ and $G/\Theta_l \simeq K_l$. Hence, there must exist a first suffix j, $0 < j \leq l$, such that G/Θ_j is not k-colorable. By the choice of j, this implies that G/Θ_j is $(k + 1)$-colorable since we can simply color B_j by the $(k + 1)$-st color, and hence by Lemma 7.4.2, G is $(k + 1)$-colorable. (Just compose the two onto homomorphisms $G \to G/\Theta_j \to K_{k+1}$.) $\qquad\square$

Exercise 3.22 shows that an interpolation theorem similar to that of complete coloring does not hold good for the b-coloring.

Exercise 4.1. Let $f : G \to H$ be a graph homomorphism and let $x, y \in V(G)$. Prove $d_H(x, y) \leq d_G(x, y)$.

Exercise 4.2. Assume that there exists a homomorphism from G onto C_k, where k is odd. Show that G must contain an odd cycle. Show by means of an example that a similar statement need not hold good if k is even.

Exercise 4.3. Prove that there exists a homomorphism from C_{2l+1} to C_{2k+1} if and only if $l \leq k$.

7.5 Triangle-Free Graphs

Definition 7.5.1. A graph G is *triangle-free* if G contains no K_3.

Remark 7.5.2. Triangle-free graphs cannot contain a $K_k, k \geq 3$, either. It is obvious that if a graph G contains a clique of size k, then $\chi(G) \geq k$. However, the converse is not true. That is, if the chromatic number of G is large, then G need not contain a clique of large size. The construction of triangle-free k-chromatic graphs, for $k \geq 3$, was raised in the middle of the 20th century. In answer to this question, Mycielski [144] developed an interesting graph transformation known as the *Mycielskian* of a graph.

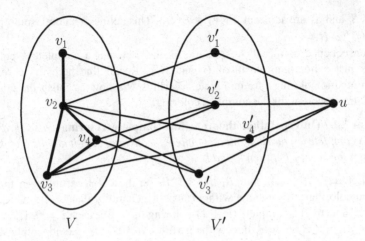

Fig. 7.5 $\mu(K_{1,3} + e)$

Definition 7.5.3. Let G be a finite simple connected graph with vertex set $V = V(G)$ and edge set $E = E(G)$. The *Mycielskian* $\mu(G)$ of G is defined as follows: The vertex set $V(\mu(G))$ of $\mu(G)$ is the disjoint union $V \cup V' \cup \{u\}$, where $V' = \{x' : x \in V\}$ and the edge set of $\mu(G)$ is $E(\mu(G)) = E \cup \{x'y : xy \in E\} \cup \{x'u : x' \in V'\}$.

We denote $V(\mu(G))$ by the triad $\{V, V', u\}$. For $x \in V$, we call $x' \in V'$, the twin of x in $\mu(G)$, and vice versa, and u, the root of $\mu(G)$. Figure 7.5 displays the Mycielskian $\mu(K_{1,3} + e)$.

Remark 7.5.4. The following facts about $\mu(G)$, where G is of order n and size m, are obvious:

  (i) $|V(\mu(G))| = 2n + 1$.
 (ii) For each $v \in V$, $d_{\mu(G)}(v) = 2d_G(v)$.
(iii) For each $v' \in V'$, $d_{\mu(G)}(v') = d_G(v) + 1$.
(iv) $d_{\mu(G)}(u) = n$.

We now establish some basic results concerning the Mycielskian.

Theorem 7.5.5. $\chi(\mu(G)) = \chi(G) + 1$.

Proof. Assume that $\chi(G) = k$. Consider a proper (vertex) k-coloring c of G using the colors, say, $1, 2, \ldots, k$. We now give a proper $(k + 1)$-coloring c' for $\mu(G)$. For $v \in V$, set $c'(v) = c(v)$. For the twin $v' \in V'$, set $c'(v') = c(v)$. For the root u of $\mu(G)$, set $c'(u) = k + 1$. Then c' is a proper coloring for $\mu(G)$ using $k + 1$ colors and therefore $\chi(\mu(G)) \leq k + 1$. [c' is proper because for any edge xy', $c'(x) = c(x) \neq c(y) = c'(y')$.] We now show that it is actually $k + 1$.

Suppose $\mu(G)$ has a proper k-coloring c'' using the colors $1, 2, \ldots, k$. Assume, without loss of generality, that $c''(u) = 1$. Then for any $v' \in V', c''(v') \neq 1$. Recolor each vertex of V that has been colored by 1 in c'' by the color of its twin under c''.

Fig. 7.6 The Grötzsch graph, $\mu(C_5)$

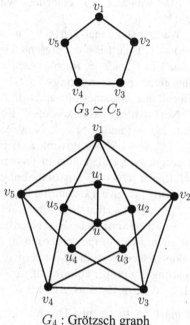

$G_3 \simeq C_5$

G_4 : Grötzsch graph

Then this gives a proper coloring of V using the $k - 1$ colors $2, 3, \dots, k$. This is impossible as $\chi(G) = k$. This proves that $\chi(\mu(G)) = k + 1 = \chi(G) + 1$. $\qquad \square$

Theorem 7.5.6. *If G triangle-free, then $\mu(G)$ is also triangle-free.*

Proof. Assume that G is triangle-free. If $\mu(G)$ contains a triangle, it can only be of the form vwz', where $v \in V$, $w \in V$, and $z' \in V'$, so that vz' and wz' are edges of $\mu(G)$. This means, by the definition of $\mu(G)$, that vz and wz are edges of G and hence vwz is a triangle in G, a contradiction. $\qquad \square$

Theorem 7.5.7 (Mycielski [144]). *For any positive integer p, there exists a triangle-free graph with chromatic number p.*

Proof. For $p = 1, 2$, the result is trivial. [For $p = 1$, take $G = K_1$, and for $p = 2$, take $G = K_2$. For $p = 3$, take $G = \mu(K_2)$. $\mu(K_2) = C_5$ is triangle-free and $\chi(C_5) = 3$.] For $p \geq 3$, by Theorems 7.5.5 and 7.5.6, the iterated Mycielskian $\mu^{p-2}(K_2) = \mu(\mu^{p-3}(K_2))$ is triangle-free and has chromatic number p. $\qquad \square$

Remark 7.5.8. The graph $\mu^2(K_2) = \mu(C_5)$ is the Grötzsch graph of Fig. 7.6.

Theorem 7.5.9. *If G is critical, then so is $\mu(G)$.*

Proof. Assume that G is k-critical. Since by Theorem 7.4.5, $\chi(\mu(G)) = k + 1$, we have to show that $\mu(G)$ is $(k + 1)$-critical.

Start with a $(k+1)$-coloring c with colors $1, 2, \ldots, k+1$ of $\mu(G)$ with vertex set $\{V, V', u\}$.

We first show that $\chi(\mu(G) - u) = k$. Without loss of generality, assume that $c(u) = 1$. Then 1 is not represented in V'. Let S be the set of vertices receiving the color 1 in V under c. Recolor each vertex v of S by the color of its twin $v' \in V'$. This gives a proper coloring of $\mu(G) - u$ using k colors and hence $\chi(\mu(G) - u) = k$. [Recall that adjacency of v and w in G implies adjacency of vw' and $v'w$ in $\mu(G)$.]

Next remove a vertex v' of V' from $\mu(G)$. Without loss of generality, assume that $c(u) = 1$ and $c(v') = 2$. Now recolor the vertices of $G - v$ by the $k - 1$ colors $3, \ldots, k, k+1$ (this is possible as G is k-critical) and recolor the vertices of $V' - v'$, if necessary, by the colors of their twins in $V - v$. Also, give color 1 to v. This coloring of $\mu(G) - v'$ misses the color 2 and gives a proper k-coloring to $\mu(G) - v'$.

Lastly, we give a k-coloring to $\mu(G) - v$, $v \in V$. Color the vertices of $G - v$ by $1, 2, \ldots, k-1$ so that the resulting coloring of $G - v$ is proper. Let A be the subset of $G - v$ whose vertices have received color 1 in this new coloring and $A' \subset V'$ denote the set of twins of the vertices in A. Now color the vertices of $(V' \backslash A') - v'$ by the colors of their twins in G, the vertices of $A' \cup \{v'\}$ by color k, and u by color 1. This coloring is a proper coloring of $\mu(G) - v$, which misses the color $k+1$ in the list $\{1, 2, \ldots, k+1\}$. Thus, $\mu(G)$ is $(k+1)$-critical. $\square$

Remark 7.5.10. Apply Theorem 7.5.12 to observe that for each $k \geq 1$, there exists a k-critical triangle-free graph. Not every k-critical graph is triangle-free; for example, the complete graph K_k $(k \geq 3)$ is k-critical but is not triangle-free.

Lemma 7.5.11. *Let* $f : G \to H$ *be a graph isomorphism of G onto H. Then* $f(N_G(x)) = N_H(f(x))$. *Further,* $G - x \simeq H - f(x)$, *and* $G - N_G[x] \simeq H - N_H[f(x)]$ *under the restriction maps of f to the respective domains.*

Proof. The proof follows from the definition of graph isomorphism. $\square$

Theorem 7.5.12 ([13]). *For connected graphs G and H, $\mu(G) \simeq \mu(H)$ if and only if $G \simeq H$.*

Proof. If $G \simeq H$, then trivially $\mu(G) \simeq \mu(H)$. So assume that G and H are connected and that $\mu(G) \simeq \mu(H)$. When $n = 2$ or 3, the result is trivial. So assume that $n \geq 4$. If G is of order n, then $\mu(G)$ and $\mu(H)$ are both of order $2n+1$, and so H is also of order n. Let $f : \mu(G) \to \mu(H)$ be the given isomorphism, where $V(\mu(G))$ and $V(\mu(H))$ are given by the triads (V_1, V_1', u_1) and (V_2, V_2', u_2), respectively.

We look at the possible images of the root u_1 of $\mu(G)$ under f. Both u_1 and u_2 are vertices of degree n. If $f(u_1) = u_2$, then by Lemma 7.5.11, $G = \mu(G) - N[u_1] \simeq \mu(H) - N[u_2] = H$.

Next we claim that $f(u_1) \notin V_2$. Suppose $f(u_1) \in V_2$. Since $d_{\mu(H)}(f(u_1)) = d_{\mu(G)}(u_1) = n$, it follows from the definition of the Mycielskian that in $\mu(H)$, $\frac{n}{2}$ neighbors of $f(u_1)$ belong to V_2 while another $\frac{n}{2}$ neighbors (the twins) belong to V_2'. (This forces n to be even.) These n neighbors of $f(u_1)$ form an independent subset of $\mu(H)$. Then $H' = \mu(H) - N_{\mu(H)}[f(u_1)] \simeq \mu(G) - N_{\mu(G)}[u_1] = G$.

Now if $x \in V_2$ is adjacent to $f(u_1)$ in $\mu(H)$, then x is adjacent to $f(u_1)'$, the twin of $f(u_1)$ belonging to V_2' in $\mu(H)$. Further, $d_{H'}(f(u_1)') = 1 = d_G(v)$, where $v \in V_1$ (the vertex set of G) corresponds to $f(u_1)'$ in $\mu(H)$. But then $d_{\mu(G)}(v) = 2$, while $d_{\mu(H)}(f(u)') = \frac{n}{2} + 1 > 2$, as $n \geq 4$. Hence, this case cannot arise.

Finally, suppose that $f(u_1) \in V_2'$. Set $f(u_1) = y'$. Then y, the twin of y' in $\mu(H)$, belongs to V_2. As $d_{\mu(G)}(u_1) = n$, $d_{\mu(H)}(y') = n$. The vertex y' has $n - 1$ neighbors in V_2, say, $x_1, x_2, \ldots, x_{n-1}$. Then $N_H(y) = \{x_1, x_2, \ldots, x_{n-1}\}$, and hence y is also adjacent to $x_1', x_2', \ldots, x_{n-1}'$ in V_2'. Further, as $N_{\mu(G)}(u_1)$ is independent, $N_{\mu(H)}(y')$ is also independent. Therefore, $H = $ star $K_{1,n-1}$ consisting of the edges $yx_1, yx_2, \ldots, yx_{n-1}$. Moreover, $G = \mu(G) - N[u_1] \simeq \mu(H) - N[y'] = $ star $K_{1,n-1}$ consisting of the edges $yx_1', yx_2', \ldots, yx_{n-1}'$. Thus, $G \simeq K_{1,n-1} \simeq H$. □

7.6 Edge Colorings of Graphs

7.6.1 The Timetable Problem

Suppose in a school there are r teachers, $T_1, T_2, \ldots, T_r$, and s classes, $C_1, C_2, \ldots, C_s$. Each teacher T_i is expected to teach the class C_j for p_{ij} periods. It is clear that during any particular period, no more than one teacher can handle a particular class and no more than one class can be engaged by any teacher. Our aim is to draw up a timetable for the day that requires only the minimum number of periods. This problem is known as the "timetable problem."

To convert this problem into a graph-theoretic one, we form the bipartite graph $G = G(T, C)$ with bipartition (T, C), where T represents the set of teachers T_i and C represents the set of classes C_j. Further, T_i is made adjacent to C_j in G with p_{ij} parallel edges if and only if teacher T_i is to handle class C_j for p_{ij} periods. Now color the edges of G so that no two adjacent edges receive the same color. Then the edges in a particular color class, that is, the edges in that color, form a matching in G and correspond to a schedule of work for a particular period. Hence, the minimum number of periods required is the minimum number of colors in an edge coloring of G in which adjacent edges receive distinct colors; in other words, it is the edge-chromatic number of G. We now present these notions as formal definitions.

Definition 7.6.1. An *edge coloring* of a loopless graph G is a function $\pi : E(G) \to S$, where S is a set of distinct colors; it is *proper* if no two adjacent edges receive the same color. Thus, a proper edge coloring π of G is a function $\pi : E(G) \to S$ such that $\pi(e) \neq \pi(e')$ whenever edges e and e' are adjacent in G, and it is a proper k-edge coloring of G if $|S| = k$.

Definition 7.6.2. The minimum k for which a loopless graph G has a proper k-edge coloring is called the *edge-chromatic number* or *chromatic index* of G. It is denoted by $\chi'(G)$. G is *k-edge-chromatic* if $\chi'(G) = k$.

Further, if an edge uv is colored by color c, we say that c is represented at both u and v. If G has a proper k-edge coloring, $E(G)$ is partitioned into k edge-disjoint matchings.

It is clear that for any (loopless) graph G, $\chi'(G) \geq \Delta(G)$ since the $\Delta(G)$ edges incident at a vertex v of maximum degree $\Delta(G)$ must all receive distinct colors. For bipartite graphs, however, equality holds.

Theorem 7.6.3 (König). *If G is a bipartite graph, $\chi'(G) = \Delta(G)$.*

Proof. The proof is by induction on the size (i.e., number of edges) m of G. The result is true for $m = 1$. Assume the result for bipartite graphs of size at most $m - 1$. Let G have m edges. Let $e = uv \in E(G)$. Then $G - e$ has [since $\Delta(G - e) \leq \Delta(G)$] a proper Δ-edge coloring, say c. Out of these Δ colors, suppose that one particular color is not represented at both u and v. Then in this coloring the edge uv can be colored with this color, and a proper Δ-edge coloring of G is obtained.

In the other case (that is, in the case in which each of the Δ colors is represented either at u or at v in $G - e$), since the degrees of u and v in $G - e$ are at most $\Delta - 1$, there exists a color out of the Δ colors that is not represented in $G - e$ at u, and similarly there exists a color not represented at v. Thus, if color j is not represented at u in c, then j is represented at v in c, and if color i is not represented at v in c, then i is represented at u in c. Since G is bipartite and u and v are not in the same parts of the bipartition, there can exist no u-v path in G in which the colors alternate between i and j.

Let P be a maximal path in $G - e$ starting from u in which the colors of the edges alternate between i and j. Interchange the colors i and j in P. This would still yield a proper edge coloring of $G - e$ using the Δ colors in which color i is not represented at both u and v. Now color the edge uv by the color i. This results in a proper Δ-edge coloring of G. $\qquad\square$

Exercise 6.1. Disprove the converse of Theorem 7.6.3 by a counterexample.

Next, we determine the chromatic index of the complete graphs.

Theorem 7.6.4. $\chi'(K_n) = \begin{cases} n - 1 & \text{if } n \text{ is even,} \\ n & \text{if } n \text{ is odd.} \end{cases}$

Proof. (Berge) Since K_n is regular of degree $n - 1$, $\chi'(K_n) \geq n - 1$.

Case 1. n is even. We show that $\chi'(K_n) \leq n - 1$ by exhibiting a proper $(n - 1)$-edge coloring of K_n. Label the n vertices of K_n as $0, 1, \ldots, n - 1$. Draw a circle with center at 0 and place the remaining $n - 1$ numbers on the circumference of the circle so that they form a regular $(n - 1)$-gon (Fig. 7.7). Then the $\frac{n}{2}$ edges $(0, 1)$, $(2, n - 1)$, $(3, n - 2), \ldots, (\frac{n}{2}, \frac{n}{2} + 1)$ form a 1-factor of K_n. These $\frac{n}{2}$ edges are the thick edges of Fig.7.7. Rotation of these edges through the angle $\frac{2\pi}{n-1}$ in succession gives $(n - 1)$ edge-disjoint 1-factors of K_n. This would account for $\frac{n}{2}(n - 1)$ edges and hence all the edges of K_n. (Actually, the above construction displays a 1-factorization of K_n when n is even.) Each 1-factor can be assigned a distinct color. Thus, $\chi'(K_n) \leq n - 1$. This proves the result in Case 1.

Fig. 7.7 Graph for proof
of Theorem 7.6.4

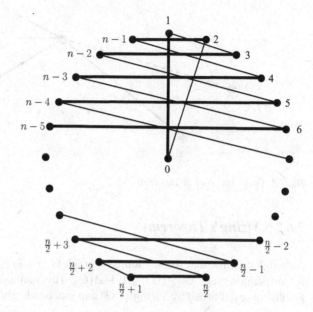

Case 2. n is odd. Take a new vertex and make it adjacent to all the n vertices of K_n. This gives K_{n+1}. By Case 1, $\chi'(K_{n+1}) = n$. The restriction of this edge coloring to K_n yields a proper n-edge coloring of K_n. Hence, $\chi'(K_n) \le n$. However, K_n cannot be edge colored properly with $n - 1$ colors. This is because the size of any matching of K_n can contain no more than $\frac{n-1}{2}$ edges, and hence $n - 1$ matchings of K_n can contain no more than $\frac{(n-1)^2}{2}$ edges. But K_n has $\frac{n(n-1)}{2}$ edges. Thus, $\chi'(K_n) \ge n$, and hence $\chi'(K_n) = n$. $\qquad\square$

Exercise 6.2. Show that a Hamiltonian cubic graph is 3-edge-chromatic.

Exercise 6.3. Show that the Petersen graph is 4-edge-chromatic.

Exercise 6.4. Show that the Herschel graph (see Fig. 5.4) is 4-edge-chromatic.

Exercise 6.5. Determine the edge-chromatic number of the Grötzsch graph (Fig. 7.6).

Exercise 6.6. Show that a simple cubic graph with a cut edge is 4-edge-chromatic.

Exercise 6.7. Describe a proper k-edge coloring of a k-regular bipartite graph.

Exercise 6.8. Show that any bipartite graph G of maximum degree Δ is a subgraph of a Δ-regular bipartite graph. Hence, furnish an alternative proof of Theorem 7.6.3, using Exercise 6.7.

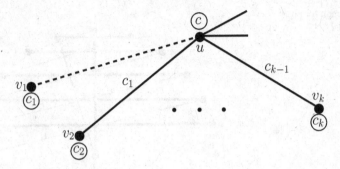

Fig. 7.8 Graph for proof of Theorem 7.6.5

7.6.2 Vizing's Theorem

Although it is true that for any loopless graph G, $\chi'(G) \geq \Delta(G)$, it turns out that for any simple graph G, $\chi'(G) \leq 1 + \Delta(G)$. This major result in edge coloring of graphs was established by Vizing [183] and independently by Gupta [81].

Theorem 7.6.5 (Vizing-Gupta). *For any simple graph G, $\Delta(G) \leq \chi'(G) \leq 1 + \Delta(G)$.*

Proof. In a proper edge coloring of G, $\Delta(G)$, colors are to be used for the edges incident at a vertex of maximum degree in G. Hence, $\chi'(G) \geq \Delta(G)$.

We now prove that $\chi'(G) \leq 1 + \Delta$, where $\Delta = \Delta(G)$.

If G is not $(1 + \Delta)$-edge-colorable, choose a subgraph H of G with a maximum possible number of edges such that H is $(1 + \Delta)$-edge-colorable. We derive a contradiction by showing that there exists a subgraph H_0 of G that is $(1 + \Delta)$-edge-colorable and has one edge more than H.

By our assumption, G has an edge $uv_1 \notin E(H)$. Since $d(u) \leq \Delta$, and $1 + \Delta$ colors are being used in H, there is a color c that is not represented at u (i.e., not used for any edge of H incident at u). For the same reason, there is a color c_1 not represented at v_1. (See Fig. 7.8, where the color not represented at a particular vertex is enclosed in a circle and marked near the vertex.)

There must be an edge, say uv_2 of H, colored c_1; otherwise, uv_1 can be assigned the color c_1, and $H \cup (uv_1)$, which has one edge more than H, would have a proper $(1 + \Delta)$-edge coloring. Again, there is a color, say c_2, not represented at v_2. Then as above, there is an edge uv_3 colored c_2 and there is a color, say c_3, not represented at v_3.

In this way, we construct a sequence of edges $\{uv_1, uv_2, \ldots, uv_k\}$ such that color c_i is not represented at vertex v_i, $1 \leq i \leq k$, and the edge uv_{j+1} receives the color c_j, $1 \leq j \leq k - 1$ (see Fig. 7.8).

Suppose at some stage, say the rth stage, where $1 \leq r \leq k$, c (the missing color at u) is not represented at v_r. We then "cascade" (i.e., shift in order) the colors $c_1, \ldots, c_{r-1}$ from $uv_2, uv_3, \ldots, uv_r$ to $uv_1, uv_2, \ldots, uv_{r-1}$. Under this new coloring,

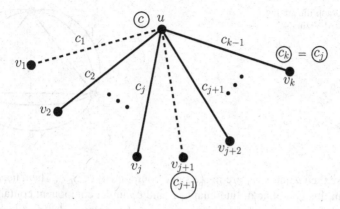

Fig. 7.9 Another graph for proof of Theorem 7.6.5

c is not represented both at u and at v_r, and therefore we can color uv_r with c. This yields a proper $(1 + \Delta)$-edge coloring to $H \cup (uv_1)$, contradicting the choice of H. Hence, we may assume that c is represented at each of the vertices $v_1, v_2, \ldots, v_k$.

Now we need to know why the sequence of edges uv_i, $1 \leq i \leq k$, had stopped. There are two possible reasons. Either there is no edge incident to u that is colored c_k, or the color $c_k = c_j$ for some $j < k-1$ and so has already been represented at u. Note that the sequence must stop at some finite stage since $d(u)$ is finite; however, it may as well stop before all the edges incident to u are exhausted.

If c_k is not represented at u in H, then we can cascade as before so that uv_i gets color c_i, $1 \leq i \leq k - 1$, and then color uv_k with color c_k. Once again, we have a contradiction to our assumption on H.

Thus, we must have $c_k = c_j$ for come $j < k - 1$. In this case, cascade the colors $c_1, c_2, \ldots, c_j$ so that uv_i has color c_i, $1 \leq i \leq j$, and leave uv_{j+1} uncolored (Fig. 7.9). Let $S = (H \cup (uv_1)) - uv_{j+1}$. Then S and H have the same number of edges.

Now consider S_{cc_j}, the subgraph of S defined by the edges of S with colors c and c_j. Clearly, each component of S_{cc_j} is either an even cycle or a path in which the adjacent edges alternate with colors c and c_j.

Now, c is represented at each of the vertices $v_1, v_2, \ldots, v_k$, and in particular at v_{j+1} and v_k. But c_j is not represented at v_{j+1} and v_k, since we have just moved c_j to uv_j, and $c_j = c_k$ is not represented at v_k. Hence in S_{cc_j}, the degrees of v_{j+1} and v_k are both equal to 1. Moreover, c_j is represented at u, but c is not. Therefore, u also has degree 1 in S_{cc_j}. As each component of S_{cc_j} is either a path or an even cycle, not all of u, v_{j+1}, and v_k can be in the same component of S_{cc_j} (since a nontrivial path has only two vertices of degree 1).

If u and v_{j+1} are in different components of S_{cc_j}, interchange the colors c and c_{j+1} in the component containing v_{j+1}. Then c is not represented at both u and v_{j+1}, and so we can color the edge uv_{j+1} with c. This gives a $(1 + \Delta)$-edge coloring to the graph $S \cup (uv_{j+1})$.

Fig. 7.10 Graph illustrating
the generalized Vizing's
theorem

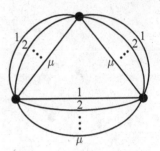

Suppose then u and v_{j+1} are in the same components of S_{cc_j}. Then, necessarily, v_k is not in this component. Interchange c and c_j in the component containing v_k. In this case, *further* cascade the colors so that uv_i has color c_i, $1 \leq i \leq k - 1$. Now color uv_k with color c.

Thus, we have extended our edge coloring of S with $1 + \Delta$ colors to one more edge of G. This contradiction proves that $H = G$, and thus $\chi'(G) \leq 1 + \Delta$.   □

Actually, Vizing proved a more general result than the one given above. Let G be any loopless graph and let μ denote the maximum number of edges joining two vertices in G. Then the generalized Vizing's theorem states that $\Delta \leq \chi' \leq \Delta + \mu$. This theorem is the best possible in that there are graphs with $\chi' = \Delta + \mu$. For example, let G be the graph of Fig. 7.10. Since any two edges of G are adjacent, $\chi' = m(G) = 3\mu = \Delta + \mu$. For a proof of the generalized Vizing's theorem, see Yap [194].

Definition 7.6.6. Graphs for which $\chi' = \Delta$ are called *Class* 1 graphs and those for which $\chi' = 1 + \Delta$ are called *Class* 2 graphs.

Example 7.6.7. Bipartite graphs are of class 1 (see Theorem 7.6.3), whereas the Petersen graph (see Exercise 6.3) and any simple cubic graph with a cut edge (see Exercise 6.6) are of class 2.

For details relating to graphs of class 1 and class 2, see [62, 194].

Exercise 6.9. Let G be a simple Δ-edge-chromatic critical graph [i.e., G is of class 1 and for every edge e of G, $\chi'(G - e) < \chi'(G)$]. Prove that if $uv \in E(G)$, then $d(u) + d(v) \geq \Delta + 2$.

We now return to the timetable problem. Following are some examples of such a problem.

Problem 1. In a social health checkup scheme, specialist physicians are to visit various health centers. Given the places each physician has to visit and also the time interval of his or her visit, how can we fit in an itinerary? The assumption is that each health center can accommodate only one doctor at a time.

Problem 2. Mobile laboratories are to visit various schools in a city. Given the places each lab has to visit and also the time interval (period) of visits in a day, how can we fit in a timetable for the laboratories?

Problem 3. In an educational institution, as is well known, teachers have to instruct various classes. Given the various classes each teacher has to instruct in a day, how can we fit in a timetable? It is presumed that a teacher can teach only one class at a time and that each class could be taught by only one teacher at a time!

We shall now discuss Problem 3. Let $x_1, x_2, \ldots, x_n$ denote the teachers and $y_1, y_2, \ldots, y_m$ the classes. Let t_{ij} denote the number of periods for which teacher x_i has to meet class y_j. How can we draw up a timetable? If there are constraints on the availability of classrooms, what is the minimum number of periods required to implement a timetable? If the number of periods in a day is specified, what is the minimum number of rooms required to implement the timetable? All these problems could be analyzed by using a suitable graph.

Let $G(T, C)$ be a bipartite graph formed with $T = \{x_1, x_2, \ldots, x_p\}$ and $C = \{y_1, y_2, \ldots, y_q\}$ as the bipartition and in which there are t_{ij} parallel edges with x_i and y_j as their common ends. If T denotes the set of teachers and C the set of classrooms, a teaching assignment for a period determines a matching in the bipartite graph G. Conversely, any matching in G corresponds to a teaching assignment for one period. The edges of G could be partitioned into Δ edge-disjoint matchings (see Theorem 7.6.3). Corresponding to the Δ matchings, a Δ-period timetable can be drawn up.

Let N be the total number of periods to be taught by all teachers put together. Then, on average, N/Δ classes are to be taught per period. Hence, at least $\lceil N/\Delta \rceil$ rooms are necessary to implement a Δ-period timetable. We present below a method for drawing up such a timetable. For this, we need Lemma 7.6.8.

Lemma 7.6.8. *Let M and N be disjoint matchings of a graph G with $|M| > |N|$. Then there are disjoint matchings M' and N' of G with $|M'| = |M| - 1$ and $|N'| = |N| + 1$ and with $M' \cup N' = M \cup N$.*

Proof. Consider the subgraph $H = G[M \cup N]$. Each component of H is either an even cycle or a path with edges alternating between M and N. Since $|M| > |N|$, some path component P of H must have its initial and terminal edges in M. Let $P = v_0 e_1 v_1 e_2 v_2 \ldots e_{2r+1} v_{2r+1}$.

Now set

$$M' = (M \setminus \{e_1, e_3, \ldots, e_{2r+1}\}) \cup \{e_2, e_4, \ldots, e_{2r}\}$$

and

$$N' = (N \setminus \{e_2, e_4, \ldots, e_{2r}\}) \cup \{e_1, e_3, \ldots, e_{2r+1}\}.$$

Then M' and N' are disjoint matchings of G satisfying the conditions of the lemma. $\qquad \square$

Theorem 7.6.9. *If G is a bipartite graph (with m edges), and if $m \geq t \geq \Delta$, then there exist t disjoint matchings $M_1, M_2, \ldots, M_t$ of G such that*

$$E = M_1 \cup M_2 \cup \ldots \cup M_t.$$

Fig. 7.11 Bipartite graph
corresponding to Problem 1

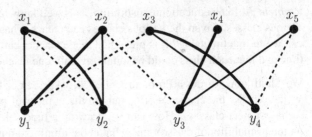

and, for $1 \le i \le t,$

$$\lfloor m/t \rfloor \le |M_i| \le \lceil m/t \rceil .$$

(In other words, any connected bipartite graph G is equitably t-edge-colorable,
where $m \ge t \ge \Delta$.)

Proof. By Theorem 7.6.3, $\chi' = \Delta$. Hence, $E(G)$ can be partitioned into Δ
matchings $M_1', M_2', \ldots, M_\Delta'$. So for $t \ge \Delta$, there exist disjoint matchings
$M_1', M_2', \ldots, M_t'$, where $M_i' = \emptyset$ for $\Delta + 1 \le i \le t$, and

$$E = M_1' \cup M_2' \cup \ldots \dot{M}_t'.$$

Now repeatedly apply Lemma 7.6.8 to pairs of matchings that differ by more
than one in size. This would eventually result in matchings $M_1, M_2, \ldots, M_t$ of G
satisfying the condition stated in the theorem. $\square$

Coming back to our timetable problem, if the number of rooms available, say r,
is less than N/Δ (so that $N/r > \Delta$), then the number of periods is to be
correspondingly increased. Hence, starting with an edge partition of $E(G)$ into
matchings $M_1', M_2', \ldots, M_\Delta'$, we apply Lemma 7.6.8 repeatedly to get an edge
partition of $E(G)$ into disjoint matchings $M_1, M_2, \ldots, M_{\lceil N/r \rceil}$. This partition gives
a $\lceil N/r \rceil$-period timetable that uses r rooms.

Illustration. The teaching assignments of five professors, x_1, x_2, x_3, x_4, x_5, in the
mathematics department of a particular university are given by the following array:

	I Year	II Year	III Year	IV Year
	y_1	y_2	y_3	y_4
x_1	1	2	–	–
x_2	1	1	1	–
x_3	1	–	–	2
x_4	–	–	1	–
x_5	–	–	1	1

The bipartite graph G corresponding to the above problem is shown in Fig. 7.11.
Each of the sets of edges drawn by the ordinary lines, dashed lines, and thick lines

Table 7.1 Timetable

		Period		
		I	II	III
Professor:	x_1	y_1	y_2	y_2
	x_2	y_2	y_3	y_1
	x_3	y_4	y_1	y_4
	x_4	$-$	$-$	y_3
	x_5	y_3	y_4	$-$

gives a matching in G. The three matchings cover the edges of G. Hence, they can be the basis of a three-period timetable. The corresponding timetable is given in Table 7.1.

In each period, four classes are to be met. Hence, at least four rooms are needed to implement this timetable. Here $\Delta = 3$ and $N = 12$. Consequently, G could be covered by three matchings each containing $\lfloor 12/3 \rfloor$ or $\lceil 12/3 \rceil$ edges, that is, exactly four edges. This gives the edge partition

$$M' = \{M'_1, M'_2, M'_3\},$$

where

$$M'_1 = \{x_1 y_1, x_2 y_2, x_3 y_4, x_5 y_3\},$$
$$M'_2 = \{x_1 y_2, x_2 y_3, x_3 y_1, x_5 y_4\},$$

and

$$M'_3 = \{x_1 y_2, x_2 y_1, x_3 y_4. x_4 y_3\}.$$

Now, take $M'' = \{M'_1, M'_2, M'_3, M'_4 = \emptyset\}$, and apply Lemma 7.6.8. This gives an edge partition $M = \{M_1, M_2, M_3, M_4\}$, where $M_1 = \{x_1 y_1, x_2 y_2, x_3 y_4\}$, $M_2 = \{x_1 y_2, x_2 y_3, x_5 y_4\}$, $M_3 = \{x_2 y_1, x_3 y_4, x_4 y_3\}$, and $M_4 = \{x_5 y_3, x_3 y_1, x_1 y_2\}$. The above partition yields a four-period timetable using three rooms.

7.7 Snarks

A consequence of the Vizing–Gupta theorem is that *if G is a simple cubic graph*, $\chi'(G) = 3$ *or* 4. By Exercise 6.6, *if G is a simple cubic graph with a cut edge*, $\chi'(G) = 4$. So the natural question is: Are there 2-edge-connected, simple cubic graphs that are 4-edge-chromatic? Such graphs are important in their own right, since their existence is related to the four-color problem (see Chap. 8). The search for such graphs has led to the study of snarks.

Definition 7.7.1. A *snark* is a cyclically 4-edge-connected cubic graph of girth at least 5 that has chromatic index 4.

Exercise 7.1. Prove that no snark can be Hamiltonian.

Fig. 7.12 Graph for proof of
Theorem 7.7.2

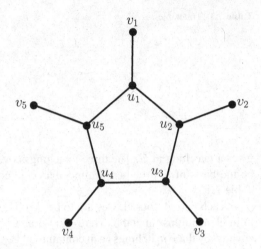

Clearly, the Petersen graph is a snark. In fact, Theorem 7.7.2 is an interesting result.

Theorem 7.7.2. *The Petersen graph P is the smallest snark and it is the unique snark on* 10 *vertices.*

Proof. Let G be a snark and A a cyclical edge cut of G. Then $G - A$ has two components, each having a cycle of length at least 5 (since G is of girth at least 5). Hence, $|V(G)| \geq 10$. Thus, P is a smallest snark since $|V(P)| = 10$.

We now show that any snark G on 10 vertices must be isomorphic to P. Let A be a cyclical edge cut of G. If $|A| = 4$, then each component of $G - A$ is a 5-cycle. But this will not account for all the 15 edges of G. If $|A| > 5$, then $|E(G)| > 5 + 5 + 5 = 15$, a contradiction. Hence, $|A| = 5$, and let $A = \{u_i v_i :$ $1 \leq i \leq 5\}$. Then $G - A$ consists of two 5-cycles. Let one of these cycles be $\{u_1, u_2, u_3, u_4, u_5\}$. Let v_i be the third neighbor of u_i not belonging to the set $\{u_1, u_2, u_3, u_4, u_5\}$ for each i. If $v_1 v_2$ or $v_1 v_5$ is an edge of G, then G contains a 4-cycle (see Fig. 7.12).

Since G is cubic, $v_1 v_3 \in E(G)$ and $v_1 v_4 \in E(G)$. Similarly, $v_2 v_4$, $v_2 v_5$, and $v_3 v_5$ are edges of G and hence $G \simeq P$.   □

The construction of snarks is not easy. In 1975, Isaacs constructed two infinite classes of snarks. Prior to that, only four kinds of snarks were known: (1) the Petersen graph on 10 vertices, (2) Blanuša's graphs on 18 vertices, (3) Szekeres' graph on 50 vertices, and (4) Blanche Descartes' graph on 210 vertices.

7.8   Kirkman's Schoolgirl Problem

Kirkman's schoolgirl problem was introduced in 1850 by Reverend Thomas J. Kirkman as "query 6" in page 48 of the *Ladies and Gentlemen's Diary*. The problem is the following: A teacher would like to take 15 schoolgirls out for a walk,

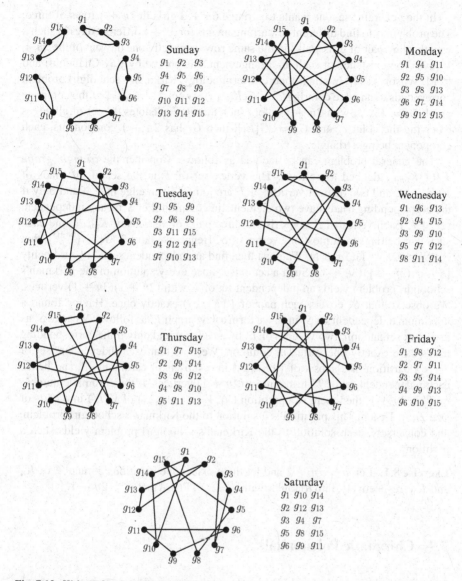

Fig. 7.13 Kirkman's schoolgirl problem

the girls being arranged in five rows of three. The teacher would like to ensure equal chances of friendship between any two girls. Hence, it is desirable to find different row arrangements for the seven days of the week such that any pair of girls walk in the same row on exactly one day of the week.

Kirkman's 15-schoolgirl problem has a solution. In fact, one of the possible schedules is given in Fig. 7.13.

In the general case, one wants to arrange $6n + 3$ girls in $2n + 1$ rows of three. The problem is to find different row arrangements for $3n + 1$ different days in such a way that any pair of girls walks in the same row on exactly one day out of the $3n + 1$ days. The existence of such an arrangement was proved by Ray-Chaudhuri and R. M. Wilson [164]. In graph-theoretic terminology, Kirkman's schoolgirl problem corresponds to an edge coloring $\mathscr{C} : E(K_{6n+3}) \to \{c_1, \ldots, c_{3n+1}\}$ of the complete graph $G = K_{6n+3}$ with $3n + 1$ colors such that if E_i denotes the set of all edges receiving the color c_i and $G_i = G[E_i]$, then G_i has $2n + 1$ components, each component being a triangle.

The general problem can be tackled as follows: Consider the *triangle graph* T of K_{6n+3} defined as follows: The vertex set of T is the set of all triads of $V(K_{6n+3})$, and two distinct vertices of T are joined by an edge in T if and only if the corresponding triads have two elements in common. Let S be any independent set of T. Each vertex of S gives rise to three pairs of vertices of K_{6n+3}, and each such pair belongs to at most one vertex of S. Hence, we have $3 |S| \leq \binom{6n+3}{2}$, that is, $|S| \leq (2n + 1)(3n + 1)$. We must then find an independent set S' of cardinality $|S'| = (2n + 1)(3n + 1)$. Such a set exists since every solution of the Kirkman's schoolgirl problem yields an independent set of T with $(2n + 1)(3n + 1)$ vertices. We observe that S' covers each pair of $V(K_{6n+3})$ exactly once. Having found a maximum independent set S' in T, we form a new graph T' as follows: We take S' as its vertex set and join two vertices of T' by an edge if and only if the corresponding triads have exactly one vertex in common. We note that each independent set of T' is a partition of a subset of $V(K_{6n+3})$ into subsets of cardinality 3, and hence each independent set of T' has at most $(2n + 1)$ vertices. If the chromatic number of T' is $3n + 1$, then there is a partition $(V_1, V_2, \ldots, V_{3n+1})$ of $V(T')$ into parts of size $2n + 1$ each. This partition is a solution to the Kirkman's schoolgirl problem, and conversely, each solution to the Kirkman's schoolgirl problem yields such a partition.

Exercise 8.1. Let $m \geq n + 2$ and let there exist edge partitions $\mathscr{F}$ and $\mathscr{G}$ of K_n and K_m, respectively, into triangles with $\mathscr{F} \subset \mathscr{G}$. Prove that $m \geq 2n + 1$.

7.9   Chromatic Polynomials

In 1946, Birkhoff and Lewis [23] introduced the chromatic polynomial of a graph in their attempt to tackle the four-color problem (see Chap. 8) through algebraic techniques.

For a graph G and a given set of λ colors, the function $f(G; \lambda)$ is defined to be the number of ways of (vertex) coloring G properly using the λ colors. Hence, $f(G; \lambda) = 0$ when G has no proper λ-coloring. Clearly, the minimum λ for which $f(G; \lambda) > 0$ is the chromatic number $\chi(G)$ of G.

It is easy to see that $f(K_n; \lambda) = \lambda(\lambda - 1) \ldots (\lambda - n + 1)$ for $\lambda \geq n$. This is because any vertex of K_n can be colored by any one of the given λ colors. After coloring a vertex of K_n, a second vertex of K_n can be colored by any one of the remaining $(\lambda - 1)$ colors, and so on. In particular, $f(K_3; \lambda) = \lambda(\lambda - 1)(\lambda - 2)$. Also, $f(K_n^c; \lambda) = \lambda^n$.

Let $e = uv$ be any edge of G. Recall (see Sect. 4.3, Chap. 4) that the graph $G \circ e$ is obtained from G by contracting the edge e. Theorem 7.9.1 presents a simple reduction formula to compute $f(G; \lambda)$.

Theorem 7.9.1. *Let G be any graph. Then $f(G; \lambda) = f(G - e; \lambda) - f(G \circ e; \lambda)$ for any edge e of G.*

Proof. $f(G - e; \lambda)$ denotes the number of proper colorings of $G - e$ using λ colors. Hence, it is the sum of the number of proper colorings of $G - e$ in which u and v receive the same color and the number of proper colorings of $G - e$ in which u and v receive distinct colors. The former number is $f(G \circ e; \lambda)$, and the latter number is $f(G; \lambda)$. $\qquad\square$

Exercise 9.1. If G and H are disjoint graphs, show that

$$f(G \cup H; \lambda) = f(G; \lambda) f(H; \lambda).$$

Theorem 7.9.1 could be used recursively to determine $f(G; \lambda)$ for graphs of small size by taking the given graph on n vertices as G and successively deleting edges until we end up with the totally disconnected graph K_n^c. It can also be determined by taking the given graph as $G - e$ and recursively adding a new edge e until we end up with the complete graph K_n. For a fixed n, when $m(G)$, the number of edges of G is small, the first method is preferable, and when it is large, the second method is preferable. These two methods are illustrated for the graph C_4. [Here the function $f(G; \lambda)$ is represented by the graph itself.]

Method 1

$$= (\lambda(\lambda - 1)^2 - \{\lambda^2(\lambda - 1) - \lambda(\lambda - 1)\} - \lambda(\lambda - 1)(\lambda - 2)$$
$$= \lambda^4 - 4\lambda^3 + 6\lambda^2 - 3\lambda.$$

Method 2

$$f(C_4; \lambda) = \left(\begin{array}{c} \square \\ \underbrace{}_{G-e} \end{array} \right)$$

4

$$= \left(\begin{array}{c} \boxed{e} \\ \underbrace{}_{G} \end{array} \right) + \left(\begin{array}{c} \infty \\ \underbrace{}_{G \circ e} \end{array} \right)$$

$$= \left(\boxed{} \right) + \left(\begin{array}{c} | \\ | \end{array} \right)$$

$$= \left(\boxtimes \right) + \left(\triangleright \right) + \left(\right) + \left(\right)$$

$$= f(K_4; \lambda) + f(K_3; \lambda) + f(K_3; \lambda) + f(K_2; \lambda)$$
$$= f(K_4; \lambda) + 2f(K_3; \lambda) + f(K_2; \lambda)$$
$$= \lambda(\lambda - 1)(\lambda - 2)(\lambda - 3) + 2\lambda(\lambda - 1)(\lambda - 2) + \lambda(\lambda - 1)$$
$$= \lambda^4 - 4\lambda^3 + 6\lambda^2 - 3\lambda.$$

The function $f(C_4; \lambda)$ computed above is a monic polynomial with integer coefficients of degree $n = 4$ in which the coefficient of $\lambda^3 = -4 = -m$, the constant term is zero, and the coefficients alternate in sign. That this is the case with all such functions $f(G; \lambda)$ is the content of Theorem 7.9.2. For this reason, the function $f(G; \lambda)$ is called the *chromatic polynomial* of the graph G.

Theorem 7.9.2. *For a simple graph G of order n and size m, $f(G; \lambda)$ is a monic polynomial of degree n in λ with integer coefficients and constant term zero. In addition, its coefficients alternate in sign and the coefficient of λ^{n-1} is $-m$.*

Proof. The proof is by induction on m. If $m = 0$, G is K_n^c and $f(K_n^c; \lambda) = \lambda^n$, and if $m = 1$, G is K_2 and $f(K_2; \lambda) = \lambda^2 - \lambda$, and the statement of the theorem is trivially true in these cases. Suppose now that the theorem holds for all graphs with fewer than m edges, where $m \geq 2$. Let G be any simple graph of order n and size m, and let e be any edge of G. Both $G - e$ and $G \circ e$ (after removal of multiple edges, if necessary) are simple graphs with at most $m - 1$ edges, and hence, by the induction hypothesis,

$$f(G - e; \lambda) = \lambda^n - a_0\lambda^{n-1} + a_1\lambda^{n-2} - \ldots + (-1)^{n-1}a_{n-2}\lambda,$$

and

$$f(G \circ e; \lambda) = \lambda^{n-1} - b_1\lambda^{n-2} + \ldots + (-1)^{n-2}b_{n-2}\lambda,$$

where $a_0, \ldots, a_{n-2}; b_1, \ldots, b_{n-2}$ are nonnegative integers (so that the coefficients alternate in sign), and a_0 is the number of edges in $G - e$, which is $m - 1$. By Theorem 7.9.1, $f(G; \lambda) = f(G - e; \lambda) - f(G \circ e; \lambda)$, and hence

$$f(G;\lambda) = \lambda^n - (a_0 + 1)\lambda^{n-1} + (a_1 + b_1)\lambda^{n-2} - \ldots + (-1)^{n-1}(a_{n-2} + b_{n-2})\lambda.$$

Since $a_0 + 1 = m$, $f(G;\lambda)$ has all the stated properties.   □

Theorem 7.9.3. *A simple graph G on n vertices is a tree if and only if $f(G;\lambda) = \lambda(\lambda - 1)^{n-1}$.*

Proof. Let G be a tree. We prove that $f(G;\lambda) = \lambda(\lambda - 1)^{n-1}$ by induction on n. If $n = 1$, the result is trivial. So assume the result for trees with at most $n - 1$ vertices, $n \geq 2$. Let G be a tree with n vertices, and e be a pendent edge of G. By Theorem 7.9.1, $f(G;\lambda) = f(G-e;\lambda) - f(G\circ e;\lambda)$. Now, $G-e$ is a forest with two component trees of orders $n-1$ and 1, and hence $f(G-e;\lambda) = (\lambda(\lambda-1)^{n-2})\lambda$ (see Exercise 9.1). Since $G \circ e$ is a tree with $n - 1$ vertices, $f(G \circ e;\lambda) = \lambda(\lambda - 1)^{n-2}$. Thus, $f(G;\lambda) = (\lambda(\lambda - 1)^{n-2})\lambda - \lambda(\lambda - 1)^{n-2} = \lambda(\lambda - 1)^{n-1}$.

Conversely, assume that G is a simple graph with $f(G;\lambda) = \lambda(\lambda - 1)^{n-1} = \lambda^n - (n - 1)\lambda^{n-1} + \ldots + (-1)^{n-1}\lambda$. Hence, by Theorem 7.9.2, G has n vertices and $n - 1$ edges. Further, the last term, $(-1)^{n-1}\lambda$, ensures that G is connected (see Exercise 9.2). Hence, G is a tree (see Theorem 4.2.4).   □

Remark 7.9.4. Theorem 7.9.3 shows that the chromatic polynomial of a graph G does not fix the graph uniquely up to isomorphism. For example, even though the graphs $K_{1,3}$ and P_4 are not isomorphic, they have the same chromatic polynomial, namely, $\lambda(\lambda - 1)^3$.

Exercise 9.2. If G has ω components, show that λ^ω is a factor of $f(G;\lambda)$.

Exercise 9.3. Show that there exists no graph with the following polynomials as chromatic polynomial (i) $\lambda^5 - 4\lambda^4 + 8\lambda^3 - 4\lambda^2 + \lambda$; (ii) $\lambda^4 - 3\lambda^3 + \lambda^2$; (iii) $\lambda^7 - \lambda^6 + 1$.

Exercise 9.4. Find a graph G whose chromatic polynomial is $\lambda^5 - 6\lambda^4 + 11\lambda^3 - 6\lambda^2$.

Exercise 9.5. Show that for the cycle C_n of length n, $f(C_n;\lambda) = (\lambda - 1)^n + (-1)^n(\lambda - 1)$, $n \geq 3$.

Exercise 9.6. Show that for any graph G, $f(G \vee K_1;\lambda) = \lambda f(G;\lambda-1)$, and hence prove that $f(W_n;\lambda) = \lambda(\lambda - 2)^n + (-1)^n\lambda(\lambda - 2)$.

Notes

A good reference for graph colorings is the book by Jensen and Toft [116]. The book by Fiorini and Wilson [62] concentrates on edge colorings. Theorem 7.5.7 (Mycielski's theorem) has also been proved independently by Blanche Descartes [50] as well as by Zykov [195]. For a complete description of graph homomorphisms, see [105].

The proof of Brooks' theorem given in this chapter is based on the proof given by Fournier [67] (see also references [27] and [106]).

The term "snark" was given to the snark graph by Martin Gardner after the unusual creature that is described in Lewis Carroll's poem, T*he Hunting of the Snark*. A detailed account of the snarks, including their constructions, can be found in the interesting book by Holton and Sheehan [106].

Chapter 8
Planarity

8.1 Introduction

The study of planar and nonplanar graphs and, in particular, the several attempts
to solve the *four-color conjecture* have contributed a great deal to the growth of
graph theory. Actually, these efforts have been instrumental to the development of
algebraic, topological, and computational techniques in graph theory.

In this chapter, we present some of the basic results on planar graphs. In
particular, the two important characterization theorems for planar graphs, namely,
Wagner's theorem (same as the Harary–Tutte theorem) and Kuratowski's theorem,
are presented. Moreover, the nonhamiltonicity of the Tutte graph on 46 vertices (see
Fig. 8.28 and also the front wrapper) is explained in detail.

8.2 Planar and Nonplanar Graphs

Definition 8.2.1. A graph G is *planar* if there exists a drawing of G in the plane in
which no two edges intersect in a point other than a vertex of G, where each edge
is a Jordan arc (that is, a simple arc). Such a drawing of a planar graph G is called
a *plane representation* of G. In this case, we also say that G has been embedded in
the plane. A *plane graph* is a planar graph that has already been embedded in the
plane.

Example 8.2.2. There exist planar as well as nonplanar graphs. In Fig. 8.1, a planar
graph and two of its plane representations are shown. Note that all trees are planar
as also are cycles and wheels. The Petersen graph is nonplanar (a proof of this result
is given later in this chapter.).

Before proceeding further, let us recall here the celebrated Jordan curve theorem.
If J is any closed Jordan curve in the plane, the complement of J (with respect

R. Balakrishnan and K. Ranganathan, *A Textbook of Graph Theory,*
Universitext, DOI 10.1007/978-1-4614-4529-6_8,
© Springer Science+Business Media New York 2012

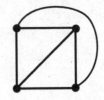

Planar graph K_4 Two plane embeddings of K_4

Fig. 8.1 A planar graph with two plane embeddings

Fig. 8.2 Arc connecting
point x in int J with point y
in ext J

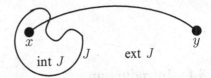

to the plane) is partitioned into two disjoint open connected subsets of the plane, one of which is bounded and the other unbounded. The bounded subset is called the *interior* of J and is denoted by int J. The unbounded subset is called the *exterior* of J and is denoted by ext J. The *Jordan curve theorem* (of topology) states that if J is any closed Jordan curve in the plane, any arc joining a point of int J and a point of ext J must intersect J at some point (see Fig. 8.2) (the proof of this result, although intuitively obvious, is tedious).

Let G be a plane graph. Then the union of the edges (as Jordan arcs) of a cycle C of G form a closed Jordan curve, which we also denote by C. A plane graph G divides the rest of the plane (i.e., plane minus the edges and vertices of G), say π, into one or more faces, which we define below. We define an equivalence relation $\sim$ on π.

Definition 8.2.3. We say that for points A and B of π, $A \sim B$ if and only if there exists a Jordan arc from A to B in π. Clearly, $\sim$ is an equivalence relation on π. The equivalence classes of the above equivalence relation are called the *faces* of G.

Remark 8.2.4. 1. We claim that a connected graph is a tree if and only if it has only one face. Indeed, since there are no cycles in a tree T, the complement of a plane embedding of T in the plane is connected (in the above sense), and hence a tree has only one face. Conversely, it is clear that if a connected plane graph has only one face, then it must be a tree.
2. Any plane graph has exactly one unbounded face. The unbounded face is also referred to as the exterior face of the plane graph. All other faces, if any, are bounded. Figure 8.3 represents a plane graph with seven faces.

The distinction between bounded and unbounded faces of a plane graph is only superfluous, as there exists a plane representation G_1 of a plane graph G in which any specified face of G_1 becomes the unbounded face, as is shown below. (This of

Fig. 8.3 A plane graph with seven faces

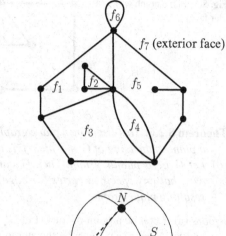

Fig. 8.4 Stereographic projection of the sphere S from N

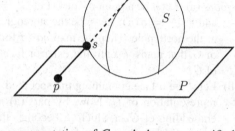

course means that there exists a plane representation of G such that any specified vertex or edge belongs to the unbounded face.) We consider embeddings of a graph on a sphere. A graph is *embeddable on a sphere* S if it can be drawn on the surface of S so that its edges intersect only at its vertices. Such a drawing, if it exists, is called an embedding of G on S. Embeddings on a sphere are called *spherical embeddings*. What we have given here is only a naive definition. For a more rigorous description of spherical embeddings, see [79].

To prove the next theorem, we need to recall the notion of stereographic projection. Let S be a sphere resting on a plane P so that P is a tangent plane to S. Let N be the "north pole," the point on the sphere diametrically opposite the point of contact of S and P. Let the straight line joining N and a point s of $S \backslash \{N\}$ meet P at p. Then the mapping $\eta : S \backslash \{N\} \to P$ defined by $\eta(s) = p$ is called the *stereographic projection* of S from N (see Fig. 8.4).

Theorem 8.2.5. *A graph is planar if and only if it is embeddable on a sphere.*

Proof. Let a graph G be embeddable on a sphere and let G' be a spherical embedding of G. The image of G' under the stereographic projection η of the sphere from a point N of the sphere not on G' is a plane representation of G on P. Conversely, if G'' is a plane embedding of G on a plane P, then the inverse of the stereographic projection of G'' on a sphere touching the plane P gives a spherical embedding of G. $\square$

Fig. 8.5 Plane graph with
four faces

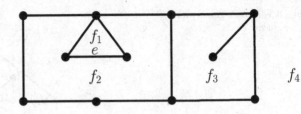

Fig. 8.5 Plane graph with four faces

Theorem 8.2.6. *(a) Let G be a plane graph and f be a face of G. Then there exists
a plane embedding of G in which f is the exterior face.*

*(b) Let G be a planar graph. Then G can be embedded in the plane in such a
way that any specified vertex (or edge) belongs to the unbounded face of the
resulting plane graph.*

Proof. (a) Let n be a point of int f. Let $G' = \sigma(G)$ be a spherical embedding of G
and let $N = \sigma(n)$. Let η be the stereographic projection of the sphere with N
as the north pole. Then the map $\eta\sigma$ (σ followed by η) gives a plane embedding
of G that maps f onto the exterior face of the plane representation $(\eta\sigma)(G)$
of G.

(b) Let f be a face containing the specified vertex (respectively, edge) in a plane
representation of G. Now, by part (a) of the theorem, there exists a plane
embedding of G in which f becomes the exterior face. The specified vertex
(respectively, edge) then becomes a vertex (respectively, edge) of the new
unbounded face.   □

Remark 8.2.7. 1. Let G be a connected plane graph. Each edge of G belongs to one
or two faces of G. A cut edge of G belongs to exactly one face, and conversely,
if an edge belongs to exactly one face of G, it must be a cut edge of G. An edge
of G that is not a cut edge belongs to exactly two faces and conversely.

2. The union of the vertices and edges of G incident with a face f of G is called the
boundary of f and is denoted by $b(f)$. The vertices and edges of a plane graph
G belonging to the boundary of a face of G are said to be *incident* with that face.
If G is connected, the boundary of each face is a closed walk in which each cut
edge of G is traversed twice. When there are no cut edges, the boundary of each
face of G is a closed trail in G. (See, for instance, face f_1 of Fig. 8.3.) However,
if G is a disconnected plane graph, then the edges and the vertices incident with
the exterior face will not define a trail.

3. The number of edges incident with a face f is defined as the *degree* of f. In
counting the degree of a face, a cut edge is counted twice. Thus, each edge of a
plane graph G contributes two to the sum of the degrees of the faces. It follows
that if $\mathcal{F}$ denotes the set of faces of a plane graph G, then $\sum_{f \in \mathcal{F}} d(f) = 2m(G)$,

where $d(f)$ denotes the degree of the face f.

In Fig. 8.5, $d(f_1) = 3$, $d(f_2) = 9$, $d(f_3) = 6$, and $d(f_4) = 8$.

Theorem 8.2.8 connects the planarity of G with the planarity of its blocks.

Theorem 8.2.8. *A graph G is planar if and only if each of its blocks is planar.*

Proof. If G is planar, then each of its blocks is planar, since a subgraph of a planar graph is planar. Conversely, suppose that each block of G is planar. We now use induction on the number of blocks of G to prove the result. Without loss of generality, we assume that G is connected. If G has only one block, then G is planar.

Now suppose that G has k planar blocks and that the result is true for all connected graphs having $(k-1)$ planar blocks. Choose any end block B_0 of G and delete from G all the vertices of B_0 except the unique cut vertex, say v_0, of G in B_0. The resulting connected subgraph G' of G contains $(k-1)$ planar blocks. Hence, by the induction hypothesis, G' is planar. Let $\tilde{G}'$ be a plane embedding of G' such that v_0 belongs to the boundary of the unbounded face, say f' (refer to Theorem 8.2.6). Let $\tilde{B}_0$ be a plane embedding of B_0 in f' so that v_0 is in the boundary of the exterior face of $\tilde{B}_0$. Then (by the identification of v_0 in the two embeddings), $\tilde{G}' \cup \tilde{B}_0$ is a plane embedding of G. $\square$

Remark 8.2.9. In testing for the planarity of a graph G, one may delete multiple edges and loops of G, if any. This is so because if a graph H is nonplanar, the removal of loops and parallel edges of H results in a subgraph of H, which is also nonplanar. Also, by Theorem 8.2.8, G can be assumed to be a block and hence 2-connected. If G has a vertex of degree 2, say v_0, and vv_0v' is the path formed by the two edges incident with v_0, contraction of vv_0 and deletion of the multiple edges (if any) thus formed again result in a planar graph. Let G' be the graph obtained from G by performing such contractions successively at vertices of degree 2 and deleting the resulting multiple edges. Then G is planar if and only if G' is planar. From these observations, it is clear that in designing a planarity algorithm (i.e., an algorithm to test planarity), it suffices to consider only 2-connected simple graphs with minimum degree at least 3. (For a planarity algorithm, see [49].)

Exercise 2.1. Show that every graph with at most three cycles is planar.

Exercise 2.2. Find a simple graph G with degree sequence $(4, 4, 3, 3, 3, 3)$ such that

(a) G is planar.
(b) G is nonplanar.

Exercise 2.3. Redraw the following planar graph so that the face f becomes the exterior face.

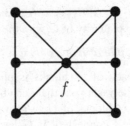

8.3   Euler Formula and Its Consequences

We have noted that a planar graph may have more than one plane representation
(see Fig. 8.1). A natural question that would arise is whether the number of faces is
the same in each such representation. The answer to this question is provided by the
Euler formula.

Theorem 8.3.1 (Euler formula).   *For a connected plane graph G, $n - m + f = 2$,*
where n, m, and f denote the number of vertices, edges, and faces of G, respectively.

Proof. We apply induction on f.

If $f = 1$, then G is a tree and $m = n - 1$. Hence, $n - m + f = 2$.

Now assume that the result is true for all plane graphs with $f - 1$ faces, $f \geq 2$,
and suppose that G has f faces. Since $f \geq 2$, G is not a tree, and hence contains a
cycle C. Let e be an edge of C. Then e belongs to exactly two faces, say f_1 and
f_2, of G and the deletion of e from G results in the formation of a single face from
f_1 and f_2 (see Fig. 8.5). Also, since e is not a cut edge of G, $G - e$ is connected.
Further, the number of faces of $G - e$ is $f - 1$. So applying induction to $G - e$, we
get $n - (m - 1) + (f - 1) = 2$, and this implies that $n - m + f = 2$. This completes
the proof of the theorem. $\square$

Below are some of the consequences of the Euler formula.

Corollary 8.3.2. *All plane embeddings of a given planar graph have the same*
number of faces.

Proof. Since $f = m - n + 2$, the number of faces depends only on n and m, and not
on the particular embedding. $\square$

Corollary 8.3.3. *If G is a simple planar graph with at least three vertices, then*
$m \leq 3n - 6$.

Proof. Without loss of generality, we can assume that G is a simple connected plane
graph. Since G is simple and $n \geq 3$, each face of G has degree at least 3. Hence,
if $\mathcal{F}$ denotes the set of faces of G, $\sum_{f \in \mathcal{F}} d(f) \geq 3f$. But $\sum_{f \in \mathcal{F}} d(f) = 2m$.
Consequently, $2m \geq 3f$, so that $f \leq \frac{2m}{3}$.

By the Euler formula, $m = n + \beta - 2$. Now $\beta \le \frac{2m}{3}$ implies that $m \le n + \left(\frac{2m}{3}\right) - 2$. This gives $m \le 3n - 6$. $\qquad\square$

The above result is not valid if $n = 1$ or 2. Also, the condition of Corollary 8.3.3 is not sufficient for the planarity of a simple connected graph as the Petersen graph shows. For the Petersen graph, $m = 15$, $n = 10$, and hence $m \le 3n - 6$, but the graph is not planar (see Corollary 8.3.7 below).

Example 8.3.4. Show that the complement of a simple planar graph with 11 vertices is nonplanar.

Solution. Let G be a simple planar graph with $n(G) = 11$. Since G is planar, $m(G) \le 3n - 6 = 27$. If G^c were also planar, then $m(G^c) \le 3n - 6 = 27$. On the one hand, $m(G) + m(G^c) \le 27 + 27 = 54$, whereas, on the other hand, $m(G) + m(G^c) = m(K_{11}) = \binom{11}{2} = 55$. Hence, we arrive at a contradiction. This contradiction proves that G^c is nonplanar. $\qquad\square$

Corollary 8.3.5. *For any simple planar graph G, $\delta(G) \le 5$.*

Proof. If $n \le 6$, then $\Delta(G) \le 5$. Hence $\delta(G) \le \Delta(G) \le 5$, proving the result for such graphs. So assume that $n \ge 7$. By Corollary 8.3.3, $m \le 3n - 6$. Now, $\delta n \le \sum_{v \in V(G)} d_G(v) = 2m \le 2(3n - 6) = 6n - 12$. Hence $n(\delta - 6) \le -12$. Consequently, $\delta - 6$ is negative, implying that $\delta \le 5$. $\qquad\square$

Recall that the *girth* of a graph G is the length of a shortest cycle in G.

Theorem 8.3.6. *If the girth k of a connected plane graph G is at least 3, then $m \le \frac{k(n-2)}{(k-2)}$.*

Proof. Let $\mathcal{F}$ denote the set of faces and β, as before, denote the number of faces of G. If $f \in \mathcal{F}$, then $d(f) \ge k$. Since $2m = \sum_{f \in \mathcal{F}} d(f)$, we get $2m \ge k\beta$. By Theorem 8.3.1, $\beta = 2 - n + m$. Hence, $2m \ge k(2 - n + m)$, implying that $m(k - 2) \le k(n - 2)$. Thus, $m \le \frac{k(n-2)}{(k-2)}$. $\qquad\square$

Corollary 8.3.7. *The Petersen graph P is nonplanar.*

Proof. The girth of the Petersen graph P is 5, $n(P) = 10$, and $m(P) = 15$. Hence, if P were planar, $15 \le \frac{5(10-2)}{5-2}$, which is not true. Hence, P is nonplanar. $\qquad\square$

Exercise 3.1. Show that every simple bipartite cubic planar graph contains a C_4.

Exercise 3.2. A nonplanar graph G is called *planar-vertex-critical* if $G - v$ is planar for every vertex v of G. Prove that a planar-vertex-critical graph must be 2-connected.

Exercise 3.3. Verify Euler's formula for the plane graph P.

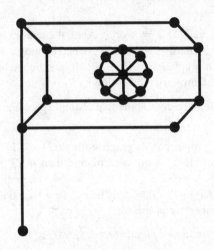

Exercise 3.4. Let G be a simple plane cubic graph having eight faces. Determine $n(G)$. Draw two such graphs that are nonisomorphic.

Exercise 3.5. Prove that if G is a simple connected planar bipartite graph, then $m \leq 2n - 4$, where $n \geq 3$.

Exercise 3.6. Prove that a simple planar graph (with at least four vertices) has at least four vertices each of degree 5 at most.

Exercise 3.7. If G is a nonplanar graph, show that it has either five vertices of degree at least 4, or six vertices of degree at least 3.

Exercise 3.8. Prove that a simple planar graph with minimum degree at least five contains at least 12 vertices. Give an example of a simple planar graph on 12 vertices with minimum degree 5.

Exercise 3.9. Show that there is no 6-connected planar graph.

Exercise 3.10. Let G be a plane graph of order n and size m in which every face is bounded by a k-cycle. Show that $m = \frac{k(n-2)}{(k-2)}$.

Definition 8.3.8. A graph G is *maximal planar* if G is planar, but for any pair of nonadjacent vertices u and v of G, $G + uv$ is nonplanar.

Remark 8.3.9. Any planar graph is a spanning subgraph of a maximal planar graph. Indeed, if $\tilde{G}$ is a plane embedding of a planar graph G with at least three vertices, and if $e = uv$ is a cut edge of $\tilde{G}$ embedded in a face f of $\tilde{G}$, it is clear that there exists a vertex w on the boundary of f such that the edge uw or vw can be drawn in f so that either $\tilde{G} + (vw)$ or $\tilde{G} + (uw)$ is also a plane graph (see Fig. 8.6a). Further, if C_0 is any cycle bounding a face f_0 of a plane graph H, then edges can be drawn in int C_0 without crossing each other so that f_0 is divided into triangles (see Fig. 8.6b).

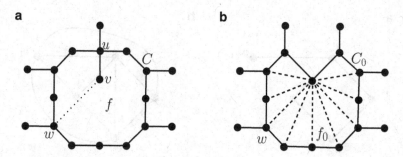

Fig. 8.6 Procedure to get maximal planar graphs

Definition 8.3.10. A *plane triangulation* is a plane graph in which each of its faces is bounded by a triangle. A plane triangulation of a plane graph G is a plane triangulation H such that G is a spanning subgraph of H.

Remark 8.3.11. Remark 8.3.9 shows that a plane embedding of a simple maximal planar graph is a plane triangulation.

Note that any simple plane graph is a subgraph of a simple maximal plane graph and hence is a spanning subgraph of some plane triangulation. Thus, to any simple plane graph G that is not already a plane triangulation, we can add a set of new edges to obtain a plane triangulation. The set of new edges thus added need not be unique.

Figure 8.7a is a simple plane graph G and Fig. 8.7b is a plane triangulation of G; Fig. 8.7c is a plane triangulation of G isomorphic to the graph of Fig. 8.7b having only straight-line edges. (A result of Fáry [60] states that every simple planar graph has a plane embedding in which each edge is a straight line.)

Exercise 3.11. Embed the 3-cube Q_3 (see Exercise 4.4 of Chap. 5) in a maximal planar graph having the same vertex set as Q_3. Count the number of new edges added.

Exercise 3.12. Prove that for a simple maximal planar graph on $n \geq 3$ vertices, $m = 3n - 6$.

Exercise 3.13. Use Exercise 3.12 to show that for any simple planar graph, $m \leq 3n - 6$.

Exercise 3.14. Show that every plane triangulation of order $n \geq 4$ is 3-connected.

Exercise 3.15. Let G be a maximal planar graph with $n \geq 4$. Let n_i denote the number of vertices of degree i in G. Then prove that $3n_3 + 2n_4 + n_5 = 12 + n_7 + 2n_8 + 3n_9 + 4n_{10} + \dots$. (Hint: Use the fact that $n = n_3 + n_4 + n_5 + n_6 + \dots$.)

Exercise 3.16. Generalize the Euler formula for disconnected plane graphs.

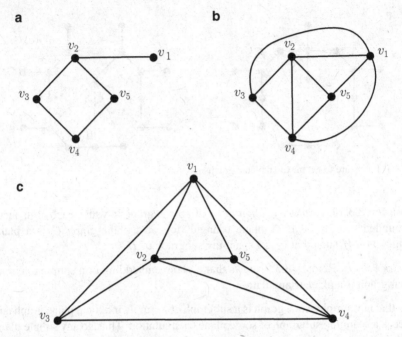

Fig. 8.7 (a) Graph G and (b), (c) are plane triangulations of G

8.4 K_5 and $K_{3,3}$ are Nonplanar Graphs

In this section we prove that K_5 and $K_{3,3}$ are nonplanar. These two graphs are basic in Kuratowski's characterization of planar graphs (see Theorem 8.7.5 given later in this chapter). For this reason, they are often referred to as the two *Kuratowski graphs*.

Theorem 8.4.1. K_5 *is nonplanar.*

First proof. This proof uses the Jordan curve theorem. Assume the contrary, namely, K_5 is planar. Let v_1, v_2, v_3, v_4, and v_5 be the vertices of K_5 in a plane representation of K_5. The cycle $C = v_1 v_2 v_3 v_4 v_1$ (as a closed Jordan curve) divides the plane into two faces, namely, the interior and the exterior of C. The vertex v_5 must belong either to int C or to ext C. Suppose that v_5 belongs to int C (a similar proof holds if v_5 belongs to ext C). Draw the edges $v_5 v_1$, $v_5 v_2$, $v_5 v_3$, and $v_5 v_4$ in int C. Now there remain two more edges $v_1 v_3$ and $v_2 v_4$ to be drawn. None of these can be drawn in int C, since it is assumed that K_5 is planar. Thus, $v_1 v_3$ lies in ext C. Then one of v_2 and v_4 belongs to the interior of the closed Jordan curve $C_1 = v_1 v_5 v_3 v_1$ and the other to its exterior (see Fig. 8.8). Hence, $v_2 v_4$ cannot be drawn without violating planarity. □

Fig. 8.8 Graph for first proof
of Theorem 8.4.1

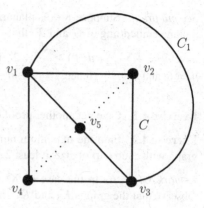

Fig. 8.9 Graph for first proof
of Theorem 8.4.3

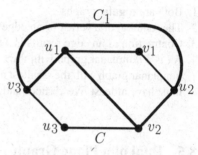

Remark 8.4.2. The first proof of Theorem 8.4.1 shows that all the edges of K_5 except one can be drawn in the plane without violating planarity. Hence for any edge e of K_5, $K_5 - e$ is planar.

Second proof. If K_5 were planar, it follows from Theorem 8.3.6 that $10 \leq \frac{3(5-2)}{(3-2)}$, which is not true. Hence K_5 is nonplanar. □

Theorem 8.4.3. $K_{3,3}$ *is nonplanar.*

First proof. The proof is by the use of the Jordan curve theorem. Suppose that $K_{3,3}$ is planar. Let $U = \{u_1, u_2, u_3\}$ and $V = \{v_1, v_2, v_3\}$ be the bipartition of $K_{3,3}$ in a plane representation of the graph. Consider the cycle $C = u_1v_1u_2v_2u_3v_3u_1$. Since the graph is assumed to be planar, the edge u_1v_2 must lie either in the interior of C or in its exterior. For the sake of definiteness, assume that it lies in int C (a similar proof holds if one assumes that the edge u_1v_2 lies in ext C). Two more edges remain to be drawn, namely, u_2v_3 and u_3v_1. None of these can be drawn in int C without crossing the edge u_1v_2. Hence, both of them are to be drawn in ext C. Now draw u_2v_3 in ext C. Then one of v_1 and u_3 belongs to the interior of the closed Jordan curve $C_1 = u_1v_2u_2v_3u_1$ and the other to the exterior of C_1 (see Fig. 8.9). Hence, the edge v_1u_3 cannot be drawn without violating planarity. This shows that $K_{3,3}$ is nonplanar. □

Second proof. Suppose $K_{3,3}$ is planar. Let f be the number of faces of $G = K_{3,3}$ in a plane embedding of G and $\mathcal{F}$, the set of faces of G. As the girth of $K_{3,3}$ is 4, we have $m = \frac{1}{2}\sum_{f \in \mathcal{F}} d(f) \geq \frac{4f}{2} = 2f$. By Theorem 8.3.1, $n - m + f = 2$. For $K_{3,3}$, $n = 6$, and $m = 9$. Hence, $f = 2 + m - n = 5$. Thus, $9 \geq 2.5 = 10$, a contradiction. □

Exercise 4.1. Give yet another proof of Theorem 8.4.3.

Exercise 4.2. Find the maximum number of edges in a planar complete tripartite graph with each part of size at least 2.

Remark 8.4.4. As in the case of K_5, for any edge e of $K_{3,3}$, $K_{3,3} - e$ is planar. Observe that the graphs K_5 and $K_{3,3}$ have some features in common.

1. Both are regular graphs.
2. The removal of a vertex or an edge from each graph results in a planar graph.
3. Contraction of an edge results in a planar graph.
4. K_5 is a nonplanar graph with the smallest number of vertices, whereas $K_{3,3}$ is a nonplanar graph with the smallest number of edges. (Hence, any nonplanar graph must have at least five vertices and nine edges.)

8.5   Dual of a Plane Graph

Let G be a plane graph. One can form out of G a new graph H in the following way. Corresponding to each face f of G, take a vertex f^* and corresponding to each edge e of G, take an edge e^*. Then edge e^* joins vertices f^* and g^* in H if and only if edge e is common to the boundaries of faces f and g in G. (It is possible that f may be the same as g.) The graph H is then called the *dual* (or more precisely, the geometric dual) of G (see Fig. 8.10). If e is a cut edge of G embedded in face f of G, then e^* is a loop at f^*. H is a planar graph and there exists a natural way of embedding H in the plane. Vertex f^*, corresponding to face f, is placed in face f of G. Edge e^*, joining f^* and g^*, is drawn so that e^* crosses e once and only once and crosses no other edge. This procedure is illustrated in Fig. 8.11. This embedding is the canonical embedding of H. H with this canonical embedding is denoted by G^*. Any two embeddings of H, as described above, are isomorphic.

The definition of the dual implies that $m(G^*) = m(G)$, $n(G^*) = f(G)$, and $d_{G^*}(f^*) = d_G(f)$, where $d_G(f)$ denotes the degree of the face f of G.

From the manner of construction of G^*, it follows that

(i) An edge e of a plane graph G is a cut edge of G if and only if e^* is a loop of G^*, and it is a loop of G if and only if e^* is a cut edge of G^*.

(ii) G^* is connected whether G is connected or not (see graphs G and G^* of Fig. 8.12).

The canonical embedding of the dual of G^* is denoted by G^{**}. It is easy to check that G^{**} is isomorphic to G if and only if G is connected. Graph isomorphism

Fig. 8.10 A plane graph G and its dual H

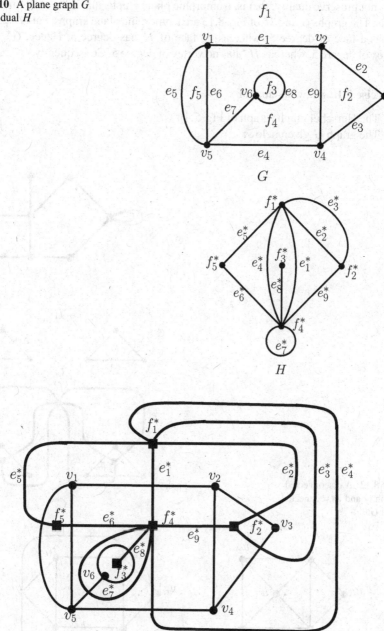

Fig. 8.11 Procedure for drawing the dual graph

does not preserve duality; that is, isomorphic plane graphs may have nonisomorphic duals. The graphs G and H of Fig. 8.13 are isomorphic plane graphs, but $G^* \not\simeq H^*$. G has a face of degree 5, whereas no face of H has degree 5. Hence, G^* has a vertex of degree 5, whereas H^* has no vertex of degree 5. Consequently, $G^* \not\simeq H^*$.

Exercise 5.1. Draw the dual of

(i) The Herschel graph (graph of Fig. 5.4).
(ii) The graph G given below:

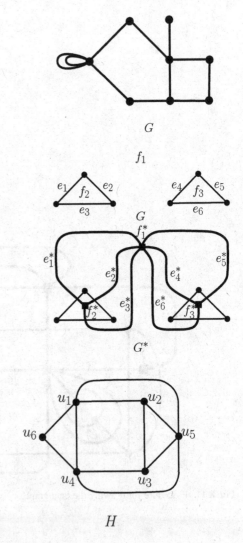

Fig. 8.12 A disconnected graph G and its (connected) dual G^*

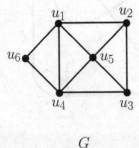

G

H

Fig. 8.13 Isomorphic graphs G and H for which $G^* \not\simeq H^*$

Exercise 5.2. A plane graph G is called *self-dual* if $G \simeq G^*$. Prove the following:

(i) All wheels W_n ($n \geq 3$) are self-dual.
(ii) For a self-dual graph, $2n = m + 2$.

Exercise 5.3. Construct two infinite families of self-dual graphs.

8.6   The Four-Color Theorem and the Heawood Five-Color Theorem

What is the minimum number of colors required to color the world map of countries so that no two countries having a common boundary receive the same color? This simple-looking problem manifested itself into one of the most challenging problems of graph theory, popularly known as the four-color conjecture (4CC).

The geographical map of the countries of the world is a typical example of a plane graph. An assignment of colors to the faces of a plane graph G so that no two faces having a common boundary containing at least one edge receive the same color is a *face coloring* of G. The face-chromatic number $\chi^*(G)$ of a plane graph G is the minimum k for which G has a face coloring using k colors. The problem of coloring a map so that no two adjacent countries receive the same color can thus be transformed into a problem of face coloring of a plane graph G. The face coloring of G is closely related to the vertex coloring of the dual G^* of G. The fact that two faces of G are adjacent in G if and only if the corresponding vertices of G^* are adjacent in G^* shows that G is k-face-colorable if and only if G^* is k-vertex-colorable.

It was young Francis Guthrie who conjectured, while coloring the district map of England, that four colors were sufficient to color the world map so that adjacent countries receive distinct colors. This conjecture was communicated by his brother to De Morgan in 1852. Guthrie's conjecture is equivalent to the statement that any plane graph is 4-face-colorable. The latter statement is equivalent to the conjecture: Every planar graph is 4-vertex-colorable.

After the conjecture was first published in 1852, many attempted to settle it. In the process of settling the conjecture, many equivalent formulations of this conjecture were found. Assaults on the conjecture were made using such varied branches of mathematics as algebra, number theory, and finite geometries. The solution found the light of the day when Appel, Haken, and Koch [8] of the University of Illinois established the validity of the conjecture in 1976 with the aid of computers (see also [6,7]). The proof includes, among other things, 10^{10} units of operations, amounting to a staggering 1200 hours of computer time on a high-speed computer available at that time.

Although the computer-oriented proof of Appel, Haken, and Koch settled the conjecture in 1976 and has stood the test of time, a theoretical proof of the four-color problem is still to be found.

Even though the solution of the 4CC has been a formidable task, it is rather easy to establish that every planar graph is 6-vertex-colorable.

Theorem 8.6.1. *Every planar graph is 6-vertex-colorable.*

Proof. The proof is by induction on n, the number of vertices of the graph. The result is trivial for planar graphs with at most six vertices. Assume the result for planar graphs with $n - 1$, $n \geq 7$, vertices. Let G be a planar graph with n vertices. By Corollary 8.3.5, $\delta(G) \leq 5$, and hence G has a vertex v of degree at most 5. By hypothesis, $G - v$ is 6-vertex-colorable. In any proper 6-vertex coloring of $G - v$, the neighbors of v in G would have used only at most five colors, and hence v can be colored by an unused color. In other words, G is 6-vertex-colorable. $\square$

It involves some ingenious arguments to reduce the upper bound for the chromatic number of a planar graph from 6 to 5. The upper bound 5 was obtained by Heawood [103] as early as 1890.

Theorem 8.6.2 (Heawood's five-color theorem). *Every planar graph is 5-vertex-colorable.*

Proof. The proof is by induction on $n(G) = n$. Without loss of generality, we assume that G is a connected plane graph. If $n \leq 5$, the result is clearly true. Hence, assume that $n \geq 6$ and that any planar graph with fewer than n vertices is 5-vertex-colorable. G being planar, $\delta(G) \leq 5$ by Corollary 8.3.5, and so G contains a vertex v_0 of degree not exceeding 5. By the induction hypothesis, $G - v_0$ is 5-vertex-colorable.

If $d(v_0) \leq 4$, at most four colors would have been used in coloring the neighbors of v_0 in G in a 5-vertex coloring of $G - v_0$. Hence, an unused color can then be assigned to v_0 to yield a proper 5-vertex coloring of G.

If $d(v_0) = 5$, but only four or fewer colors are used to color the neighbors of v_0 in a proper 5-vertex coloring of $G - v_0$, then also an unused color can be assigned to v_0 to yield a proper 5-vertex coloring of G.

Hence assume that the degree of v_0 is 5 and that in every 5-coloring of $G - v_0$, the neighbors of v_0 in G receive five distinct colors. Let v_1, v_2, v_3, v_4, and v_5 be the neighbors of v_0 in a cyclic order in a plane embedding of G. Choose some proper 5-coloring of $G - v_0$ with colors, say, $c_1, c_2, \ldots, c_5$. Let $\{V_1, V_2, \ldots, V_5\}$ be the color partition of $G - v_0$, where the vertices in V_i are colored c_i, $1 \leq i \leq 5$. Assume further that $v_i \in V_i$, $1 \leq i \leq 5$.

Let G_{ij} be the subgraph of $G - v_0$ induced by $V_i \cup V_j$. Suppose v_i and v_j, $1 \leq i, j \leq 5$, belong to distinct components of G_{ij}. Then the interchange of the colors c_i and c_j in the component of G_{ij} containing v_i would give a recoloring of $G - v_0$ in which only four colors are assigned to the neighbors of v_0. But this is against our assumption. Hence, v_i and v_j must belong to the same component of G_{ij}. Let $P_{i,j}$

Fig. 8.14 Graph for proof
of Theorem 8.6.2

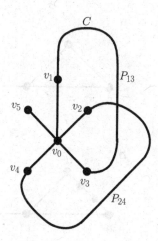

be a v_i-v_j path in G_{ij}. Let C denote the cycle $v_0 v_1 P_{13} v_3 v_0$ in G (Fig. 8.14). Then C separates v_2 and v_4; that is, one of v_2 and v_4 must lie in int C and the other in ext C. In Fig. 8.14, $v_2 \in$ int C and $v_4 \in$ ext C. Then P_{24} must cross C at a vertex of C. But this is clearly impossible since no vertex of C receives either of the colors c_2 and c_4. Hence this possibility cannot arise, and G is 5-vertex-colorable.   □

Note that the bound 4 in the inequality $\chi(G) \leq 4$ for planar graphs G is best possible since K_4 is planar and $\chi(K_4) = 4$.

Exercise 6.1. Show that a planar graph G is bipartite if and only if each of its faces is of even degree in any plane embedding of G.

Exercise 6.2. Show that a connected plane graph G is bipartite if and only if G^* is Eulerian. Hence, show that a connected plane graph is 2-face-colorable if and only if it is Eulerian.

Exercise 6.3. Prove that a Hamiltonian plane graph is 4-face-colorable and that its dual is 4-vertex-colorable.

Exercise 6.4. Show that a plane triangulation has a 3-face coloring if and only if it is not K_4. (Hint: Use Brooks' theorem.)

Remark 8.6.3. (Grötzsch): If G is a planar graph that contains no triangle, then G is 3-vertex-colorable.

8.7   Kuratowski's Theorem

Definition 8.7.1. 1. A *subdivision of an edge* $e = uv$ of a graph G is obtained by introducing a new vertex w in e, that is, by replacing the edge $e = uv$ of G by the path uwv of length 2 so that the new vertex w is of degree 2 in the resulting graph (see Fig. 8.15a).

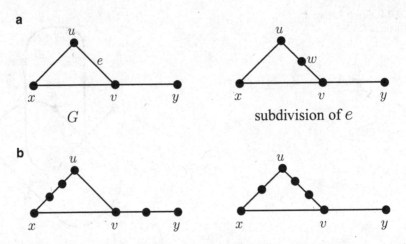

Fig. 8.15 (a) Subdivision of edge e of graph G, (b) two homeomorphs of graph G

2. A *homeomorph* or a *subdivision of a graph* G is a graph obtained from G by applying a finite number of subdivisions of edges in succession (see Fig. 8.15b). G itself is regarded as a subdivision of G.
3. Two graphs G_1 and G_2 are called *homeomorphic* if they are both homeomorphs of some graph G. Clearly, the graphs of Fig. 8.15b are homeomorphic, even though neither of the two graphs is a homeomorph of the other.

Kuratowski's theorem [129] characterizing planar graphs was one of the major breakthrough results in graph theory of the 20th century. As mentioned earlier, while examining planarity of graphs, we need only consider simple graphs since the presence of loops and multiple edges does not affect the planarity of graphs. Consequently, *a graph is planar if and only if its underlying simple graph is planar.* We therefore consider in this section only (finite) simple graphs. We recall that for any edge e of a graph G, $G - e$ is the subgraph of G obtained by deleting the edge e, whereas $G \circ e$ denotes the contraction of e. We always discard isolated vertices when edges get deleted and remove the new multiple edges when edges get contracted. More generally, for a subgraph H of G, $G \circ H$ denotes the graph obtained by the successive contractions of all the edges of H in G. The resulting graph is independent of the order of contraction. Moreover, if G is planar, then $G \circ e$ is planar; consequently, $G \circ H$ is planar. In other words, if $G \circ H$ is nonplanar for some subgraph H of G, then G is also nonplanar. Further, any two homeomorphic graphs are contractible to the same graph.

Definition 8.7.2. If $G \circ H = K$, we call K a *contraction* of G; we also say that G is *contractible* to K. G is said to be *subcontractible* to K if G has a subgraph H contractible to K. We also refer to this fact by saying that K is a *subcontraction* of G.

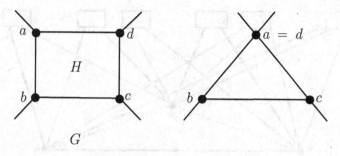

Fig. 8.16 Graph G subcontractible to triangle abc

Example 8.7.3. For instance, in Fig. 8.16, graph G is subcontractible to the triangle abc. (Take H to be the cycle $abcd$ and contract the edge ad in H. By abuse of notation, the new vertex is denoted by a or d.) We note further that if G' is a homeomorph of G, then contraction of one of the edges incident at each vertex of degree 2 in $V(G')\backslash V(G)$ results in a graph homeomorphic to G.

Our first aim is to prove the following result, which was established by Wagner [186] and, independently, by Harary and Tutte [96].

Theorem 8.7.4 ([96,186]). *A graph is planar if and only if it is not subcontractible to K_5 or $K_{3,3}$.*

As a consequence, we establish Kuratowski's characterization theorem for planar graphs.

Theorem 8.7.5 (Kuratowski [129]). *A graph is planar if and only if it has no subgraph homeomorphic to K_5 or $K_{3,3}$.*

The proofs of Theorems 8.7.4 and 8.7.5, as presented here, are due to Fournier [68]. Recall that any subgraph and any contraction of a planar graph are both planar.

Definition 8.7.6. A simple connected nonplanar graph G is *irreducible* if, for each edge e of G, $G \circ e$ is planar.

For instance, both K_5 and $K_{3,3}$ are irreducible.

Proof of theorem 8.7.4. If G has a subgraph G_0 contractible to K_5 or $K_{3,3}$, then since K_5 and $K_{3,3}$ are nonplanar, G_0 and therefore G are nonplanar.

We now prove the converse. Assume that G is a simple connected nonplanar graph. By Theorem 8.2.8, at least one block of G is nonplanar. Hence, assume that G is a simple 2-connected nonplanar graph. We now show that G has a subgraph contractible to K_5 or $K_{3,3}$.

Keep contracting edges of G (and delete the new multiple edges, if any, at each stage of the contraction) until a (2-connected) irreducible (nonplanar) graph H results. Clearly, $\delta(H) \geq 3$. Now, if e and f are any two distinct edges of G, then $(G \circ e) - f = (G - f) \circ e$. Hence, the graph H may as well be obtained by

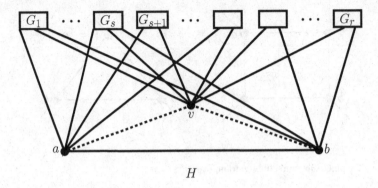

$$H$$

Fig. 8.17 Graph H for case 1 of proof of Theorem 8.7.4

deleting a set (which may be empty) of edges of G, resulting in a subgraph G_0 of G and then contracting a subgraph of G_0. We now complete the proof of the theorem by showing that H has a subgraph K homeomorphic (and hence contractible) to K_5 or $K_{3,3}$. In this case, G has the subgraph G_0, which is contractible to K_5 or $K_{3,3}$.

Let $e = ab \in E(H)$ and $H' = H - \{a, b\}$. Then H' is connected. If not, $\{a, b\}$ is a vertex cut of H. Let $G'_1, \ldots, G'_r$ be the components of H'. As H is irreducible, $H - V(G'_r)$ is planar, and there exists a plane embedding of H' in which the edge ab is in the exterior face. As G'_r is planar, G'_r can be embedded in this exterior face of H'. This would make H a planar graph, a contradiction. Thus, H' is connected.]

Case 1. H' has a cut vertex v. Let $G_1, G_2, \ldots, G_r$ $(r \geq 2)$ be the components of $H' - \{v\}$, and let $G_1, G_2, \ldots, G_s$, $0 \leq s \leq r$, be those components that are connected to both a and b. (see Fig. 8.17). If $r > s$, then each of $G_{s+1}, \ldots, G_r$ is connected to only one of a or b. Assume that G_r is connected to b and not to a. From the plane representation of $G \circ (G_{s+1} \cup \ldots \cup G_r)$, the contraction of G obtained by contracting the edges of $G_{s+1}, \ldots, G_r$, we can obtain a plane representation of H' (see Fig. 8.17). [In fact, if G_r is contracted to the vertex w_r, then as the subgraph $A_r = \langle v, b, v(G_r) \rangle$ of H' is planar, the pair of edges $\{vw_r, w_rb\}$ can be replaced by the planar subgraph A_r and so on.] Hence this case cannot arise. Consequently, $r = s$. If $r = s = 2$, the plane embeddings of $H' \circ G_1$ and $H' \circ G_2$ yield a plane embedding of H', a contradiction (see Fig. 8.18). Consequently, $r = s \geq 3$. In this case, H' contains a homeomorph of $K_{3,3}$ (see Fig. 8.19), with $\{w_1, w_2, w_3 ; a, b, v\}$ being the vertex set of $K_{3,3}$. (Other possibilities for w_1, w_2, w_3 will also yield a homeomorph of $K_{3,3}$.)

Case 2. H' is 2-connected. Then H' contains a cycle C of length at least 3. Consider a plane embedding of $H \circ e$ (where $e = ab$, as above). If c denotes the new vertex to which a and b get contracted, $(H \circ e) - c = H'$. We may therefore suppose without loss of generality that c is in the interior of the cycle C in the plane embedding of $H \circ e$.

Fig. 8.18 Plane embedding
for case 1 of proof of
Theorem 8.7.4

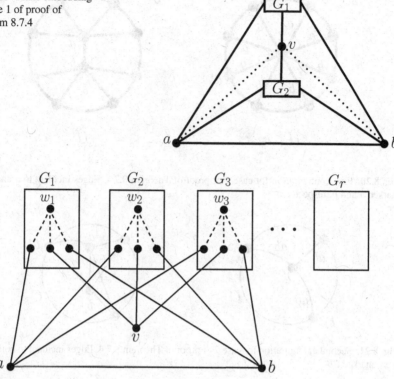

Fig. 8.19 Homeomorph for case 1 of proof of Theorem 8.7.4

Now, the edges of $H \circ e$ incident to c arise out of edges of H incident to a or b. There arise three possibilities with reference to the positions of the edges of $H \circ e$ incident to c relative to the cycle C.

(i) Suppose the edges incident to c occur so that the edges incident to a and the edges incident to b in H are consecutive around c in a plane embedding of $H \circ e$, as shown in Fig. 8.20a. Since H is a minimal nonplanar graph, the paths from c to C can only be single edges. Then the plane representation of $H \circ e$ gives a plane representation of H, as in Fig. 8.20b, a contradiction. So this possibility cannot arise.

(ii) Suppose there are three edges of $H \circ e$ incident with c, with each edge corresponding to a pair of edges of H, one incident to a and the other to b, as in Fig. 8.21a. Then H contains a subgraph contractible to K_5, as shown in Fig. 8.21b.

We are now left with only one more possibility.

(iii) There are four edges of $H \circ e$ incident to c, and they arise alternately out of edges incident to a and b in H, as in Fig. 8.22a. Then there arises in H

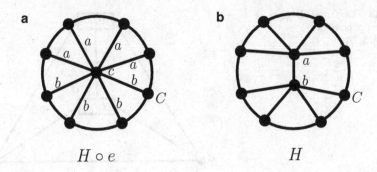

Fig. 8.20 First configuration for case 2 of proof of Theorem 8.7.4. Edges incident to a and b are marked a and b, respectively

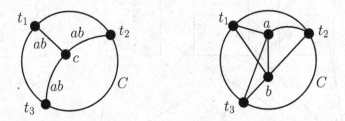

Fig. 8.21 Second configuration for case 2 of proof of Theorem 8.7.4. Edges incident to both a and b are marked ab

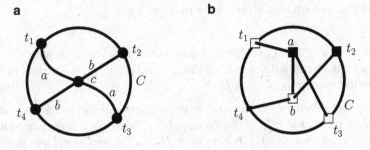

Fig. 8.22 Third configuration for case 2 of proof of Theorem 8.7.4

a homeomorph of $K_{3,3}$, as shown in Fig. 8.22b. The sets $X = \{a, t_2, t_4\}$ and $Y = \{b, t_1, t_3\}$ are the sets of the bipartition of this homeomorph of $K_{3,3}$.   □

We now proceed to prove Theorem 8.7.5.

Proof of theorem 8.7.5. The "sufficiency" part of the proof is trivial. If G contains a homeomorph of either K_5 or $K_{3,3}$, G is certainly nonplanar, since a homeomorph of a planar graph is planar.

Fig. 8.23 Graphs for proof
of Theorem 8.7.5

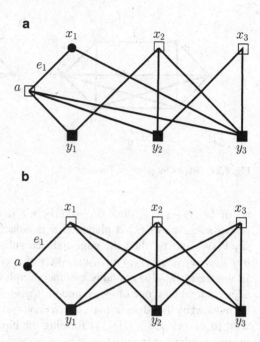

Assume that G is connected and nonplanar. Remove edges from G one after another until we get an edge-minimal connected nonplanar subgraph G_0 of G; that is, G_0 is nonplanar and for any edge e of G, $G_0 - e$ is planar. Now contract the edges in G_0 incident with vertices of degree at most 2 in some order. Let us denote the resulting graph by G_0'. Then G_0' is nonplanar, whereas $G_0' - e$ is planar for any edge e of G_0', and the minimum degree of G_0' is at least 3. We now have to show that G_0' contains a homeomorph of K_5 or $K_{3,3}$.

By Theorem 8.7.4, G_0' is subcontractible to K_5 or $K_{3,3}$. This means that G_0' contains a subgraph H that is contractible to K_5 or $K_{3,3}$. As $G_0' - e$ is planar for any edge e of G_0', $G_0' = H$. Thus, G_0' itself is contractible to K_5 or $K_{3,3}$. If G_0' is either K_5 or $K_{3,3}$, we are done. Assume now that G_0' is neither K_5 nor $K_{3,3}$. Let $e_1, e_2, \ldots, e_r$ be the edges of G_0', when contracted in order, that result in a K_5 or $K_{3,3}$.

First, let us assume that $r = 1$, so that $G_0' \circ e_1$ is either K_5 or $K_{3,3}$. Suppose that $G_0' \circ e_1 = K_{3,3}$ with $\{x_1, x_2, x_3\}$ and $\{y_1, y_2, y_3\}$ as the partite sets of vertices. Suppose that x_1 is the vertex obtained by identifying the ends of e_1. We may then take $e_1 = x_1 a$ (by abuse of notation), where a is a vertex distinct from the x_i's and y_j's (Fig. 8.23a). If a is adjacent to all of y_1, y_2 and y_3, then $\{a, x_2, x_3\}$ and $\{y_1, y_2, y_3\}$ form a bipartition of a $K_{3,3}$ in G_0'. If a is adjacent to only one or two of $\{y_1, y_2, y_3\}$ (Fig. 8.23b), then again G_0' contains a homeomorph of $K_{3,3}$.

Next, let us assume that $G_0' \circ e_1 = K_5$ with vertex set $\{v_1, v_2, v_3, v_4, v_5\}$. Suppose that v_1 is the vertex obtained by identifying the ends of e_1. As before, we may take $e_1 = v_1 a$, where $a \notin \{v_1, v_2, v_3, v_4, v_5\}$. If a is adjacent to

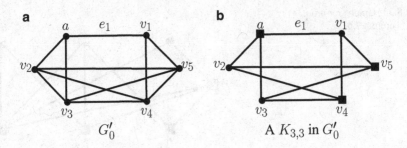

Fig. 8.24 Graphs for proof of Theorem 8.7.5

all of $\{v_2, v_3, v_4, v_5\}$, then $G_0' - v_1$ is a K_5, contradiction to the fact that any proper subgraph of G_0' is planar. If a is adjacent to only three of $\{v_2, v_3, v_4, v_5\}$, say v_2, v_3, and v_4, then the edge-induced subgraph of G_0' induced by the edges $av_1, av_2, av_3, av_4, v_1v_5, v_2v_3, v_2v_4, v_2v_5, v_3v_4, v_3v_5$, and v_4v_5 is a homeomorph of K_5. In this case, G_0' also contains a homeomorph of $K_{3,3}$. Since $d_{G_0'}(v_1) \geq 3$, v_1 is adjacent to at least one of v_2, v_3, and v_4, say v_2. Then the edge-induced subgraph of G_0' induced by the edges in $\{av_1, av_3, av_4, v_1v_2, v_2v_3, v_2v_4, v_1v_5, v_3v_5, v_4v_5\}$ is a $K_{3,3}$, with $\{a, v_4, v_5\}$ and $\{v_1, v_2, v_3\}$ forming the bipartition. We now consider the case when a is adjacent to only two of v_2, v_3, v_4 and v_5, say v_2 and v_3. Then, necessarily, v_1 is adjacent to v_4 and v_5 (since on contraction of the edge v_1a, v_1 is adjacent to v_2, v_3, v_4, and v_5). In this case G_0' also contains a $K_{3,3}$ (see Fig. 8.24b). Finally, the case when a is adjacent to at most one of v_2, v_3, v_4, and v_5 cannot arise since the degree of a is at least 3 in G_0'. Thus, in any case, we have proved that when $r = 1$, G_0' contains a homeomorph of $K_{3,3}$. The result can now easily be seen to be true by induction on r. Indeed, if $H_2 = H_1 \circ e$ and H_2 contains a homeomorph of $K_{3,3}$, then H_1 contains a homeomorph of $K_{3,3}$. $\qquad\square$

The nonplanarity of the Petersen graph (Fig. 8.25a) can be established by showing that it is contractible to K_5 (see Fig. 8.25b) or by showing that it contains a homeomorph of $K_{3,3}$ (see Fig. 8.25c).

Exercise 7.1. Prove that the following graph is nonplanar.

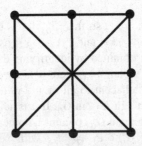

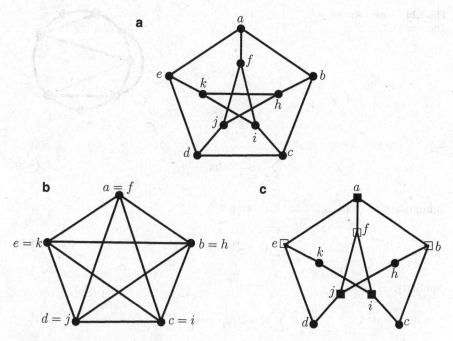

Fig. 8.25 Nonplanarity of the Petersen graph. (**a**) The Petersen graph P, (**b**) contraction of P to K_5, (**c**) A subdivision of $K_{3,3}$ in P

8.8 Hamiltonian Plane Graphs

An elegant necessary condition for a plane graph to be Hamiltonian was given by Grinberg [78].

Theorem 8.8.1. *Let G be a loopless plane graph having a Hamilton cycle C. Then $\sum_{i=2}^{n}(i-2)(\phi_i'-\phi_i'') = 0$, where ϕ_i' and ϕ_i'' are the numbers of faces of G of degree i contained in int C and ext C, respectively.*

Proof. Let E' and E'' denote the sets of edges of G contained in int C and ext C, respectively, and let $|E'| = m'$ and $|E''| = m''$. Then int C contains exactly $m' + 1$ faces (see Fig. 8.26), and so

$$\sum_{i=2}^{n} \phi_i' = m' + 1. \tag{8.1}$$

(Since G is loopless, $\phi_1' = \phi_1'' = 0$.)

Moreover, each edge in int C is on the boundary of exactly two faces in int C, and each edge of C is on the boundary of exactly one face in int C. Hence, counting the edges of all the faces in int C, we get

Fig. 8.26 Graph for proof of
Theorem 8.8.1

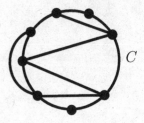

$$\sum_{i=2}^{n} i\phi_i' = 2m' + n. \qquad (8.2)$$

Eliminating m' from (8.1) and (8.2), we get

$$\sum_{i=2}^{n} (i-2)\phi_i' = n - 2. \qquad (8.3)$$

Similarly,

$$\sum_{i=2}^{n} (i-2)\phi_i'' = n - 2. \qquad (8.4)$$

Equations (8.3) and (8.4) give the required result.   □

Grinberg's condition is quite useful in that by applying this result, many plane graphs can easily be shown to be non-Hamiltonian by establishing that they do not satisfy the condition.

Example 8.8.2. The Herschel graph G of Fig. 5.4 is non-Hamiltonian.

Proof. G has nine faces and all the faces are of degree 4. Hence, if G were Hamiltonian, we must have $2(\phi_4' - \phi_4'') = 0$. This means that $\phi_4' = \phi_4''$. This is impossible, since $\phi_4' + \phi_4'' =$ (number of faces of degree 4 in G) $= 9$ is odd. Hence, G is non-Hamiltonian. (In fact, it is the smallest planar non-Hamiltonian 3-connected graph.)

Exercise 8.1. Does there exist a plane Hamiltonian graph with faces of degrees 5, 7, and 8, and with just one face of degree 7?

Exercise 8.2. Prove that the Grinberg graph given in Fig. 8.27 is non-Hamiltonian.

8.9   Tait Coloring

In an attempt to solve the four-color problem, Tait considered edge colorings of 2-edge-connected cubic planar graphs. He conjectured that every such graph was 3-edge colorable. Indeed, he could prove that his conjecture was equivalent to the

Fig. 8.27 The Grinberg graph

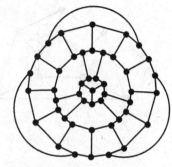

Fig. 8.28 The Tutte graph

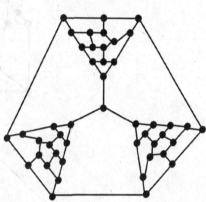

four-color problem (see Theorem 8.9.1). Tait did this in 1880. He even went to the extent of giving a "proof" of the four-color theorem using this result. Unfortunately, Tait's proof was based on the wrong assumption that any 2-edge-connected cubic planar graph is Hamiltonian. A counterexample to his assumption was given by Tutte in 1946 (65 years later). The graph given by Tutte is the graph of Fig. 8.28. It is a non-Hamiltonian cubic 3-connected (and therefore 3-edge-connected; see Theorem 3.3.4) planar graph. Tutte used ad hoc techniques to prove this result. (The Grinberg condition does not establish this result.)

We indicate below the proof of the fact that the Tutte graph of Fig. 8.28 is non-Hamiltonian. The graphs G_1 to G_5 mentioned below are shown in Fig. 8.29.

It is easy to check that there is no Hamilton cycle in the graph G_1 containing both of the edges e_1 and e_2. Now, if there is a Hamilton cycle in G_2 containing both of the edges e_1' and e_2', then there will be a Hamilton cycle in G_1 containing e_1 and e_2. Hence there is no Hamilton cycle in G_2 containing e_1' and e_2'. In $G_3 - e'$, u and w are vertices of degree 2. Hence if $G_3 - e'$ were Hamiltonian, then in any Hamilton cycle of $G_3 - e'$, both the edges incident to u as well as both the edges incident to w must be consecutive. This would imply that G_2 has a Hamilton cycle containing e_1' and e_2', which is not the case. Consequently, any Hamilton cycle of G_3 must contain the edge e'. It follows that there exists no Hamilton path from x to y in $G_3 - w$. A redrawing of $G_3 - w$ is the graph G_4. It is called the "Tutte triangle." The Tutte

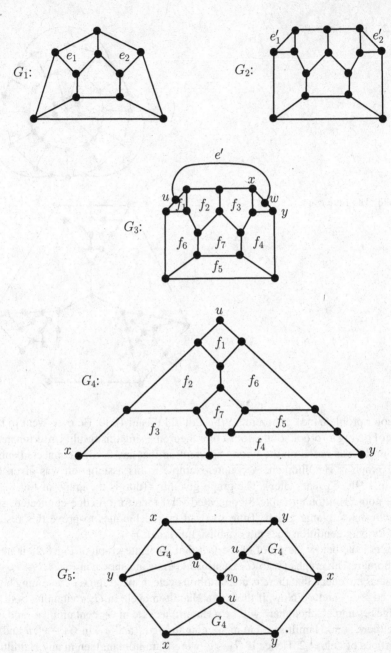

Fig. 8.29 Graphs G_1 to G_5

Fig. 8.30 Graph for proof of
(i) ⇒ (ii) in Theorem 8.9.1

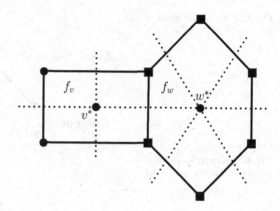

graph (Fig. 8.28) contains three copies of G_4 together with a vertex v_0. It has been
redrawn as graph G_5 of Fig. 8.29. Suppose G_5 is Hamiltonian with a Hamilton cycle
C. If we describe C starting from v_0, it is clear that C must visit each copy of G_4
exactly once. Hence, if C enters a copy of G_4, it must exit that copy through x or
y after visiting all the other vertices of that copy. But this means that there exists a
Hamilton path from y to x (or from x to y) in G_4, a contradiction. Thus, the Tutte
graph G_5 is non-Hamiltonian.

We now give the proof of Tait's result. Recall that by Vizing–Gupta's theorem
(Theorem 7.5.5), every simple cubic graph has chromatic index 3 or 4. A 3-edge
coloring of a cubic planar graph is often called a *Tait coloring*.

Theorem 8.9.1. *The following statements are equivalent:*

(i) All plane graphs are 4-vertex-colorable.
(ii) All plane graphs are 4-face-colorable.
*(iii) All simple 2-edge-connected cubic planar graphs are 3-edge-colorable (i.e.,
    Tait colorable).*

Proof.   (i) ⇒ (ii). Let G be a plane graph. Let G^* be the dual of G (see Sect. 8.4).
Then, since G^* is a plane graph, it is 4-vertex-colorable. If v^* is a
vertex of G^*, and f_v is the face of G corresponding to v^*, assign to
f_v the color of v^* in a 4-vertex coloring of G^*. Then, by the definition
of G^*, it is clear that adjacent faces of G will receive distinct colors.
(See Fig. 8.30, in which f_v and f_w receive the colors of v^* and w^*,
respectively.) Thus, G is 4-face-colorable.

(ii) ⇒ (iii). Let G be a plane embedding of a 2-edge-connected cubic planar
graph. By assumption, G is 4-face-colorable. Denote the four colors
by $(0,0)$, $(1,0)$, $(0,1)$, and $(1,1)$, the elements of the ring $\mathbb{Z}_2 \times \mathbb{Z}_2$.
If e is an edge of G that separates the faces, say f_1 and f_2, color e
with the color given by the sum (in $\mathbb{Z}_2 \times \mathbb{Z}_2$) of the colors of f_1 and f_2.
Since G has no cut edge, each edge is the common boundary of exactly

Fig. 8.31 Graph for proof of
(ii) $\Rightarrow$ (iii) in Theorem 8.9.1

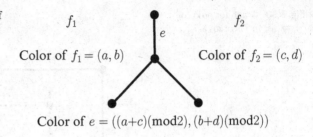

Color of $f_1 = (a, b)$ Color of $f_2 = (c, d)$

Color of $e = ((a+c)(\text{mod}2), (b+d)(\text{mod}2))$

Fig. 8.32 Graph for proof of
(iii) $\Rightarrow$ (i) for Theorem 8.9.1

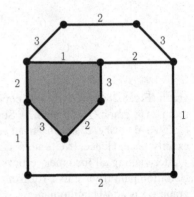

two faces of G. This gives a 3-edge coloring of G using the colors
$(1, 0)$, $(0, 1)$, and $(1, 1)$, since the sum of any two distinct elements of
$\mathbb{Z}_2 \times \mathbb{Z}_2$ is not $(0, 0)$ (see Fig. 8.31).

(iii) $\Rightarrow$ (i) Let G be a planar graph. We want to show that G is 4-vertex-colorable.
We may assume without loss of generality that G is simple. Let $\tilde{G}$ be
a plane embedding of G. Then $\tilde{G}$ is a spanning subgraph of a plane
triangulation T, (see Sect. 8.2), and hence it suffices to prove that T is
4-vertex-colorable.

Let T^* be the dual of T. Then T^* is a 2-edge-connected cubic plane graph. By
our assumption, T^* is 3-edge-colorable using, for example, the colors c_1, c_2, and c_3.
Since T^* is cubic, each of the above three colors is represented at each vertex of T^*.
Let T_{ij}^* be the edge subgraph of T^* induced by the edges of T^* which have been
colored using the colors c_i and c_j. Then T_{ij}^* is a disjoint union of even cycles, and
thus it is 2-face-colorable. But each face of T^* is the intersection of a face of T_{12}^*
and a face of T_{23}^* (see Fig. 8.32). Now the 2-face colorings of T_{12}^* and T_{23}^* induce a
4-face coloring of T^* if we assign to each face of T^* the (unordered) pair of colors
assigned to the faces whose intersection is f. Since $T^* = T_{12}^* \cup T_{23}^*$, this defines a
proper 4-face coloring of T^*. Thus, $\chi(G) = \chi(\tilde{G}) \le \chi(T) = \chi^*(T^*) \le 4$, and G
is 4-vertex-colorable. (Recall that $\chi^*(T^*)$ is the face-chromatic number of T^*.) $\square$

Exercise 9.1. Exhibit a 3-edge coloring for the Tutte graph (see Fig. 8.28).

Notes

The proof of Heawood's theorem uses arguments based on paths in which the vertices are colored alternately by two colors. Such paths are called "Kempe chains" after Kempe [121], who first used such chains in his "proof" of the 4CC. Even though Kempe's proof went wrong, his idea of using Kempe chains and switching the colors in such chains had been effectively exploited by Heawood [103] in proving his five-color theorem (Theorem 8.6.2) for planar graphs, as well as by Appel, Haken, and Koch [8] in settling the 4CC. As the reader might notice, the same technique had been employed in the proof of Brooks' theorem (Theorem 7.3.7). Chronologically, Francis Guthrie conceived the four-color theorem in 1852 (if not earlier). Kempe's purported "proof" of the 4CC was given in 1879, and the mistake in his proof was pointed out by Heawood in 1890. The Appel–Haken–Koch proof of the 4CC was first announced in 1976. Between 1879 and 1976, graph theory witnessed an unprecedented growth along with the methods to tackle the 4CC. The reader who is interested in getting a detailed account of the four-color problem may consult Ore [152] and Kainen and Saaty [120].

Even though the Tutte graph of Fig. 8.28 shows that not every cubic 3-connected planar graph is Hamiltonian, Tutte himself showed that every 4-connected planar graph is Hamiltonian [180].

Chapter 9
Triangulated Graphs

9.1 Introduction

Triangulated graphs form an important class of graphs. They are a subclass of the class of perfect graphs and contain the class of interval graphs. They possess a wide range of applications. We describe later in this chapter an application of interval graphs in phasing the traffic lights at a road junction.

We begin with the definition of perfect graphs.

9.2 Perfect Graphs

For a simple graph G, we have the following parameters:

$\chi(G)$: The chromatic number of G
$\omega(G)$: The clique number of G (= the order of a maximum clique of G)
$\alpha(G)$:   The independence number of G
$\theta(G)$: The clique covering number of G (= the minimum number of cliques of G that cover the vertex set of G).

For instance, for the graph G of Fig. 9.1, $\chi(G) = \omega(G) = 4$ and $\alpha(G) = \theta(G) = 4$.

A minimum set of cliques that covers $V(G)$ is $\{\{1\}, \{2\}, \{3, 4, 5, 6\}, \{7, 8, 9\}\}$. In any proper vertex coloring of G, the vertices of any clique must receive distinct colors. Hence it is clear that $\chi(G) \geq \omega(G)$. Further, if A is any independent set of G, any clique of a clique cover of G can contain at most one vertex of A. Hence, to cover the $\alpha(G)$ vertices of a maximum independent set of G, at least $\alpha(G)$ distinct cliques of G are needed. Therefore, $\theta(G) \geq \alpha(G)$.

If G is an odd cycle $C_{2n+1}, n \geq 2$, $\chi(G) = 3, \omega(G) = 2, \theta(G) = n + 1$, and $\alpha(G) = n$. Hence, for such a $G, \chi(G) > \omega(G)$, and $\theta(G) > \alpha(G)$. Moreover,

R. Balakrishnan and K. Ranganathan, *A Textbook of Graph Theory*,
Universitext, DOI 10.1007/978-1-4614-4529-6_9,
© Springer Science+Business Media New York 2012

Fig. 9.1 Graph G

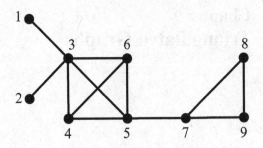

$A \subset V(G)$ is an independent set of vertices of G if and only if A induces a clique in G^c. Therefore, for any simple graph G,

$$\chi(G) = \theta(G^c), \text{ and}$$

$$\alpha(G) = \omega(G^c). \tag{9.1}$$

Definition 9.2.1. Let G be a simple graph. Then

(i) G is χ-*perfect* if and only if for every $A \subseteq V(G)$, $\chi(G[A]) = \omega(G[A])$.
(ii) G is α-*perfect* if and only if for every $A \subseteq V(G)$, $\alpha(G[A]) = \theta(G[A])$.

Remark 9.2.2. 1. By (9.1) above, it is clear that a graph is χ-perfect if and only if its complement is α-perfect.
2. Berge [19] conjectured that the concepts of χ-perfectness and α-perfectness are equivalent for any simple graph. This was shown to be true by Lovász [134] (and independently by Fulkerson [69]). This result is often referred to in the literature as the *perfect graph theorem*.

Theorem 9.2.3 (Perfect graph theorem). *For a simple graph G, the following statements are equivalent:*

(i) G is χ-perfect.
(ii) G is α-perfect.
(iii) $\alpha(G[A])\,\omega(G[A]) \geq |A|$ for every $A \subseteq V(G)$.

In view of the perfect graph theorem, there is no need to distinguish between α-perfectness and χ-perfectness; hence, graphs that satisfy any one of these three equivalent conditions can be referred to as merely *perfect graphs*. In particular, this means that a simple graph G is perfect if and only if its complement is perfect. For a proof of the perfect graph theorem, see [76] or [134].

Remark 9.2.4. If G is perfect, by what is mentioned above, G cannot contain an odd hole, that is, an induced odd cycle C_{2n+1}, $n \geq 2$; likewise, by (9.1), G cannot contain an odd antihole, that is, an induced C_{2n+1}^c, $n \geq 2$. Equivalently, if G is perfect, then G can contain neither $C_{2n+1}, n \geq 2$ nor its complement as an induced subgraph. The converse of this result is the celebrated "*strong perfect graph conjecture*" of Berge, settled affirmatively by Chudnovsky et al. [36] (see notes at the end of this chapter).

9.3 Triangulated Graphs

Definition 9.3.1. A simple graph G is called *triangulated* if every cycle of length at least four in G has a chord, that is, an edge joining two nonconsecutive vertices of the cycle (see Fig. 9.2). For this reason, triangulated graphs are also called *chordal graphs* and sometimes *rigid circuit graphs*.

A graph is *weakly triangulated* if it contains neither a chordless cycle of length at least 5 nor the complement of such a cycle as an induced subgraph. Note that any triangulated graph is weakly triangulated.

Remark 9.3.2. It is clear that the property of a graph being triangulated is hereditary; that is, if G is triangulated, then every induced subgraph of G is also triangulated.

Definition 9.3.3. A vertex v of a graph G is a *simplicial vertex* of G if the closed neighborhood $N_G[v]$ of v in G induces a clique in G.

Example 9.3.4. In Fig. 9.2a, the vertices u_1, u_2, u_3, and u_4 are simplicial, whereas v_1, v_2, v_3, and v_4 are not.

Triangulated graphs can be recognized by the presence of a perfect vertex elimination scheme.

Definition 9.3.5. A *perfect vertex elimination scheme* (or, briefly, a *perfect scheme*) of a graph G is an ordering $\{v_1, v_2, \ldots, v_n\}$ of the vertex set of G in such a way that, for $1 \leq i \leq n$, v_i is a simplicial vertex of the subgraph induced by $\{v_i, v_{i+1}, \ldots, v_n\}$ of G.

Example 9.3.6. For the graph of Fig. 9.2a, $\{u_1, u_2, u_3, u_4, v_4, v_2, v_1, v_3\}$ is a perfect scheme.

Remark 9.3.7. Any vertex of degree 1 is trivially simplicial. Hence, any tree has a perfect vertex elimination scheme. Also, any tree is trivially triangulated. It turns out that these facts can be generalized to assert that any triangulated graph has

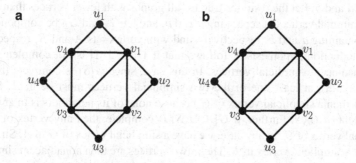

Fig. 9.2 (a) Triangulated and (b) nontriangulated graphs

a perfect vertex elimination scheme. (Based on this, Fulkerson and Gross [70] gave a "good algorithm" to test for triangulated graphs, namely, repeatedly locate a simplicial vertex and remove it from the graph until there is left out a single vertex and the graph is triangulated, or else at some stage no simplicial vertex exists and the graph is not triangulated.) Before we establish the above result, we need another characterization of triangulated graphs. This result is due to Hajnal and Surányi [89] and also due to Dirac [56].

Lemma 9.3.8. *A graph G is triangulated if and only if every minimal vertex cut of G is a clique.*

Proof. Assume that G is triangulated and that S is a minimal vertex cut of G. Let a and b be vertices in distinct components, say G_A and G_B, respectively, of $G \backslash S$. Now every vertex x of S must be adjacent to some vertex of G_A, since if x is adjacent to no vertex of G_A, then $G \backslash (S \backslash x)$ is disconnected and this would contradict the minimality of S. Similarly, x is adjacent to some vertex of G_B. Hence for any pair $x, y \in S$, there exist paths $P_1 : xa_1 \ldots a_r y$ and $P_2 : xb_1 \ldots b_s y$, with each $a_i \in G_A$ and each $b_j \in G_B$. Let us assume further that the a_i's and b_j's have been so chosen that these x-y paths are of least length. Then $xa_1 \ldots a_r y b_s b_{s-1} \ldots b_1 x$ is a cycle whose length is at least 4, and so it must have a chord. But such a chord cannot be of the form $a_i a_j$ or $b_k b_\ell$ in view of the minimality of the length of P_1 and P_2. Nor can it be $a_i b_j$ for some i and j, as a_i and b_j belong to a distinct component of $G \backslash S$. Hence, it can be only xy. Thus, every pair x, y in S is adjacent, and S is a clique.

Conversely, assume that every minimal vertex cut of G is a clique. Let $axby_1y_2 \ldots y_r a$ be a cycle C of length ≥ 4 in G. If ab were not a chord of C, denote by S a minimal vertex cut that puts a and b in distinct components of $G \backslash S$. Then S must contain x and y_j for some j. By hypothesis, S is a clique, and hence $xy_j \in E(G)$, and xy_j is a chord of C. Thus, G is triangulated. $\square$

Lemma 9.3.9. *Every triangulated graph G has a simplicial vertex. Moreover, if G is not complete, it has two nonadjacent simplicial vertices.*

Proof. The lemma is trivial either if G is complete or if G has just two or three vertices. Assume therefore that G is not complete, so that G has two nonadjacent vertices a and b. Let the result be true for all graphs with fewer vertices than G. Let S be a minimal vertex cut separating a and b, and let G_A and G_B be components of $G \backslash S$ containing a and b, respectively, and with vertex sets A and B, respectively. By the induction hypothesis, it follows that if $G[A \cup S]$ is not complete, it has two nonadjacent simplicial vertices. In this case, since $G[S]$ is complete (refer to Lemma 9.3.8), at least one of the two simplicial vertices must be in A. Such a vertex is then a simplicial vertex of G because none of its neighbors is in any other component of $G \backslash S$. Further, if $G[A \cup S]$ is complete, then any vertex of A is a simplicial vertex of G. In any case, we have a simplicial vertex of G in A. Similarly, we have a simplicial vertex in B. These two vertices are then nonadjacent simplicial vertices of G. $\square$

We are now ready to prove the second characterization theorem of triangulated graphs.

Theorem 9.3.10. *A graph G is triangulated if and only if it has a perfect vertex elimination scheme.*

Proof. The result is obvious for graphs with at most three vertices. So assume that G is a triangulated graph with at least four vertices. Assume that every triangulated graph with fewer vertices than G has a perfect vertex elimination scheme. By Lemma 9.3.9, G has a simplicial vertex v. Then $G \setminus v$ has a perfect vertex elimination scheme. Then v followed by a perfect scheme of $G \setminus v$ gives a perfect scheme of G.

Conversely, assume that G has a perfect scheme, say $\{v_1, v_2, \ldots, v_n\}$. Let C be a cycle of length ≥ 4 in G. Let j be the first suffix with $v_j \in V(C)$. Then $V(C) \subseteq G[\{v_j, v_{j+1}, \ldots, v_n\}]$ and, since v_j is simplicial in $G[\{v_j, v_{j+1}, \ldots, v_n\}]$, the neighbors of v_j in C form a clique in G, and hence C has a chord. Thus, G is triangulated. $\qquad\square$

Theorem 9.3.11. *A triangulated graph is perfect.*

Proof. The result is clearly true for triangulated graphs of order at most 4. So assume that G is a triangulated graph of order at least 5. We apply induction. Assume that the theorem is true for all graphs having fewer vertices than G. If G is disconnected, we can consider each component of G individually. So assume that G is connected. By Lemma 9.3.9, G contains a simplicial vertex v. Let u be a vertex adjacent to v in G. Since v is simplicial in G (and so in $G-u$), $\theta(G-u) = \theta(G)$. By the induction hypothesis, $G - u$ is triangulated and therefore perfect and therefore $\theta(G - u) = \alpha(G - u)$. Hence (see Exercise 7.4), $\theta(G) = \theta(G - u) = \alpha(G - u) \leq \alpha(G)$. This together with the fact that $\theta(G) \geq \alpha(G)$ implies that $\theta(G) = \alpha(G)$. The proof is complete since by the induction assumption, for any proper subset A of $V(G)$, the subgraph $G[A]$ is triangulated and therefore perfect. $\qquad\square$

9.4 Interval Graphs

One of the special classes of triangulated graphs is the class of interval graphs.

Definition 9.4.1. An *interval graph* G is the intersection graph of a family of intervals of the real line. This means that for each vertex v of G, there corresponds an interval $J(v)$ of the real line such that $uv \in E(G)$ if and only if $J(u) \cap J(v) \neq \emptyset$.

Figure 9.3 displays a graph G and its interval representation.

Remark 9.4.2. 1. Interval graphs occur in a natural manner in various applications. In genetics, the Benzer model [18] deals with the conditions under which two subsets of the fine structure inside a gene overlap. In fact, one can tell when they overlap on the basis of mutation data. Is this overlap information consistent with

Fig. 9.3 (a) Graph G; (b) its interval representation

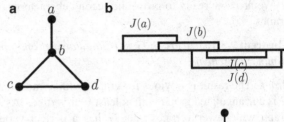

Fig. 9.4 Example of a graph that is not an interval graph

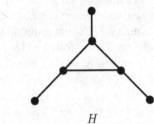

H

the hypothesis that the fine structure inside the gene is linear? The answer is "yes" if the graph defined by the overlap information is an interval graph.

2. It is clear that the intervals may be taken as either open or closed.
3. The cycle C_4 is not an interval graph. In fact, if $V(C_4) = \{a, b, c, d\}$ and if ab, bc, cd, and da are the edges of C_4, then $J(a) \cap J(b) \neq \emptyset$, $J(b) \cap J(c) \neq \emptyset$, $J(c) \cap J(d) \neq \emptyset$, and $J(d) \cap J(a) \neq \emptyset$ imply that either $J(a) \cap J(c) \neq \emptyset$ or $J(b) \cap J(d) \neq \emptyset$ [i.e., $ac \in E(G)$ or $bd \in E(G)$], which is not the case. Hence, an interval graph cannot contain C_4 as an induced subgraph. For a similar reason, it can be checked that the graph H of Fig. 9.4 is not an interval graph.

Recall that an *orientation* of a graph G is an assignment of a direction to each edge of G. Hence, an orientation of G converts G into a directed graph. As mentioned in Chap. 2, an orientation is *transitive* if, when (a, b) and (b, c) are arcs in the orientation, then (a, c) is also an arc in the orientation.

Lemma 9.4.3. *If G is an interval graph, G^c has a transitive orientation.*

Proof. Let $J(a)$ denote the interval that represents the vertex a of the interval graph G. Let $ab \in E(G^c)$ and $bc \in E(G^c)$ so that $ab \notin E(G)$ and $bc \notin E(G)$. Hence, $J(a) \cap J(b) = \emptyset$, and $J(b) \cap J(c) = \emptyset$. Now, introduce an orientation for the edges of G^c by orienting an edge xy of G^c from x to y if and only if $J(x)$ lies to the left of $J(y)$. Then $J(a)$ lies to the left of $J(b)$ and $J(b)$ lies to the left of $J(c)$, and therefore $J(a)$ lies to the left of $J(c)$. Hence, whenever (a, b) and (b, c) are arcs in the defined orientation, arc (a, c) also belongs to this orientation. Thus, G^c has a transitive orientation. $\square$

Gilmore and Hoffman [73] have shown that the above two properties (Remark 3 of 9.4.2 and Lemma 9.4.3) characterize interval graphs.

Theorem 9.4.4. *A graph G is an interval graph if and only if G does not contain C_4 as an induced subgraph and G^c admits a transitive orientation.*

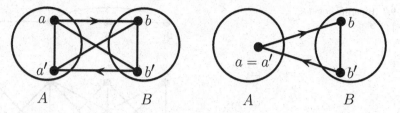

Fig. 9.5 Graph for proof of first condition of Theorem 9.4.4

Fig. 9.6 Ordering of
maximal cliques

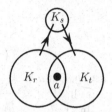

Proof. We have just seen the necessity of these two conditions. We now prove their sufficiency. Assume that G has no induced C_4 and that G^c has a transitive orientation. We look at the set of maximal cliques of G and introduce a linear ordering on it. If A and B are two distinct maximal cliques of G, then for any $a \in A$, there exists $b \in B$ with $ab \notin E(G)$ and therefore $ab \in E(G^c)$. (Otherwise, $G[A \cup B]$ would be a clique of G properly containing both A and B, a contradiction, since A and B are maximal cliques in G.) If ab has the orientation from a to b in the transitive orientation of G^c, we set $A < B$. This ordering is well defined in that if $a' \in A$ and $b' \in B$ with $a'b' \in E(G^c)$, then $a'b'$ must be oriented from a' to b' in G^c (see Fig. 9.5).

To see this, first assume that $a \neq a'$ and $b \neq b'$ and that edge $a'b'$ is oriented from b' to a' in G^c. Then at least one of the edges ab' and $a'b$ must be an edge of G^c. Otherwise, the edges aa', $a'b$, bb', and $b'a$ induce a C_4 in G, a contradiction. Suppose then that $a'b \in E(G^c)$. Then if $a'b$ is oriented from a' to b in G^c, by the transitivity of the orientation in G^c, $b'b \in E(G^c)$, a contradiction. A similar argument applies when ba', ab', or $b'a$ is an oriented arc of G^c. The cases when $a = a'$ or $b = b'$ can also be treated similarly. Thus, if one arc of G^c goes from A to B, then all the arcs between A and B go from A to B in G^c. In this case, we set $A < B$. Since the number of maximal cliques of G is finite, and any two maximal cliques can be ordered by "$<$," we obtain a linear ordering of the set of maximal cliques of G, say, $K_1 < K_2 < \ldots < K_p$.

We now claim that if a vertex a of G belongs to K_r and K_t, where $K_r < K_t$, then it also belongs to K_s, where $K_r < K_s < K_t$ (see Fig. 9.6).

Suppose $a \notin K_s$. First note that there exists some vertex b in K_s such that b is nonadjacent to a. If not, $K_s \vee \{a\}$ would be a clique properly containing K_s, a contradiction. But then, since $K_r < K_s$, the edge ab of G^c must be oriented from

Fig. 9.7 Bipartite graph
$B(G)$ of G

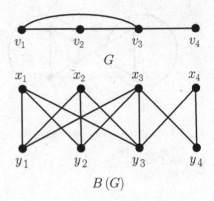

a to b. But $a \in K_t$, and this means that $K_t < K_s$, a contradiction. Thus, $a \in K_s$ as well.

In $\{1, 2, \ldots, p\}$, let i be the smallest and j be the greatest numbers such that $a \in K_i$ and $a \in K_j$. We now define the interval $J(a) =$ the closed interval $[i, j]$. Then $J(a) \cap J(b) \neq \emptyset$ if and only if there exists a positive integer k such that $k \in J(a) \cap J(b)$. But this can happen if and only if both a and b are in K_k [i.e., if and only if $ab \in E(G)$]. Thus, G is an interval graph. $\square$

9.5 Bipartite Graph $B(G)$ of a Graph G

Given a graph G, we define the associated bipartite graph $B(G)$ as follows: Let $V(G) = \{v_1, v_2, \ldots, v_n\}$. Corresponding to $V(G)$, take disjoint sets $X = \{x_1, x_2, \ldots, x_n\}$ and $Y = \{y_1, y_2, \ldots, y_n\}$ and form the bipartite graph $B(G)$ by taking X and Y as sets of the bipartition of the vertex set of $B(G)$. Adjacency in $B(G)$ is defined by setting $x_i y_i \in E(B(G))$ for every i, $1 \leq i \leq n$, and for $i \neq j$, x_i is adjacent to y_j in $B(G)$ if and only if $v_i v_j \in E(G)$ (Fig. 9.7).

Our next theorem relates the chordal nature of a graph G with that of the bipartite graph $B(G)$. Since a bipartite graph has no odd cycles and a 4-cycle of a bipartite graph cannot have a chord, a *bipartite graph* is defined to be *chordal* if each of its cycles of length at least 6 has a chord.

Theorem 9.5.1. *If the bipartite graph $B(G)$ formed out of G is chordal, then G is chordal.*

Proof. Let $C = v_1 v_2 \ldots v_p v_1$ be any cycle of G of length $p \geq 4$. If p is odd, take C' to be the cycle $x_1 y_2 x_3 y_4 \ldots x_p y_p x_1$, while if p is even, take C' to be the cycle $x_1 y_2 x_3 y_4 \ldots x_{p-1} y_p x_p y_1 x_1$ in $B(G)$. As $B(G)$ is chordal and C' is of length at least 6, C' has a chord in $B(G)$. Such a chord can only be of the form $x_i y_j$, where $|i - j| \geq 2$. This means that $v_i v_j$ is a chord of C. Thus, G is chordal. $\square$

9.6 Circular Arc Graphs

Circular arc graphs are similar to interval graphs except that the $J(a)$'s are now taken to be arcs of a particular circle. Consider an interval graph G. Since the number of intervals $J(a)$, $a \in V(G)$, is finite, there are real numbers m and M such that $J(a) \subseteq (m, M)$ for every $a \in V(G)$. Consequently, identification of m and M (i.e., conversion of the closed interval $[m, M]$ into a circle by the identification of m and M) makes G a circular arc graph. Thus, every interval graph is a circular arc graph. Clearly, the converse is not true. However, if there exists a point p on the circle that does not belong to any arc $J(a)$, then the circle can be cut at p and the circular arc graph can be made into an interval graph.

9.7 Exercises

7.1 If e is an edge of a cycle of a triangulated graph G, show that e belongs to a triangle of G.

7.2 What are the simplicial vertices of the triangulated graph of Fig. 9.2a?

7.3 Give a perfect elimination scheme for the triangulated graph of Fig. 9.2a.

7.4 If v is a simplicial vertex of a triangulated graph G, and $vu \in E(G)$, prove that $\theta(G - u) = \theta(G)$.

7.5 Let $t(G)$ denote the smallest positive integer k such that G^k is triangulated. Determine $t(C_n)$, $n \geq 4$.

7.6 Prove G and G^c are triangulated if and only if G does not contain C_4, C_4^c, or C_5 as an induced subgraph. Hence, or otherwise, show that C_n^c, $n \geq 5$ is not triangulated.

7.7 Prove that $L(G)$ is triangulated if and only if every block of G is either K_2 or K_3. Hence, show that the line graph of a tree is triangulated.

7.8 Let $K(G)$ and $L(G)$ denote, respectively, the clique graph and the line graph of a graph G. [$K(G)$ is defined as the intersection graph of the family of maximal cliques of G; i.e., the vertices of $K(G)$ are the maximal cliques of G, and two vertices of $K(G)$ are adjacent in $K(G)$ if and only if the corresponding maximal cliques of G have a nonempty intersection.] Then prove or disprove

  (i) G is triangulated $\Rightarrow K(G)$ is triangulated

  (ii) $K(G)$ is triangulated $\Rightarrow G$ is triangulated

  (iii) $L(G)$ is triangulated $\Rightarrow G$ is triangulated

  (iv) G is triangulated $\Rightarrow L(G)$ is triangulated

7.9 Show by means of an example that an even power of a triangulated graph need not be triangulated.

7.10 Prove the following by means of a counterexample: G is chordal need not imply that $B(G)$ is chordal.

7.11 Draw the interval graph of the family of intervals below and display a transitive orientation for G^c.

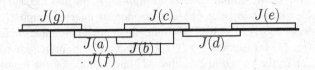

7.12 If G is cubic and if G does not contain an odd cycle of length at least 5 as an induced subgraph, prove that G is perfect. (Hint: Use Brooks' theorem.)

7.13 Show that every bipartite graph is perfect.

7.14 For a bipartite graph G, prove that $\chi(G^c) = \omega(G^c)$.

7.15 Give an example of a triangulated graph that is not an interval graph.

7.16 Give an example of a perfect graph that is not triangulated.

7.17 Show that a 2-connected triangulated graph with at least four vertices is locally connected. Hence, show that a 2-connected triangulated $K_{1,3}$-free graph is Hamiltonian. (See reference [149].)

7.18 Show by means of an example that a 2-connected triangulated graph need not be Hamiltonian.

7.19 Show that the line graph of a 2-edge-connected triangulated graph is Hamiltonian.

7.20 Give an example of a circular arc graph that is not an interval graph.

7.21 * Show that a graph G is perfect if and only if every induced subgraph G' of G contains an independent set that meets all the maximum cliques of G'.

7.22 Let $\{v_1, v_2, \ldots, v_n\}$ be a simplicial ordering of the vertices of a chordal graph G. Let

$$d_i = \deg(v_i) \text{ in the subgraph } \langle v_i, v_{i+1}, \ldots, v_n \rangle \text{ of } G.$$

Prove that the chromatic polynomial of G is given by $\prod_{i=1}^{n}(t - d_i)$. Hence show that $\chi(G) = \max_{1 \le i \le n} \{1 + d_i\}$. (This shows that the roots of the chromatic polynomial of a chordal graph are nonnegative integers.)

9.8   Phasing of Traffic Lights at a Road Junction

We present an application of interval graphs to the problem of phasing of traffic lights at a road junction. The problem is to install traffic lights at a road junction in such a way that traffic flows smoothly and efficiently at the junction.

We take a specific example and explain how our problem could be tackled. Figure 9.8 displays the various traffic streams, namely, a, b, $\ldots$, g, that meet at the Main Guard Gate road junction at Tiruchirappalli, Tamil Nadu (India).

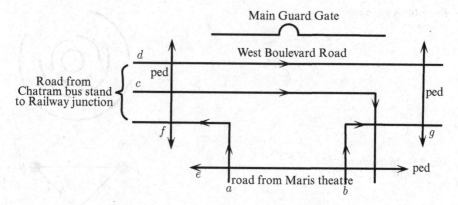

Fig. 9.8 Traffic streams at a road junction (ped = pedestrian crossing)

Fig. 9.9 Compatibility graph
of Fig. 9.8

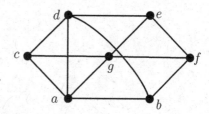

   Certain traffic streams may be termed "compatible" if their simultaneous flow
would not result in any accidents. For instance, in Fig. 9.8, streams a and d are
compatible, whereas b and g are not. The phasing of lights should be such that when
the green lights are on for two streams, they should be compatible. We suppose that
the total time for the completion of green and red lights during one cycle is two
minutes.
   We form a graph G whose vertex set consists of the traffic streams in question,
and we make two vertices of G adjacent if and only if the corresponding streams
are compatible. This graph is the compatibility graph corresponding to the problem
in question. The compatibility graph of Fig. 9.8 is shown in Fig. 9.9.
   We take a circle and assume that its perimeter corresponds to the total cycle
period, namely, 120 seconds. We may think that the duration when a given traffic
stream gets green light corresponds to an arc of this circle. Hence, two such arcs
of the circle can overlap only if the corresponding streams are compatible. The
resulting circular arc graph may not be the compatibility graph because we do
not demand that two arcs intersect whenever they correspond to compatible flows.
(There may be two compatible streams, but they need not get green light at the
same time.) However, the intersection graph H of this circular arc graph will be a
spanning subgraph of the compatibility graph.
   The efficiency of our phasing may be measured by minimizing the total red light
time during a traffic cycle, that is, the total waiting time for all the traffic streams

Fig. 9.10 A green light
assignment

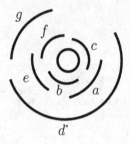

Fig. 9.11 Intersection graph
for Fig. 9.10

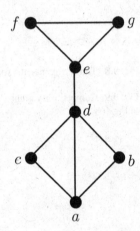

during a cycle. For the sake of concreteness, we may assume that at the time of
starting, all lights are red. This would ensure that H is an interval graph (see the last
sentence of Sect. 9.5 on circular arc graphs).

Figure 9.10 gives a feasible green light assignment whose corresponding intersection graph H is given in Fig. 9.11. The maximal cliques of H are $K_1 =
\{a, b, d\}$, $K_2 = \{a, c, d\}$, $K_3 = \{d, e\}$, and $K_4 = \{e, f, g\}$. Since H is
an interval graph, by Theorem 9.4.4, H^c has a transitive orientation. A transitive
orientation of H^c is given in Fig. 9.12.

Since (b, c), (c, e), and (d, f) are arcs of H^c, and since $b \in K_1, c \in
K_2, d, e \in K_3$, and $f \in K_4$, etc., we have

$$K_1 < K_2 < K_3 < K_4$$

in the consecutive ordering of the maximal cliques of H. Each clique K_i, $1 \le i \le 4$,
corresponds to a phase during which all streams in that clique receive green lights.
We then start a given traffic stream with green light during the first phase in which
it appears, and we keep it green until the last phase in which it appears. Because of
the consecutiveness of the ordering of the phases K_i, this gives an arc on the clock
circle. In phase 1, traffic streams a, b, and d receive a green light; in phase 2, a, c,
and d receive a green light, and so on.

Fig. 9.12 Transitive
orientation of H^c

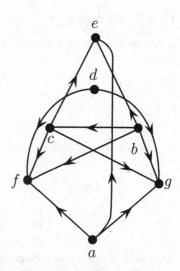

Suppose we assign to each phase K_i a duration d_i. Our aim is to determine the
d_i's (≥ 0) so that the total waiting time is minimum. Further, we may assume that
the minimum green light time for any stream is 20 seconds. Traffic stream a gets a
red light when the phases K_3 and K_4 receive a green light. Hence, a's total red light
time is $d_3 + d_4$. Similarly, the total red light times of traffic streams b, c, d, e, f,
and g, respectively, are $d_2 + d_3 + d_4$; $d_1 + d_3 + d_4$; d_4; $d_1 + d_2$; $d_1 + d_2 + d_3$; and
$d_1 + d_2 + d_3$. Therefore, the total red light time of all the streams in one cycle is
$Z = 4d_1 + 4d_2 + 4d_3 + 3d_4$. Our aim is to minimize Z subject to $d_i \geq 0$; $1 \leq i \leq 4$,
and $d_1 + d_2 \geq 20$; $d_1 \geq 20$, $d_2 \geq 20$, $d_1 + d_2 + d_3 \geq 20$, $d_3 + d_4 \geq 20$, $d_4 \geq$
20, $d_3 \geq 0$ and $d_1 + d_2 + d_3 + d_4 = 120$. (The condition $d_1 + d_2 \geq 20$ signifies
that the green light time that stream a receives, namely, the sum of the green light
times of phases K_1 and K_2, is at least 20. A similar reasoning applies to the other
inequalities. The last condition gives the total cycle time.) An optimal solution to
this problem is $d_1 = 80$, $d_2 = 20$, $d_3 = 0$, and $d_4 = 20$ and min $Z = 480$ (in
seconds). But this is not the end of our problem. There are other possible circular
arc graphs. Figures 9.13a,b give another feasible green light arrangement and its
corresponding intersection graph. With respect to this graph, min $Z = 500$ seconds.
Thus, we have to exhaust all possible circular arc graphs and then take the least of
all the minima thus obtained. The phasing that corresponds to this least value would
then be the best phasing of the traffic lights. (For the above particular problem, this
minimum value is 480 seconds.)

Notes

Exercise 7.9 shows that an even power of a triangulated graph need not be
triangulated. However, an odd power of a triangulated graph is triangulated [11].

Fig. 9.13 (a) Another green
light arrangement; (b)
corresponding intersection
graph

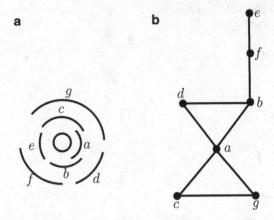

Moreover, if G^k is triangulated, then so is G^{k+2} [130], and consequently, if G and G^2 are triangulated, then so are all the powers of G.

A simple graph G is called *Berge* if it contains neither an odd cycle of length at least 5 nor its complement as an induced subgraph. The *strong perfect graph conjecture* asserted that a graph G is perfect if it is Berge. This conjecture was proposed by Claude Berge in 1960 and was settled affirmatively by Maria Chudnovsky, Neil Robertson, Paul Seymour, and Robin Thomas in 2002 [36]. The authors show that every Berge graph is in one of the four classes of perfect graphs—basic, 2-join, M-join, and balanced skew partition. Earlier the conjecture was proved to be true for several classes of graphs: (i) $K_{1,3}$-free graphs [154]; (ii) $(K_4 - e)$-free graphs [156]; (iii) K_4-free graphs [178]; (iv) bull-free, that is, ◿▵◺-free graphs [40] (v) triangulated graphs (see Theorem 9.3.11); (vi) weakly triangulated graphs [102] and so on.

Perfect graphs were first discovered by Berge in 1958–1959. Their importance is both theoretical (because of their bearing on graph coloring problems) and practical (because of their applications to perfect communication channels, operations research, optimization of municipal services, etc.).

Four books that give a very good account of perfect graphs are references [19, 21, 20, 76]. In addition to the classes of perfect graphs mentioned above, there are also other known classes of perfect graphs, for instance, wing-triangulated graphs and, more generally, strict quasi-parity graphs. For details, see reference [107]. Our discussion on the phasing of traffic lights is based on Roberts [166], which also contains some other applications of perfect graphs.

Chapter 10
Domination in Graphs

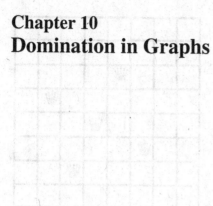

10.1 Introduction

"Domination in graphs" is an area of graph theory that has received a lot of attention in recent years. It is reasonable to believe that "domination in graphs" has its origin in "chessboard domination." The "queen domination" problem asks: What is the minimum number of queens required to be placed on an 8×8 chessboard so that every square not occupied by any of these queens will be dominated (that is, can be attacked) by one of these queens? Recall that a queen can move horizontally, vertically, and diagonally on the chessboard. The answer to the above question is 5. Figure 10.1 gives one set of dominating locations for the five queens.

10.2 Domination in Graphs

The concept of chessboard domination can be extended to graphs in the following way:

Definition 10.2.1. Let G be a graph. A set $S \subseteq V$ is called a *dominating set* of G if every vertex $u \in V \backslash S$ has a neighbor $v \in S$. Equivalently, every vertex of G is either in S or in the neighbor set $N(S) = \bigcup_{v \in S} N(v)$ of S in G. A vertex u is said to be *dominated by* a vertex $v \in G$ if either $u = v$ or $uv \in E(G)$.

Definition 10.2.2. A *γ-set* of G is a minimum dominating set of G, that is, a dominating set of G whose cardinality is minimum. A dominating set S of G is *minimal* if S properly contains no dominating set S' of G.

Definition 10.2.3. The *domination number* of G is the cardinality of a minimum dominating set (that is, γ-set) of G; it is denoted by $\gamma(G)$.

R. Balakrishnan and K. Ranganathan, *A Textbook of Graph Theory,*
Universitext, DOI 10.1007/978-1-4614-4529-6_10,
© Springer Science+Business Media New York 2012

Fig. 10.1 Queen domination

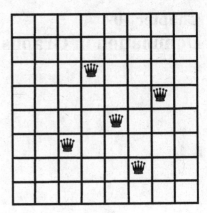

Fig. 10.2 Petersen graph P
for which $\gamma(P) = 3$

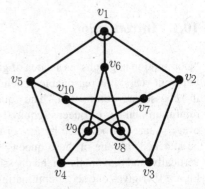

Example 10.2.4. For the Petersen graph P, $\gamma(P) = 3$. In Fig. 10.2, $\{v_1, v_8, v_9\}$ is a γ-set of P while the set $\{v_1, v_2, v_3, v_4, v_5\}$ is a minimal dominating set of P.

The study of domination was formally initiated by Ore [151]. A comprehensive introduction to "domination in graphs" is given in the first volume of the two-volume book by Haynes, Hedetniemi, and Slater [100, 101]. The next three theorems are due to Ore [151]. Given a dominating set S of G, when is S a minimal dominating set? This question is answered in Theorem 10.2.5 below.

Theorem 10.2.5. *Let S be a dominating set of a graph G. Then S is a minimal dominating set of G if and only if for each vertex u of S, one of the following two conditions holds:*

(i) u is an isolated vertex of $G[S]$, the subgraph induced by S in G.
(ii) There exists a vertex $v \in V \backslash S$ such that u is the only neighbor of v in S.

Proof. Suppose that S is a minimal dominating set of G. Then for each vertex u of S, $S \backslash \{u\}$ is not a dominating set of G. Hence, there exists $v \in V \backslash (S \backslash \{u\})$ such that v is dominated by no vertex of $S \backslash \{u\}$. If $v = u$, then u is an isolated vertex of $G[S]$, and hence condition (i) holds. If $v \neq u$, as S is a dominating set of G, condition (ii) holds.

Conversely, assume that S is not a minimal dominating set of G. Then there exists a vertex $u \in S$, such that $S \setminus \{u\}$ is also a dominating set of G. Hence, u is dominated by some vertex of $S \setminus \{u\}$. This means that u is adjacent to some vertex of $S \setminus \{u\}$, and hence u is not an isolated vertex of $G[S]$. Moreover, if v is any vertex of $V \setminus S$, then v is adjacent to some vertex of $S \setminus \{u\}$. Hence, neither condition (i) nor condition (ii) holds. $\square$

Notice that in Example 10.2.4, the set $S = \{v_1, v_2, v_3, v_4, v_5\}$ is a minimal dominating set of the Petersen graph P in which no vertex is an isolated vertex of $P[S]$ and for each $i = 1, 2, \ldots, 5$, v_i is the only vertex of S that is adjacent to v_{i+5}.

Theorem 10.2.5 suggests the following definition.

Definition 10.2.6. Let S be a dominating set of a graph G, and $u \in S$. The *private neighborhood of u relative to S in G* is the set of vertices which are in the closed neighborhood of u, but not in the closed neighborhood of any vertex in $S \setminus \{u\}$.

Thus, the private neighborhood $P_N(u, S)$ of u with respect to S is given by $P_N(u, S) = N[u] \setminus (\bigcup_{v \in S \setminus \{u\}} N[v])$.

Note that $u \in P_N(u, S)$ if and only if u is an isolated vertex of $G[S]$ in G.

Theorem 10.2.5 can now be restated as follows:

Theorem 10.2.5'. *A dominating set S of a graph G is a minimal dominating set of G if and only if $P_N(u, S) \neq \emptyset$ for every $u \in S$.*

Corollary 10.2.7. *Let G be a graph having no isolated vertices. If S is a minimal dominating set of G, then $V \setminus S$ is a dominating set of G.*

Proof. As S is a minimal dominating set, by Theorem 10.2.5', $P_N(u, S) \neq \emptyset$ for every $u \in S$. This means that for every $u \in S$, there exists $v \in V \setminus S$ such that $uv \in E(G)$, and consequently, $V \setminus S$ is a dominating set of G. $\square$

Corollary 10.2.8. *Let G be a graph of order $n \geq 2$. If $\delta(G) \geq 1$, then $\gamma(G) \leq \frac{n}{2}$.*

Proof. As $\delta(G) \geq 1$, G has no isolated vertices. If S is a minimal dominating set of G, by Corollary 10.2.7, both S and $V \setminus S$ are dominating sets of G. Certainly, at least one of them is of cardinality at most $\frac{n}{2}$. $\square$

Corollary 10.2.9. *If G is a connected graph of order $n \geq 2$, $\gamma(G) \leq \frac{n}{2}$.*

Proof. As G is connected and $n \geq 2$, G has no isolated vertices. Now apply Corollary 10.2.8. $\square$

We note that the conclusion in Corollary 10.2.9 would remain valid even if G is disconnected as long as no component of G is a K_1.

10.3 Bounds for the Domination Number

In this section, we present lower and upper bounds for the domination number $\gamma(G)$.
We first make two observations:

Observation 10.3.1. (i) A vertex v dominates $N(v)$ and $|N(v)| \le \Delta(G)$.
(ii) Let v be any vertex of G. Then $V \setminus N(v)$ is a dominating set of G.

These two observations yield the following lower and upper bounds for $\gamma(G)$.

Theorem 10.3.2. *For any graph G,* $\left\lceil \frac{n}{1+\Delta(G)} \right\rceil \le \gamma(G) \le n - \Delta(G)$.

Proof. By Observation 10.3.1 (i), for any vertex v of G, the vertices of $N[v]$
will be dominated by v. To cover all the vertices of G, at least $\left\lceil \frac{n}{1+\Delta(G)} \right\rceil$ closed
neighborhoods are required. This gives the lower bound.
    By Observation 10.3.1 (ii), $\gamma(G) \le |V \setminus N(v)|$ for each $v \in V(G)$. The minimum
is attained on the right when $|N(v)| = \Delta(G)$. Hence, $\gamma(G) \le n - \Delta(G)$.   □.
    The lower bound in Theorem 10.3.2 is due to Walikar, Acharya, and Sampathku-
mar [187], while the upper bound is due to Berge [19].

10.4 Bound for the Size m in Terms of Order n and Domination Number $\gamma(G)$

In this section, we present a basic result of Vizing [184], which bounds m (the size
of G) in terms of n (the order of G) and $\gamma = \gamma(G)$.

Theorem 10.4.1 (Vizing [184]). *Let G be a graph of order n, size m, and
domination number γ. Then*

$$ m \le \left\lfloor \frac{1}{2}(n - \gamma)(n - \gamma + 2) \right\rfloor. \tag{10.1} $$

Proof. If $\gamma = 1$, $\frac{1}{2}(n - \gamma)(n - \gamma + 2) = \frac{1}{2}(n^2 - 1)$, while the maximum value for
$m = \frac{1}{2}n(n - 1)$ (when $G = K_n$), and the result is true. If $\gamma = 2$, $\frac{1}{2}(n - \gamma)(n - \gamma + 2) = \frac{1}{2}n(n - 2)$. Now when $\gamma = 2$ by Theorem 10.3.2, $\Delta \le n - 2$ and
$m \le \frac{1}{2}n(n - 2)$ (by Euler's theorem), and the result is true. Thus, the result is true
for $\gamma = 1$ and 2. We now assume that $\gamma \ge 3$. We apply induction on n. Let G be
a graph of order n, size m, and $\gamma \ge 3$. If v is a vertex of maximum degree Δ of G,
again by Theorem 10.3.2, $|N(v)| = \Delta \le n - \gamma$, and hence $\Delta = n - \gamma - r$, where
$0 \le r$ (see Fig. 10.3).

Let $S = V \setminus N[v]$. Then

$$ |S| = |V| - |N(v)| - 1 = n - (n - \gamma - r) - 1 = \gamma + r - 1. \tag{10.2} $$

Fig. 10.3 The set S in the
proof of Theorem 10.4.1

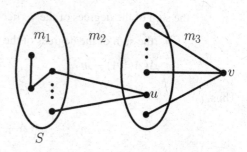

Let m_1 be the size of $G[S]$, m_2 be the number of edges between S and $N(v)$, and m_3 be the size of $G[N[v]]$. Clearly, $m = m_1 + m_2 + m_3$. If D is a γ-set of $G[S]$, then $D \cup \{v\}$ is a dominating set of G. Hence,

$$\gamma(G) = \gamma \leq |D| + 1. \tag{10.3}$$

By the induction hypothesis, this implies, by virtue of (10.2) and (10.3), that

$$m_1 \leq \left\lfloor \frac{1}{2}(|S| - |D|)(|S| - |D| + 2) \right\rfloor$$
$$\leq \left\lfloor \frac{1}{2}[(\gamma + r - 1) - (\gamma - 1)][(\gamma + r - 1) - (\gamma - 1) + 2] \right\rfloor \quad \text{(by (10.3))}$$
$$= \frac{1}{2}r(r + 2). \tag{10.4}$$

If $u \in N(v)$, then $(S \backslash N(u)) \cup \{u, v\}$ is a dominating set of G. Therefore,

$$\gamma \leq |S \backslash N(u)| + 2$$
$$= |S| - |S \cap N(u)| + 2$$
$$\leq (\gamma + r - 1) - |S \cap N(u)| + 2 \text{ (by (10.2))}.$$

This is turn implies that for each vertex $u \in N(v)$, $|S \cap N(u)| \leq r + 1$. Consequently,

$$m_2 = \text{ the number of edges between } N(v) \text{ and } S$$
$$\leq |N(v)|(r + 1)$$
$$= \Delta(r + 1). \tag{10.5}$$

Now the sum of the degrees of the vertices of $N[v] \leq (\Delta + 1)\Delta$. As there are m_2 edges between $N(v)$ and S, the sum of the degrees of the vertices of $N[v]$ in $G[N[v]]$

$= $ (the sum of the degrees of the vertices of $N[v]$ in G) $- m_2$

$\leq \Delta(\Delta + 1) - m_2$.

Thus,

$$m_3 \leq \frac{1}{2}[\Delta(\Delta + 1) - m_2]. \tag{10.6}$$

From (10.4), (10.5), and (10.6), we get

$$
\begin{aligned}
m = m_1 + m_2 + m_3 \\
&\leq \frac{1}{2}r(r+2) + m_2 + \frac{1}{2}[\Delta(\Delta+1) - m_2] \\
&= \frac{1}{2}r(r+2) + \frac{1}{2}[\Delta(\Delta+1) + m_2] \\
&\leq \frac{1}{2}r(r+2) + \frac{1}{2}[\Delta(\Delta+1) + \Delta(r+1)](by\,(10.5)) \\
&\leq \frac{1}{2}(n-\gamma-\Delta)(n-\gamma-\Delta+2) + \frac{1}{2}[\Delta(\Delta+1) + \Delta(r+1)](as\,\Delta = n-\gamma-r) \\
&= \frac{1}{2}(n-\gamma)(n-\gamma+2) - \frac{\Delta}{2}[(n-\gamma-\Delta+2) + (n-\gamma-\Delta) - \Delta - (\Delta+1)(r+1)] \\
&= \frac{1}{2}(n-\gamma)(n-\gamma+2) - \frac{\Delta}{2}[(n-\gamma+2) + (n-\gamma) - \Delta - (\Delta+1) - (r+1)] \\
&= \frac{1}{2}(n-\gamma)(n-\gamma+2) - \frac{\Delta}{2}[(\Delta+r+2) + (\Delta+r) - \Delta - (\Delta+1) - (r+1)] \\
&= \frac{1}{2}(n-\gamma)(n-\gamma+2) - \frac{\Delta}{2}r \\
&\leq \frac{1}{2}(n-\gamma)(n-\gamma+2)(as\ r \geq 0). \qquad \square
\end{aligned}
$$

The bound given in Theorem 10.4.1 is sharp. In other words, there are graphs G for which

$$m = \frac{1}{2}(n-\gamma)(n-\gamma+2). \tag{10.7}$$

Example 10.4.2 (Vizing [184]). Let H_t be the graph obtained from K_t by removing the edges of a minimum edge cover (that is, the smallest number of edges containing all the vertices of K_t) $\mathscr{C}$. We construct for any positive integer $n \geq 2$, a graph G of order n with domination number γ satisfying (10.7).

Case (i). $\gamma = 2$. Take $t = n - 2$ and $G = H_n$. Now $\mathscr{C}$ has $\lceil \frac{n-2}{2} \rceil$ edges. Hence, $m = m(G) = \{\frac{1}{2}(n-2)(n-3) - \lceil\frac{n-2}{2}\rceil\} + 2(n-2) = \lfloor\frac{1}{2}(n-2)n\rfloor = \lfloor\frac{1}{2}(n-\gamma)(n-\gamma+2)\rfloor$.

Case (ii.) $\gamma > 2$. Take $t = n - \gamma + 2$ and $G = H_t \cup K^c_{\gamma-2}$. Then $\gamma(G) = 2 + (\gamma - 2) = \gamma$, $|V(G)| = (n - \gamma) + 2 + (\gamma - 2) = n$, and

$$
\begin{aligned}
m &= \left\{ \frac{(n-\gamma)(n-\gamma-1)}{2} - \left\lceil \frac{n-\gamma}{2} \right\rceil \right\} + 2(n-\gamma) \\
&= \frac{(n-\gamma)(n-\gamma+3)}{2} - \left\lceil \frac{n-\gamma}{2} \right\rceil \\
&= \frac{1}{2}(n-\gamma)(n-\gamma+2).
\end{aligned}
$$

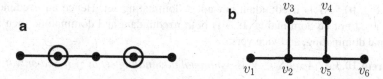

Fig. 10.4 (a) $\gamma(G) = 2 = i(G)$ (b) $\gamma(G) = 2$ while $i(G) = 3$

10.5 Independent Domination and Irredundance

Definition 10.5.1. A subset S of the vertex set of a graph G is an *independent dominating set* of G if S is both an independent and a dominating set. The *independent domination number* $i(G)$ of G is the minimum cardinality of an independent dominating set of G.

It is clear that $\gamma(G) \leq i(G)$ for any graph G. For the path P_5, $\gamma(P_5) = i(P_5) = 2$, (see Fig. 10.4(a)) while for the graph G of Fig. 10.4(b), $\gamma(G) = 2$ and $i(G) = 3$. In fact, $\{v_2, v_5\}$ is a γ-set for G, while $\{v_1, v_3, v_5\}$ is a minimum independent dominating set of G.

Theorem 10.5.2. *Every maximal independent set of a graph G is a minimal dominating set.*

Proof. Let S be a maximal independent set of G. Then S must be a dominating set of G. If not, there exists a vertex $v \in V \backslash S$ that is not dominated by S, and so $S \cup \{v\}$ is an independent set of G, violating the maximality of S. Further, S must be a minimal dominating set of G. If not, there exists a vertex u of S such that $T = S \backslash \{u\}$ is also a dominating set of G. This means, as $u \notin T$, u has a neighbor in T and hence S is not independent, a contradiction. □

Definition 10.5.3. A set $S \subset V(G)$ is called *irredundant* if every vertex v of S has at least one private neighbor.

Fig. 10.5 S is an irredundant
but not a dominating set

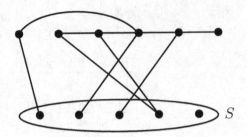

This definition means that either v is an isolated vertex of $G[S]$ or else v has a private
neighbor in $V \setminus S$; that is, there exists at least one vertex $w \in V \setminus S$ that is adjacent
only to v in S.

In Fig. 10.5, S is irredundant but not a dominating set. Hence an irredundant
set S need not be dominating. If S is both irredundant and dominating, then it is
minimal dominating, and vice versa.

Theorem 10.5.4. *A set* $S \subset V$ *is a minimal dominating set of G if and only if S is
both dominating and irredundant.*

Proof. Assume that S is both a dominating and an irredundant set of G. If S were
not a minimal dominating set, there exists $v' \in S$ such that $S \setminus \{v'\}$ is also a
dominating set. But as S is irredundant, v' has a private neighbor w' (may be equal
to v'). Since w' has no neighbor in $S \setminus v'$, $S \setminus \{v'\}$ is not a dominating set of G. Thus,
S is a minimal dominating set of G.

The proof of the converse is similar.   □

We define below a few more well-known graph parameters:

(i) The minimum cardinality of a maximal irredundant set of a graph G is known
    as the *irredundance number* and is denoted by $ir(G)$.
(ii) The maximum cardinality of an irredundant set is known as the *upper
    irredundance number* and is denoted by $IR(G)$.
(iii) The maximum cardinality of a minimal dominating set is known as the *upper
    domination number* and is denoted by $\Gamma(G)$.

From our earlier results and definitions, one can prove the following result of
Cockeyne, Hedetniemi and Miller [45].

Theorem 10.5.5 ([45]). *For any graph G, the following inequality chain holds:*

$$ir(G) \leq \gamma(G) \leq i(G) \leq \beta_0(G) \leq \Gamma(G) \leq IR(G).$$

Proof. Exercise.   □

10.6 Exercises

6.1. If G is a graph of diameter 2, show that $\gamma(G) \le \delta(G)$.

6.2. If D is a dominating set of a graph G, show that D meets every closed neighborhood of G.

6.3. Show that for any edge e of a graph G, $\gamma(G) \le \gamma(G - e) \le \gamma(G) + 1$, and that for any vertex v of G, $\gamma(G) - 1 \le \gamma(G - v)$.

6.4. If $d_1 \ge d_2 \ge \ldots \ge d_n$ is the degree sequence of a graph G, prove that

$$\gamma(G) \ge \min\{k : k + (d_1 + \cdots + d_k) \ge n\}.$$

6.5. For any graph G, prove that $\gamma(G) \le \chi(G^c)$.

6.6. Show that every minimal dominating set in a graph G is a maximal irredundant set of G.

6.7. Prove that an independent set is maximal independent if and only if it is dominating and independent.

6.8. Prove: If $\gamma(G^c) \ge 3$, then $\text{diam}(G) \le 2$.

6.9. Prove: If G is connected, then $\left\lceil \frac{\text{diam}(G)+1}{3} \right\rceil \le \gamma(G)$.

6.10. For any graph G, $n - m \le \gamma(G) \le n + 1 - \sqrt{1 + 2m}$. Prove further that $\gamma(G) = n - m$ if and only if G is a forest in which each component is a star. [Hint: To establish the upper bound, use Vizing's theorem (Theorem 10.4.1).]

6.11. Give the proof of Theorem 10.5.5.

6.12. For the graph G of Fig. 10.6, determine the six parameters given in Theorem 10.5.5.

Fig. 10.6 Refer Question 6.12

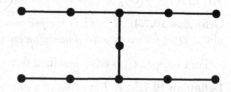

6.13. Exhibit a graph (different from the graph of Fig. 10.4b) for which no minimum dominating set is independent.

10.7 Vizing's Conjecture

All graphs considered in this section are simple. In this section we present Vizing's conjecture on the domination number of the Cartesian product of two graphs. In 1963, Vizing [182] proposed the problem of determining a lower bound for the domination number of the Cartesian product of two graphs. Five years later, in 1968, he presented it as a conjecture [185]. In the same year, E. Cockayne included it in his survey article [44]. This conjecture is one of the major unsolved problems in graph theory.

Conjecture 10.7.1 (Vizing [185]). For any two graphs G and H, $\gamma(G \,\square\, H) \geq \gamma(G)\gamma(H)$.

In what follows, we dwell upon some partial results toward this conjecture as well as some of the techniques that have been adopted in attempts to tackle this conjecture.

We write $G \leq H$ to denote the fact that G is a spanning subgraph of H. By definition (see Chap. 1),

$G \boxtimes H \leq G[H]$,

$G \,\square\, H \leq G \boxtimes H$, and

$G \times H \leq G \boxtimes H$.

It is clear that if $G \leq H$, then $\gamma(G) \geq \gamma(H)$. Consequently, we have the following result.

Theorem 10.7.2. *For any two graphs G and H,*

$\gamma(G[H]) \leq \gamma(G \boxtimes H) \leq min\{\gamma(G \,\square\, H), \gamma(G \times H)\}.$

In the absence of a proof of Vizing's conjecture, what is normally done is to fix one of the two graphs, say G, and allow the other graph H to vary and see if the conjecture 10.7.1 holds for all graphs H. Since the Cartesian product is commutative, it is immaterial as to which of the two graphs is fixed and which is varied. With this in view, we make the following definition:

Definition 10.7.3. A graph G is said to *satisfy Vizing's conjecture if and only if* $\gamma(G \,\square\, H) \geq \gamma(G)\gamma(H)$ *for every graph H*.

Definition 10.7.4. A graph G is *edge-maximal with respect to domination if* $\gamma(G + uv) < \gamma(G)$ *for every pair of nonadjacent vertices u, v of G*.

For example, C_4 is edge-maximal since $\gamma(C_4 + e) = 1 < 2 = \gamma(C_4)$.

Definition 10.7.5. A 2-*packing* of a graph G is a set P of vertices of G such that $N[x] \cap N[y] = \emptyset$ for every pair of (distinct) vertices x, y of P. The 2-*packing number* $\rho(G)$ of a graph G is the largest cardinality of a 2-packing of G. In other words, $\rho(G)$ is the maximum number of pairwise disjoint closed neighborhoods of G.

Before we set out to prove some theorems relating to Vizing's conjecture, we point out that in the relation $\gamma(G \square H) \geq \gamma(G)\gamma(H)$, both equality and strict inequality are possible.

For instance, $\gamma(C_4 \square P_3) = 3 > 2 \times 1 = \gamma(C_4)\gamma(P_3)$ (see Fig. 1.28), while $\gamma(C_4 \square P_2) = 2 = \gamma(C_4)\gamma(P_2)$.

Most of the results supporting Vizing's conjecture are of the following two types:

(i) If H is a graph related to G in some way, and if G satisfies Vizing's conjecture, then H also does.

(ii) Let $\mathscr{P}$ be a graph property. If G satisfies $\mathscr{P}$, then G satisfies Vizing's conjecture.

First, we present two results (Lemmas 10.7.6 and 10.7.7) that come under the first category.

Lemma 10.7.6. *Let $K \leq G$ such that $\gamma(K) = \gamma(G)$. If G satisfies Vizing's conjecture, then K also does.*

Proof. The graph K is obtained from G by removing edges of G (if $K = G$, there is nothing to prove). Let $e \in E(G)\backslash E(K)$. Then $K \leq G - e \leq G$. Hence, $\gamma(K) \geq \gamma(G-e) \geq \gamma(G)$. By hypothesis, $\gamma(K) = \gamma(G)$. Hence $\gamma(G-e) = \gamma(G)$, and since $(G - e)\Box H \leq G\Box H$, we have

$$\gamma((G - e)\Box H) \geq \gamma(G\Box H)$$
$$\geq \gamma(G)\,\gamma(H)\text{(by hypothesis)}$$
$$= \gamma(G - e)\,\gamma(H).$$

Hence, $G - e$ also satisfies Vizing's conjecture. Now start from $G - e$ and delete edges in succession until the resulting graph is K. Thus, K also satisfies Vizing's conjecture. $\square$

Lemma 10.7.6 is about edge deletion. We now consider vertex deletion.

Lemma 10.7.7. *Let $v \in V(G)$ such that $\gamma(G - v) < \gamma(G)$. If G satisfies Vizing's conjecture, then so does $G - v$.*

Proof. The inequality $\gamma(G - v) < \gamma(G)$ means that $\gamma(G - v) = \gamma(G) - 1$. Set $K = G - v$ so that $\gamma(K) = \gamma(G) - 1$. Suppose the result is false. Then there exists a graph H such that

$$\gamma(K\Box H) < \gamma(K)\,\gamma(H).$$

Let A be a γ-set of $K\Box H$ and B a γ-set of H. (Recall that a γ-set stands for a minimum dominating set.) Set $D = A \cup \{(v, b) : b \in B\} = A \cup (\{v\} \times B)$. Then D is a dominating set of $G\Box H$. But then, as the sets A and $\{v\} \times B$ are disjoint,

$$\gamma(G\Box H) \leq |D| = |A| + |(\{v\} \times B)| = |A| + |B|$$
$$= \gamma(K\Box H) + \gamma(H)$$
$$< \gamma(K)\,\gamma(H) + \gamma(H)$$
$$= \gamma(H)\,(\gamma(K) + 1)$$
$$= \gamma(H)\,\gamma(G),$$

and this contradicts the hypothesis that G satisfies Vizing's conjecture. $\square$

We next present a lower bound (Theorem 10.7.8) and an upper bound (Theorem 10.7.10) for $\gamma(G\square H)$.

Theorem 10.7.8 (El-Zahar and Pareek [59]). $\gamma(G\square H) \geq \min\{|V(G)|, |V(H)|\}$.

Proof. Let $V(G) = \{u_1, u_2, \ldots, u_p\}$ and $V(H) = \{v_1, v_2, \ldots, v_q\}$. We have to prove that $\gamma(G\square H) \geq \min\{p, q\}$. Suppose D is a dominating set of $G\square H$ with

$$|D| < \min\{p, q\}. \tag{10.8}$$

Then $|D| < p$ and $|D| < q$.

Recall that the G-fibers of $G\square H$ are pairwise disjoint. A similar statement applies for the H-fibers of $G\square H$ as well. In view of (10.8), D does not meet all the G-fibers nor does it meet all the H-fibers. Hence, there exist a G-fiber, say G_y, with $y \in V(H)$, and a H-fiber, say H_x, with $x \in V(G)$, which are both disjoint from D. Now any vertex that dominates (x, y) must belong either to G_y or to H_x. But this is not the case as D is disjoint from both G_y and H_x. This contradicts the fact that D is a dominating set of $G\square H$. Thus, $|D| \geq \min\{p, q\}$. □

Corollary 10.7.9 (Rall [99]). *Let H be an arbitrary graph. Then there exists a positive integer $r = r(H)$ such that if G is any graph with $\gamma(G) \leq r$ and $|V(G)| \geq |V(H)|$, then $\gamma(G\square H) \geq \gamma(G)\gamma(H)$.*

Proof. Recall that $\gamma(H) \leq \frac{1}{2}|V(H)|$ (Corollary 10.2.7). Let $c = \frac{\gamma(H)|}{|V(H)|}$, and $r = \lfloor \frac{1}{c} \rfloor$. Then

$$\gamma(G\square H) \geq \min\{|V(G)|, |V(H)|\}\text{(by Theorem 10.7.8)}$$

$$= |V(H)|\text{(since by hypothesis } |V(G)| \geq |V(H)|)$$

$$= \frac{\gamma(H)|}{c}$$

$$\geq r\,\gamma(H)$$

$$\geq \gamma(G)\,\gamma(H) \text{ (since } \gamma(G) \leq r).$$ □

Theorem 10.7.10 (Vizing [182]). *For any two graphs G and H,*

$$\gamma(G\square H) \leq \min\{\gamma(G)|V(H)|, \gamma(H)|V(G)|\}.$$

Proof. Let D be a γ-set for G. Then $\bigcup_{v \in D} H_v$ is a dominating set for $G\square H$. To see this, consider any vertex (x, y) of $G\square H$. As $x \in V(G)$, and D is a dominating set of G, either $x \in D$ or there exists $v \in D$ with $vx \in E(G)$. Hence, either $(x, y) \in D \times V(H)$ or (v, y) dominates (x, y) in $G\square H$. Thus, $\bigcup_{v \in D} H_v$ is a dominating set of $G\square H$. Further, $|D| = \gamma(G)$, and so

$$\gamma(G \square H) \le \Big| \bigcup_{v \in D} H_v \Big| = |D| \, |V(H)| = \gamma(G) \, |V(H)|.$$

Similarly, $\gamma(G \square H) \le \gamma(H) \, |V(G)|$. $\qquad\qquad\qquad\qquad\qquad\qquad\quad\Box$

Lemma 10.7.11. *If D is any dominating set of $G \square H$ and x is any vertex of G, then $|D \cap (N[x] \times V(H))| \ge \gamma(H)$.*

Proof. Let h be any vertex of H so that (x, h) is an arbitrary vertex of the fiber $H_x = \{x\} \square H$. Let $N_G[x] = \{x, u_1, \dots, u_k\} \subseteq V(G)$, and let $p : (N[x] \times V(H)) \to H_x$ be the projection map defined by $p((x, h)) = (x, h)$ and $p((u_i, h)) = (x, h)$ for $i = 1, \dots, k$. Since D is a dominating set of $G \square H$, D must meet each closed neighborhood in $G \square H$ (see Exercise 6.2.). Now the closed neighborhood of (x, h) in $G \square H$ is $N[(x, h)] = \{(x, h)\} \cup (\{x\} \times N(h)) \cup (N(x) \times \{h\})$, and therefore D must contain (x, h) or a vertex either of the form (x, h'), where $h' \in N(h) \subset V(H)$ or of the form (u_i, h), $i = 1, \dots, k$. Now $p((x, h)) = (x, h)$ and $p((x, h')) = (x, h')$, while $p((u_i, h)) = (x, h)$. Thus, $p(D \cap (N[x] \times V(H)))$ dominates H_x and so $|D \cap (N[x] \times V(H))| \ge |p(D \cap (N[x] \times V(H)))| \ge \gamma(H_x) = \gamma(H)$ (as $H_x \simeq H$). $\qquad\Box$

Next consider a set $\mathscr{S}$ of pairwise disjoint closed neighborhoods of G. Let H be any graph, and D, a dominating set of $G \square H$. Then D must meet every star in $\mathscr{S}$. Hence if $N[x] \in \mathscr{S}$, then by Lemma 10.7.11,

$$|D \cap (N[x] \times V(H))| \ge \gamma(H).$$

As this is true for each of the closed neighborhoods in $\mathscr{S}$, we have

$$|D| \ge |\mathscr{S}| \gamma(H). \qquad\qquad\qquad\qquad (10.9)$$

Recall that $\rho(G)$ denotes the maximum number of pairwise disjoint closed neighborhoods in the graph G. Replacing $|\mathscr{S}|$ by $\rho(G)$ in (10.9), we get the following result of Jacobson and Kinch.

Theorem 10.7.12 (Jacobson and Kinch [112]). *For any two graphs G and H, $\gamma(G \square H) \ge \max\{\rho(G) \gamma(H), \rho(H) \gamma(G)\}$.*

Proof. Replacing $|\mathscr{S}|$ by $\rho(G)$ in (10.9), we get

$$\gamma(G \square H) \ge \rho(G) \gamma(H).$$

In a similar manner,

$$\gamma(H \square G) \ge \rho(H) \gamma(G).$$

The result now follows from the fact that $G \square H \simeq H \square G$. $\qquad\qquad\Box$

Fig. 10.7 A graph G
satisfying Vizing's conjecture

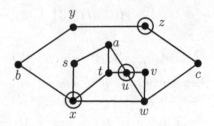

If G is a graph for which $\rho(G) = \gamma(G)$, then Theorem 10.7.12 implies that $\gamma(G \square H) \geq \gamma(G)\,\gamma(H)$. In other words, Vizing's conjecture is true for such graphs G. Now Meir and Moon [139] have shown that for any tree T, $\rho(T) = \gamma(T)$. Hence, Vizing's conjecture is true for all trees. This fact was first proved by Barcalkin and German [15]. We state this result as a corollary.

Corollary 10.7.13 ([15]). *If T is any tree, then T satisfies Vizing's conjecture.* □

Example 10.7.14. Let G be the graph of Fig. 10.7 The set $D = \{x, u, z\}$ is a γ-set for G so that $\gamma(G) = 3$. Moreover, the closed neighborhoods $N[a]$, $N[b]$, and $N[c]$ are pairwise disjoint in G. Hence, by (10.9) or by Theorem 10.7.12, $|D| \geq 3\,\gamma(H) = \gamma(G)\,\gamma(H)$, and hence G satisfies Vizing's conjecture.

10.8 Decomposable Graphs

In 1979, Barcalkin and German [15] showed that Vizing's conjecture is true for a very large class of graphs. Their result was published in Russian and remained unnoticed until 1991. The result of Barcalkin–German is on the validity of Vizing's conjecture for any decomposable graph. So we now give the definition of a decomposable graph.

Definition 10.8.1. A graph G is called *decomposable* if its vertex set can be partitioned into $\gamma(G)$ subsets with each part inducing a clique (that is a complete subgraph) of G.

Figure 10.8 displays a decomposable graph with $\gamma = 2$.

Theorem 10.8.2 (Barcalkin and German [15]). *If a graph G is decomposable, then G satisfies Vizing's conjecture.*

However, the converse of Theorem 10.8.2 is false. For instance, the graph of Fig. 10.9 is not decomposable, but it satisfies Vizing's conjecture (as $\rho = \gamma = 2$).

Theorem 10.8.2, when taken in conjunction with Lemma 10.7.6, yields the following result.

Theorem 10.8.3. *If $H \leq G$, $\gamma(H) = \gamma(G)$ and G is decomposable, then H satisfies Vizing's conjecture.*

Fig. 10.8 A decomposable
graph with $\gamma = 2$

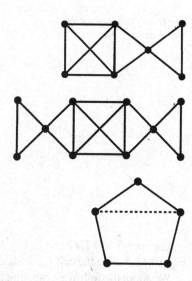

Fig. 10.9 Graph not
decomposable but satisfies
Vizing's conjecture

Fig. 10.10 Graph G of
Example 10.8.4

Example 10.8.4. C_5 satisfies Vizing's conjecture. This is because if G stands for
the graph of Fig. 10.10, then $C_5 \leq G$, $\gamma(C_5) = \gamma(G) = 2$ and G is decomposable
as its vertex set can be partitioned into the cliques K_3 and K_2.

Proof of Barcalkin–German theorem. Our proof is based on [30].

Let G be a decomposable graph with $\gamma(G) = k$, and let $\mathscr{C} = (C_1, \ldots, C_k)$
be a partition of $V(G)$ into cliques. Let $\{C_{i_1}, \ldots, C_{i_p}\}$ be a set of $p < k$ cliques
belonging to $\mathscr{C}$. Suppose that S is a smallest set of vertices in $G \setminus (C_{i_1} \cup \ldots \cup C_{i_p})$
which dominates (all the vertices of) $C_{i_1} \cup \ldots \cup C_{i_p}$. In other words, S is a set of
vertices of G outside $C_{i_1} \cup \ldots \cup C_{i_p}$ dominating the latter. Suppose further that
$C_{j_1}, \ldots, C_{j_q}$ are those cliques of $\mathscr{C}$ that have a nonempty intersection with S so that
$(\bigcup_{t=1}^{q} C_{j_t}) \cap S = S$. We then claim the following:

$$\text{Claim}: \quad \sum_{t=1}^{q} (|(C_{j_t} \cap S)| - 1) \geq p. \tag{10.10}$$

Since any vertex of a clique will dominate that clique, the vertices in S will
dominate the $p + q$ cliques $C_{i_1}, \ldots, C_{i_p}; C_{j_1}, \ldots, C_{j_q}$. Hence, if $|S| < p + q$, then
all the cliques of $\mathscr{C}$ will be dominated by $|S| + (k - p - q) < k$ vertices; equivalently,
$\gamma(G) < k$, a contradiction. This contradiction proves our claim.

We now complete the proof of the theorem. Let D be a minimum dominating set
of $G \square H$. The main idea of the proof is that each vertex from D will get a label
from 1 to k, and for each label i, the projections to H of the vertices from D that
are labeled i form a dominating set of H. This means that there are at least $\gamma(H)$
vertices in D that are labeled i, $1 \leq i \leq k$, and this implies that $|D| \geq k\gamma(H) = \gamma(G)\gamma(H)$.

Fig. 10.11 The partition of $G \square H$ into G-cells

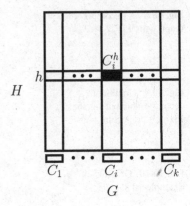

For each $h \in V(H)$ and i, $1 \leq i \leq k$, we call $C_i^h = V(C_i) \times \{h\}$ a *G-cell* (see Fig. 10.11, where the cell C_i^h is shaded).

We adopt the following labeling procedure: If a G-cell C_i^h contains a vertex from D, then one of the vertices from $D \cap C_i^h$ is given the label i. Hence, in the projection to H, h will also get the label i. Note that we have not yet determined the labels of the remaining vertices in $D \cap C_i^h$, if any.

Fix an arbitrary vertex $h \in V(H)$. We need to prove that for an arbitrary i, $1 \leq i \leq k$, there exists a vertex from D, labeled by i, that is projected to the neighborhood of h.

There are two cases. First, if there exists a vertex of D in $V(C_i) \times N[h]$, then by our labeling procedure, there will be a vertex in $N[h]$ to which the label i is projected, and so this case is settled.

The second case is that there is no vertex of D in $V(C_i) \times N[h]$, and we call such C_i^h a *missing G-cell for h*. Let $C_{i_1}^h, \ldots, C_{i_p}^h$ be the missing G-cells for h. Now by the definition of the Cartesian product, the missing G-cells for h must be dominated within the G-fiber G^h. Here there must be vertices in $D \cap G^h$ that dominate $C_{i_1}^h \cup \ldots \cup C_{i_p}^h$. Let $C_{j_1}^h, \ldots, C_{j_q}^h$ be the G-cells in G^h that intersect D. Since G^h is isomorphic to G, by inequality (10.10) we have

$$\sum_{t=1}^{q} (|C_{j_t}^h \cap D| - 1) \geq p.$$

Thus, there are enough additional vertices in $D \cap G^h$ (that have not been already labeled) so that for each missing G-cell C_i^h, the label i can be given to one of the vertices in $C_{j_t}^h \cap D$, where $|C_{j_t}^h \cap D| \geq 2$ (so that C_{j_t} has at least one unlabeled vertex of D at this stage). Hence, in this case, the label i will be projected to h. This concludes the proof. $\square$

We now present two applications of Barcalkin–German theorem.

Corollary 10.8.5. *If $\gamma(G) = 1$, then G satisfies Vizing's conjecture.*

Proof. If $\gamma(G) = 1$, G is a spanning subgraph of the complete graph K_n. As $\gamma(K_n) = 1$, the corollary follows. [Any complete graph satisfies Vizing's conjecture as $\gamma(K_n \square H) = \gamma(H)$.]

Corollary 10.8.6. *If $\gamma(G) = 2$, then G satisfies Vizing's conjecture.*

Proof. Let G' be the graph obtained from G by adding edges so that G' is edge-maximal and $\gamma(G') = 2$. We prove that G' is decomposable. This would mean, by virtue of Theorem 10.8.3, that G satisfies Vizing's conjecture.

Let Q_1 and Q_2 be disjoint cliques of G' such that $|V(Q_1)| + |V(Q_2)|$ is maximum. We claim that $|V(Q_1)| + |V(Q_2)| = |V(G')| (= |V(G)|)$.

If not, there exists a vertex v of G, with $v \notin V(Q_1) \cup V(Q_2)$. v cannot be adjacent to all the vertices of Q_1 [else the subgraph induced by $V(Q_1) \cup \{v\}$ is a clique Q_1' of G with $|V(Q_1')| + |V(Q_2)| = |V(Q_1)| + |V(Q_2)| + 1$, a contradiction]. Similarly, v cannot be adjacent to all the vertices of Q_2. Let w be a vertex of Q_1 which is nonadjacent to v in G'. As G' is edge-maximal with respect to γ, $\gamma(G' + vw) = 1$. Hence, $G' + vw$ has a single vertex that forms a minimum dominating set of $G' + vw$. Such a vertex cannot be v (as v is not adjacent to all the vertices of Q_2). Hence, it can only be w (as only w dominates v in $G' + vw$). This means that w is adjacent to all the other vertices in $V(G') \cup \{v\}$, and in particular to all vertices of Q_2. Let A be the set of vertices in Q_1 not adjacent to v in G'. Then each vertex of A is adjacent to all the vertices of Q_2. Then $(Q_1 \setminus A) \cup \{v\}$ and $Q_2 \cup A$ span cliques in G' whose union contains one vertex more than $Q_1 \cup Q_2$. This contradiction proves that $|V(Q_1)| + |V(Q_2)| = |V(G')|$, and so G' is decomposable. $\qquad\square$

The preceding theorems gave graphs G for which Vizing's conjecture holds. However, a result providing a general lower bound for $\gamma(G \square H)$ for all graphs G and H was given by Clark and Suen [42], stating that $\gamma(G \square H) \geq \frac{1}{2}\gamma(G)\gamma(H)$ for all graphs G and H. Another interesting result concerning Vizing's conjecture is the following result of Clark, Ismail, and Suen [43]. If G and H are both δ-regular graphs, then with only a few possible exceptions, Vizing's conjecture holds for the graph $G \square H$.

While so much is known about the domination number of the Cartesian product of two graphs, not much is known with regard to other products. We now present two easy results on the direct product.

10.9 Domination in Direct Products

Theorem 10.9.1. *Let G_1 and G_2 be graphs without isolated vertices. Then*

$$\gamma(G_1 \times G_2) \leq 4\gamma(G_1)\gamma(G_2).$$

The proof of Theorem 10.9.1 uses Lemma 10.9.2, which is an immediate consequence of Theorem 10.2.5'.

Lemma 10.9.2 ([160]). *Let D be a γ-set of a graph G, (that is, a minimum dominating set) without isolated vertices. Then there exists a matching in $E(D, V \setminus D)$ that saturates all the vertices of D.*

Proof of theorem 10.9.1. Let D_1 and D_2 be γ-sets for G_1 and G_2, respectively. Let D_1' and D_2' be the matching vertex sets of D_1 and D_2, respectively, as given by Lemma 10.9.2. Then $|D_1| = |D_1'| = \gamma(G_1)$, and $|D_2| = |D_2'| = \gamma(G_2)$. Clearly,

$$(D_1 \times D_2) \cup (D_1 \times D_2') \cup (D_1' \times D_2) \cup (D_1' \times D_2')$$ is a dominating set of $G_1 \times G_2$, of cardinality $4\gamma(G_1)\gamma(G_2)$. $\qquad\square$

Definition 10.9.3. A graph G is a *split graph* if $V(G)$ can be partitioned into two subsets K and I such that the subgraph, $G[K]$, induced by K in G is a clique in G, and I is an independent subset of G.

Definition 10.9.4. A subset D of the vertex set of a graph is called a *total dominating set* of G if any vertex v of G has a neighbor in D. (In other words, D dominates not only vertices outside D but also vertices in D.) The *total dominating number*, $\gamma_t(G)$, of G is the minimum cardinality of a total dominating set of G.

Lemma 10.9.5. *For any split graph G, $\gamma_t(G) = \gamma(G)$.*

Proof. Let $G = (K|I)$ be a split graph with K, a clique of G, and I, an independent set of G. Let D be any minimum dominating set of G, and let $D \cap K = K^*$ and $D \cap I = I^*$. By Lemma 10.9.2, G contains a matching in $H = E[D, V \setminus D]$ that saturates all the vertices of D. Let $I^* = \{u_1, \ldots, u_k\}$. Then G contains matching edges $u_1v_1, \ldots, u_kv_k$, where $\{v_1, \ldots, v_k\} \subset K \setminus K^*$. Clearly, $K^* \cup \{v_1, \ldots, v_k\}$ is a minimum dominating set D_1 of G, and hence $|D_1| = \gamma(G)$. Now since K is a clique, D_1 is a total dominating set. Thus, $\gamma_t(G) \leq |D_1| = \gamma(G)$. Since $\gamma(G) \leq \gamma_t(G)$ always (as any total dominating set of G is a dominating set of G), we have $\gamma(G) = \gamma_t(G)$. $\qquad\square$

Lemma 10.9.6. *For any two graphs G_1 and G_2 with no isolates, $\gamma(G_1 \times G_2) \leq \gamma_t(G_1)\gamma_t(G_2)$.*

Proof. Let A and B be minimum total dominating sets of G_1 and G_2, respectively. Then for any $(x, y) \in V(G_1) \times V(G_2)$, there exist $a \in A$ and $b \in B$ with $(x, a) \in E(G_1)$ and $(y, b) \in E(G_2)$. Hence, (x, y) is adjacent to (a, b) in $G_1 \times G_2$. This means that $A \times B$ is a dominating set for $G_1 \times G_2$, and hence $\gamma(G_1 \times G_2) \leq |A \times B| = |A||B| = \gamma_t(G_1)\gamma_t(G_2)$. $\qquad\square$

Theorem 10.9.7. *If G_1 and G_2 are split graphs, then $\gamma(G_1 \times G_2) \leq \gamma(G_1)\gamma(G_2)$.*

Proof. The proof is an immediate consequence of Lemmas 10.9.5 and 10.9.6. $\qquad\square$

Notes

"Domination in graphs" is one of the major areas of current research in graph theory. The two-volume book by Haynes, Hedetniemi, and Slater [100, 101] is a comprehensive reference work on graph domination. Several special types of domination in graphs have been studied by researchers—strong domination, weak domination, global domination, connected domination, independent domination, and so on.

As regards Vizing's conjecture, the technique of partitioning the vertex set of a graph, adopted by Barcalkin and German [15], has been exploited in two different ways to expand the classes of graphs satisfying Vizing's conjecture. In [98], Hartnell and Rall introduce the Type χ partition, which includes the Barcalkin–German class of graphs. The second has been proposed by Brešar and Rall [29]. Chordal graphs form yet another family that satisfies Vizing's conjecture. This was first established by Aharoni and Szabó [2] by using the approach of Clark and Suen [42], who showed that $\gamma(G \Box H) \geq \frac{1}{2}\gamma(G)\gamma(H)$ for *all* graphs G and H. For further details on Vizing's conjecture, the interested reader can refer to the article by Brešar et al. [30].

Domination in graph products, other than the Cartesian product, remains an area that has still not been fully explored.

Chapter 11
Spectral Properties of Graphs

11.1 Introduction

In this chapter, we look at the properties of graphs from our knowledge of their eigenvalues. The set of eigenvalues of a graph G is known as the *spectrum* of G and denoted by $Sp(G)$. We compute the spectra of some well-known families of graphs—the family of complete graphs, the family of cycles etc. We present Sachs' theorem on the spectrum of the line graph of a regular graph. We also obtain the spectra of product graphs—Cartesian product, direct product, and strong product. We introduce Cayley graphs and Ramanujan graphs and highlight their importance. Finally, as an application of graph spectra to chemistry, we discuss the *"energy of a graph"*—a graph invariant that is widely studied these days. All graphs considered in this chapter are finite, undirected, and simple.

11.2 The Spectrum of a Graph

Let G be a graph of order n with vertex set $V = \{v_1, \ldots, v_n\}$. The *adjacency matrix* of G (with respect to this labeling of V) is the n by n matrix $A = (a_{ij})$, where

$$a_{ij} = \begin{cases} 1 & \text{if } v_i \text{ is adjacent to } v_j \text{ in } G \\ 0 & \text{otherwise.} \end{cases}$$

Thus A is a real symmetric matrix of order n. Hence,

(i) The spectrum of A, that is, the set of its eigenvalues is real.
(ii) $\mathbb{R}^n$ has an orthonormal basis of eigenvectors of A.
(iii) The sum of the entries of the ith row (column) of A is $d(v_i)$ in G.

R. Balakrishnan and K. Ranganathan, *A Textbook of Graph Theory*,
Universitext, DOI 10.1007/978-1-4614-4529-6_11,
© Springer Science+Business Media New York 2012

The spectrum of A is called the *spectrum* of G and denoted by $Sp(G)$. We note that $Sp(G)$, as defined above, depends on the labeling of the vertex set V of G. We now show that it is independent of the labeling of G. Suppose we consider a new labeling of V. Let A' be the adjacency matrix of G with respect to this labeling. The new labeling can be obtained from the original labeling by means of a permutation π of $V(G)$. Any such permutation can be effected by means of a *permutation matrix* P of order n (got by permuting the rows of I_n, the identity matrix of order n).

(For example, if $n = 3$, the permutation matrix $P = \begin{pmatrix} 0 & 1 & 0 \\ 0 & 0 & 1 \\ 1 & 0 & 0 \end{pmatrix}$ takes v_1, v_2, v_3 to v_3, v_1, v_2 respectively since $(1\ 2\ 3) \begin{pmatrix} 0 & 1 & 0 \\ 0 & 0 & 1 \\ 1 & 0 & 0 \end{pmatrix} = (3\ 1\ 2)$.

Let $P = (p_{ij})$. Now given the new labeling of V, that is, given the permutation π on $\{1, 2, \ldots, n\}$, and the vertices v_i and v_j, there exist unique α_0 and β_0, such that $\pi(i) = \alpha_0$ and $\pi(j) = \beta_0$ or equivalently $p_{\alpha_0 i} = 1$ and $p_{\beta_0 j} = 1$, while for $\alpha \neq \alpha_0$ and $\beta \neq \beta_0$, $p_{\alpha i} = 0 = p_{\beta j}$. Thus, the (α_0, β_0)th entry of the matrix $A' = PAP^{-1} = PAP^{T}$ (where P^{T} stands for the transpose of P) is

$$\sum_{k,l=1}^{n} p_{\alpha_0 k} a_{kl} p_{\beta_0 l} = a_{ij}.$$

Hence, $v_{\alpha_0} v_{\beta_0} \in E(G)$ if and only if $v_i v_j \in E(G)$. This proves that the adjacency matrix of the same graph with respect to two different labelings are similar matrices. But then similar matrices have the same spectra. $\square$

We usually arrange the eigenvalues of G in their nondecreasing order: $\lambda_1 \geq \lambda_2 \geq \ldots \geq \lambda_n$. If $\lambda_1, \ldots, \lambda_s$ are the distinct eigenvalues of G, and if m_i is the multiplicity of λ_i as an eigenvalue of G, we write

$$Sp(G) = \begin{pmatrix} \lambda_1 & \lambda_2 & \ldots & \lambda_s \\ m_1 & m_2 & \ldots & m_s \end{pmatrix}.$$

Definition 11.2.1. The *characteristic polynomial* of G is the characteristic polynomial of the adjacency matrix of G with respect to some labeling of G. It is denoted by $\chi(G; \lambda)$.

Hence, $\chi(G; \lambda) = \det(xI - A) = \det(P(xI - A)P^{-1}) = \det(xI - PAP^{-1})$ for any permutation matrix of P, and hence $\chi(G; \lambda)$ is also independent of the labeling of $V(G)$.

Definition 11.2.2. A *circulant of order n* is a square matrix of order n in which all the rows are obtainable by successive cyclic shifts of one of its rows (usually taken as the first row).

For example, the circulant with first row $(a_1\ a_2\ a_3)$ is the matrix $\begin{pmatrix} a_1 & a_2 & a_3 \\ a_3 & a_1 & a_2 \\ a_2 & a_3 & a_1 \end{pmatrix}$.

Lemma 11.2.3. *Let A be a circulant matrix of order n with first row $(a_1\, a_2 \ldots a_n)$. Then $Sp(A) = \{a_1 + a_2\omega + \cdots + a_n\omega^{n-1}\ :\ \omega = $ an nth root of unity$\} = \{a_1 + \zeta^r + \zeta^{2r} + \cdots + \zeta^{(n-1)r},\ 0 \le r \le n-1$ and $\zeta = $ a primitive nth root of unity$\}$*

Proof. The characteristic polynomial of A is the determinant $D = \det(xI - A)$. Hence,

$$
D = \begin{vmatrix}
x - a_1 & -a_2 & \cdots & -a_n \\
-a_n & x - a_1 & \cdots & -a_{n-1} \\
\vdots & \vdots & & \vdots \\
-a_2 & -a_3 & \cdots & x - a_1
\end{vmatrix}.
$$

Let C_i denote the ith column of D, $1 \le i \le n$, and ω, an nth root of unity. Replace C_1 by $C_1 + C_2\omega + \cdots + C_n\omega^{n-1}$. This does not change D. Let $\lambda_\omega = a_1 + a_2\omega + \cdots + a_n\omega^{n-1}$. Then the new first column of D is $(x - \lambda_\omega,\ \omega(x - \lambda_\omega),\ \ldots,\ \omega^{n-1}(x - \lambda_\omega))^T$, and hence $x - \lambda_\omega$ is a factor of D. This gives $D = \prod_{\omega:\omega^n=1} (x - \lambda_\omega)$, and $Sp(A) = \{\lambda_\omega\ :\ \omega^n = 1\}$. $\qquad\qquad\square$

11.3   Spectrum of the Complete Graph K_n

For K_n, the adjacency matrix A is given by $A = \begin{pmatrix} 0 & 1 & 1 & \cdots & 1 \\ 1 & 0 & 1 & \cdots & 1 \\ \vdots & \vdots & \vdots & & \vdots \\ 1 & 1 & 1 & \cdots & 0 \end{pmatrix}$, and so by Lemma 11.2.3,

$$
\lambda_\omega = \omega + \omega^2 + \cdots + \omega^{n-1}
$$

$$
= \begin{cases} n - 1 & \text{if } \omega = 1 \\ -1 & \text{if } \omega \ne 1. \end{cases}
$$

Hence, $Sp(K_n) = \begin{pmatrix} n-1 & -1 \\ 1 & n-1 \end{pmatrix}$.

11.4   Spectrum of the Cycle C_n

Label the vertices of C_n as $0, 1, 2, \ldots, n-1$ in this order. Then i is adjacent to $i \pm 1 \pmod n$. Hence,

$$
A = \begin{bmatrix}
0 & 1 & 0 & 0 & \cdots & 0 & 1 \\
1 & 0 & 1 & 0 & \cdots & 0 & 0 \\
\vdots & \vdots & \vdots & \vdots & \ddots & \vdots & \vdots \\
1 & 0 & 0 & 0 & \cdots & 1 & 0
\end{bmatrix}
$$

is the circulant with the first row $(0\ 1\ 0\ \dots\ 0\ 1)$. Again, by Lemma 11.2.3, $Sp(C_n) = \{\omega^r + \omega^{r(n-1)} : 0 \le r \le n-1$, where ω is a primitive nth root of unity$\}$. Taking $\omega = \cos\frac{2\pi}{n} + i\sin\frac{2\pi}{n}$, we get $\lambda_r = \omega^r + \omega^{r(n-1)} = (\cos\frac{2\pi r}{n} + i\sin\frac{2\pi r}{n}) + (\cos\frac{2\pi r(n-1)}{n} + i\sin\frac{2\pi r(n-1)}{n})$. This simplifies to the following:

(i) If n is odd, $Sp(C_n) = \begin{pmatrix} 2 & 2\cos\frac{2\pi}{n} & \dots & 2\cos\frac{(n-1)\pi}{n} \\ 1 & 2 & \dots & 2 \end{pmatrix}$.

(ii) If n is even, $Sp(C_n) = \begin{pmatrix} 2 & 2\cos\frac{2\pi}{n} & \dots & 2\cos\frac{(n-1)\pi}{n} & -2 \\ 1 & 2 & \dots & 2 & 1 \end{pmatrix}$.

11.4.1 Coefficients of the Characteristic Polynomial

Let G be a connected graph on n vertices, and let $\chi(G;x) = \det(xI_n - A) = x^n + a_1 x^{n-1} + a_2 x^{n-2} + \dots + a_n$ be the characteristic polynomial of G. It is easy to check that $(-1)^r a_r =$ sum of the principal minors of A of order r. (Recall that a principal minor of order r of A is the determinant minor of A common to the same set of r rows and columns.)

Lemma 11.4.1. *Let G be a graph of order n and size m, and let $\chi(G;x) = x^n + a_1 x^{n-1} + a_2 x^{n-2} + \dots + a_n$ be the characteristic polynomial of A. Then*

(i) $a_1 = 0$
(ii) $a_2 = -m$
(iii) $a_3 = -$ (twice the number of triangles in G)

Proof. (i) Follows from the fact that all the entries of the principal diagonal of A are zero.

(ii) A nonvanishing principal minor of order 2 of A is of the form $\begin{vmatrix} 0 & 1 \\ 1 & 0 \end{vmatrix}$ and its value is -1. Since any 1 in A corresponds to an edge of G, we get (ii).

(iii) A nontrivial principal minor of order 3 of A can be one of the following three types:

$$\begin{vmatrix} 0 & 1 & 0 \\ 1 & 0 & 0 \\ 0 & 0 & 0 \end{vmatrix}, \begin{vmatrix} 0 & 1 & 1 \\ 1 & 0 & 0 \\ 1 & 0 & 0 \end{vmatrix}, \begin{vmatrix} 0 & 1 & 1 \\ 1 & 0 & 1 \\ 1 & 1 & 0 \end{vmatrix}.$$

Of these, only the last determinant is nonvanishing. Its value is 2 and corresponds to a triangle in G. This proves (iii).   □

11.5   The Spectra of Regular Graphs

In this section, we look at the spectra of some regular graphs.

Theorem 11.5.1. *Let G be a k-regular graph of order n. Then*

(i)   *k is an eigenvalue of G.*
(ii) *If G is connected, every eigenvector corresponding to the eigenvalue k is a multiple of **1**, (the all 1-column vector of length n) and the multiplicity of k as an eigenvalue of G is one.*
(iii) *For any eigenvalue λ of G, $|\lambda| \le k$. (Hence $Sp(G) \subset [-k, k]$).*

Proof.   (i) We have $A\mathbf{1} = k\mathbf{1}$, and hence k is an eigenvalue of A.

(ii) Let $\mathbf{x} = (x_1, \ldots, x_n)^{\mathrm{T}}$ be any eigenvector of A corresponding to the eigenvalue k so that $A\mathbf{x} = k\mathbf{x}$. We may suppose that $\mathbf{x}$ has a positive entry (otherwise, take $-\mathbf{x}$ in place of $\mathbf{x}$) and that x_j is the largest positive entry in $\mathbf{x}$. Let $v_{i_1}, v_{i_2}, \ldots, v_{i_k}$ be the k neighbors of v_j in G. Taking the inner product of the jth row of A with $\mathbf{x}$, we get $x_{i_1} + x_{i_2} + \cdots + x_{i_k} = kx_j$. This gives, by the choice of x_j,
$x_{i_1} = x_{i_2} = \ldots = x_{i_k} = x_j$.

Now start at $v_{i_1}, v_{i_2}, \ldots, v_{i_k}$ in succession and look at their neighbors in G. As before, the entries x_p in $\mathbf{x}$ corresponding to these neighbors must all be equal to x_j. As G is connected, all the vertices of G are reachable in this way step by step. Hence $\mathbf{x} = x_j(1, 1, \ldots, 1)^{\mathrm{T}}$, and every eigenvector $\mathbf{x}$ of A corresponding to the eigenvalue k is a multiple of $\mathbf{1}$. Thus, the space of eigenvectors of A corresponding to the eigenvalue k is one-dimensional, and therefore, the multiplicity of k as an eigenvalue of G is 1.

(iii) The proof is similar to (ii). If $A\mathbf{y} = \lambda\mathbf{y}$, $\mathbf{y} \neq \mathbf{0}$, and if y_j is the entry in $\mathbf{y}$ with the largest absolute value, we see that the equation $\sum_{p=1}^{k} y_{i_p} = \lambda y_j$ implies that $|\lambda||y_j| = |\lambda y_j| = \left| \sum_{p=1}^{k} y_{i_p} \right| \le \sum_{p=1}^{k} |y_{i_p}| \le k|y_j|$. Thus, $|\lambda| \le k$. $\square$

Corollary 11.5.2. *If Δ denotes the maximum degree of G, then for any eigenvalue λ of G, $|\lambda| \le \Delta$.*

Proof. Considering a vertex v_j of maximum degree Δ, and imitating the proof of (iii) above, we get $|\lambda||y_j| \le \Delta|y_j|$. $\square$

11.5.1   The Spectrum of the Complement of a Regular Graph

Theorem 11.5.3. *Let G be a k-regular connected graph of order n with spectrum $\begin{pmatrix} k & \lambda_2 & \lambda_3 & \ldots & \lambda_s \\ 1 & m_2 & m_3 & \ldots & m_s \end{pmatrix}$. Then the spectrum of G^c, the complement of G, is given by*
$$Sp(G^c) = \begin{pmatrix} n-1-k & -\lambda_2-1 & -\lambda_3-1 & \ldots & -\lambda_s-1 \\ 1 & m_2 & m_3 & \ldots & m_s \end{pmatrix}.$$

Proof. As G is k-regular, G^c is $n - 1 - k$ regular, and hence by Theorem 11.5.1, $n - 1 - k$ is an eigenvalue of G^c. Further, the adjacency matrix of G^c is $A^c = J - I - A$, where J is the all-1 matrix of order n, I is the identity matrix of order n, and A is the adjacency matrix of G. If $\chi(\lambda)$ is the characteristic polynomial of A, $\chi(\lambda) = (\lambda - k)\chi_1(\lambda)$. By Cayley–Hamilton theorem, $\chi(A) = 0$ and hence we have $A\chi_1(A) = k\chi_1(A)$. Hence, every column vector of $\chi_1(A)$ is an eigenvector of A corresponding to the eigenvalue k. But by Theorem 11.5.1, the space of eigenvectors of A is generated by $\mathbf{1}$, G being connected. Hence, each column vector of $\chi_1(A)$ is a multiple of $\mathbf{1}$. But $\chi_1(A)$ is symmetric and hence $\chi_1(A)$ is a multiple of J, say, $\chi_1(A) = \alpha J$, $\alpha \neq 0$. Thus, J and hence $J - I - A$ are polynomials in A (remember: $A^0 = I$). Let $\lambda \neq k$ be any eigenvalue of A [so that $\chi_1(\lambda) = 0$], and Y an eigenvector of A corresponding to λ. Then

$$A^c Y = (J - I - A)Y$$

$$= \left(\frac{\chi_1(A)}{\alpha} - I - A\right)Y$$

$$= \left(\frac{\chi_1(\lambda)}{\alpha} - 1 - \lambda\right)Y \text{ (see Note 11.5.4 below)}$$

$$= (-1 - \lambda)Y.$$

Thus, $A^c Y = (-1-\lambda)Y$, and therefore $-1-\lambda$ is an eigenvalue of A^c corresponding to the eigenvalue $\lambda(\neq k)$ of A.   □

Note 11.5.4. We recall that if $f(\lambda)$ is a polynomial in λ, and $Sp(A) = \begin{pmatrix} \lambda_1 & \lambda_2 & \cdots & \lambda_s \\ m_1 & m_2 & \cdots & m_s \end{pmatrix}$, then $Sp(f(A)) = \begin{pmatrix} f(\lambda_1) & f(\lambda_2) & \cdots & f(\lambda_s) \\ m_1 & m_2 & \cdots & m_s \end{pmatrix}$.

11.5.2 Spectra of Line Graphs of Regular Graphs

We now establish Sachs' theorem, which determines the spectrum of the line graph of a regular graph G in terms of $Sp(G)$.

Let G be a labeled graph with vertex set $V(G) = \{v_1, \ldots, v_n\}$ and edge set $E(G) = \{e_1, \ldots, e_m\}$. With respect to these labelings, the incidence matrix $B = (b_{ij})$ of G, which describes the incidence structure of G, is defined as the m by n matrix $B = (b_{ij})$, where $b_{ij} = \begin{cases} 1 & \text{if } e_i \text{ is incident to } v_j, \\ 0 & \text{otherwise.} \end{cases}$

Lemma 11.5.5. *Let G be a graph of order n and size m with A and B as its adjacency and incidence matrices, respectively. Let A_L denote the adjacency matrix of the line graph of G. Then*

(i) $BB^T = A_L + 2I_m$.
(ii) *If G is k-regular, $B^T B = A + kI_n$.*

Proof. Let $A = (a_{ij})$ and $B = (b_{ij})$. We have

(i)

$$(BB^{\mathrm{T}})_{ij} = \sum_{p=1}^{n} b_{ip} b_{jp}$$

$$= \text{number of vertices } v_p \text{ that are incident to both } e_i \text{ and } e_j$$

$$= \begin{cases} 1 & \text{if } e_i \text{ and } e_j \text{ are adjacent} \\ 0 & \text{if } i \neq j \text{ and } e_i \text{ and } e_j \text{ are nonadjacent} \\ 2 & \text{if } i = j. \end{cases}$$

(ii) Proof of (ii) is similar. $\square$

Theorem 11.5.6 (Sachs' theorem).   *Let G be a k-regular graph of order n. Then $\chi(L(G); \lambda) = (\lambda + 2)^{m-n} \chi(G; \lambda + 2 - k)$, where $L(G)$ is the line graph (see Chap. 1) of G.*

Proof. Consider the two partitioned matrices U and V, each of order $n + m$ (where B stands for the incidence matrix of G):

$$U = \begin{bmatrix} \lambda I_n & -B^{\mathrm{T}} \\ 0 & I_m \end{bmatrix}, V = \begin{bmatrix} I_n & B^{\mathrm{T}} \\ B & \lambda I_m \end{bmatrix}.$$

We have

$$UV = \begin{bmatrix} \lambda I_n - B^{\mathrm{T}} B & 0 \\ B & \lambda I_m \end{bmatrix} \text{ and } VU = \begin{bmatrix} \lambda I_n & 0 \\ \lambda B & \lambda I_m - BB^{\mathrm{T}} \end{bmatrix}.$$

Now $\det(UV) = \det(VU)$ gives:

$$\lambda^m \det(\lambda I_n - B^{\mathrm{T}} B) = \lambda^n \det(\lambda I_m - BB^{\mathrm{T}}). \tag{11.1}$$

Replacement of λ by $\lambda + 2$ in (11.1) yields

$$(\lambda + 2)^{m-n} \det((\lambda + 2)I_n - B^{\mathrm{T}} B) = \det((\lambda + 2)I_m - BB^{\mathrm{T}}). \tag{11.2}$$

Hence, by Lemma 11.5.5,

$$\chi(L(G); \lambda) = \det(\lambda I_m - A_L)$$
$$= \det((\lambda + 2)I_m - (A_L + 2I_m))$$
$$= \det((\lambda + 2)I_m - BB^{\mathrm{T}})$$
$$= (\lambda + 2)^{m-n} \det((\lambda + 2)I_n - B^{\mathrm{T}} B) \quad \text{(by (11.2))}$$

$$= (\lambda + 2)^{m-n} \det((\lambda + 2)I_n - (A + kI_n)) \quad \text{(by Lemma 11.5.5)}$$
$$= (\lambda + 2)^{m-n} \det((\lambda + 2 - k)I_n - A)$$
$$= (\lambda + 2)^{m-n} \chi(G; \lambda + 2 - k).$$

$\square$

Sachs' theorem implies the following: As $\chi(G; \lambda) = \prod_1^n (\lambda - \lambda_i)$, it follows that

$$\chi(L(G); \lambda) = (\lambda + 2)^{m-n} \prod_{i=1}^n (\lambda + 2 - k - \lambda_i)$$

$$= (\lambda + 2)^{m-n} \prod_{i=1}^n \left(\lambda - (k - 2 + \lambda_i)\right).$$

Hence if $Sp(G) = \begin{pmatrix} k & \lambda_2 & \dots & \lambda_s \\ 1 & m_2 & \dots & m_s \end{pmatrix}$, then

$$Sp(L(G)) = \begin{pmatrix} 2k - 2 & k - 2 + \lambda_2 & \dots & k - 2 + \lambda_s & -2 \\ 1 & m_2 & \dots & m_s & m - n \end{pmatrix}.$$

We note that the spectrum of the Petersen graph can be obtained by using Theorems 11.5.3 and 11.5.6 (see Exercise 14.14). We use this result to prove a well-known result on the Petersen graph.

Theorem 11.5.7. *The complete graph K_{10} cannot be decomposed into (that is, expressed as an edge-disjoint union of) three copies of the Petersen graph.*

Proof (Schwenk and Lossers [169]). Assume the contrary. Suppose K_{10} is expressible as an edge-disjoint union of three copies, say, P_1, P_2, P_3, of the Petersen graph P (see Fig. 11.1). Then

$$A(K_{10}) = J - I = A(P_1) + A(P_2) + A(P_3), \tag{11.3}$$

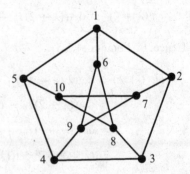

Fig. 11.1 Petersen graph P

where $A(H)$ stands for the adjacency matrix of the graph H, and J the all-1 matrix of order 10. (Note that each P_i is a spanning subgraph of K_{10}, that the number of edges of P_i, namely 15, is a divisor of the number of edges of K_{10}, namely, 45, and that the degree of any vertex v of P, namely, 3, is a divisor of the degree of v in K_{10}, namely 9.) It is easy to check that 1 is an eigenvalue of P. Further, as P is 3-regular, the all-1 column vector $\mathbf{1}$ is an eigenvector of P. Now $\mathbb{R}^{10}$ has an orthonormal basis of eigenvectors of $A(P)$ containing $\mathbf{1}$. Again, the null space of $(A(P) - 1.I) = A(P) - I$ is of dimension 5 and is orthogonal to $\mathbf{1}$. (For P with the labeling of Fig. 11.1, one can check that for $A(P) - I$, the null space is spanned by the five vectors: $(1\,0\,0\,0\,0\,1\,-1\,0\,0\,-1)^{\mathrm{T}}$, $(0\,1\,0\,0\,0\,-1\,1\,-1\,0\,0)^{\mathrm{T}}$, $(0\,0\,1\,0\,0\,0\,-1\,1\,-1\,0)^{\mathrm{T}}$, $(0\,0\,0\,1\,0\,0\,0\,-1\,1\,-1)^{\mathrm{T}}$, $(0\,0\,0\,0\,1\,-1\,0\,0\,-1\,1)^{\mathrm{T}}$) The orthogonal complement of $\mathbf{1}$ in $\mathbb{R}^{10}$ is of dimension $10 - 1 = 9$. Hence, the null spaces of $A(P_1) - I$ and $A(P_2) - I$ must have a common eigenvector $\mathbf{x}$ orthogonal to $\mathbf{1}$. Multiplying the matrices on the two sides of (11.3) to $\mathbf{x}$, we get $(J - I)\mathbf{x} = A(P_1)\mathbf{x} + A(P_2)\mathbf{x} + A(P_3)\mathbf{x}$, that is (as $J\mathbf{x} = 0$), $-\mathbf{x} = \mathbf{x} + \mathbf{x} + A(P_3)\mathbf{x}$. Thus, $A(P_3)\mathbf{x} = -3\mathbf{x}$, and this means that -3 is an eigenvalue of the Petersen graph, a contradiction (see Exercise 14.14). $\qquad\square$

Various proofs of Theorem 11.5.7 are available in literature. For a second proof, see [32].

11.6 Spectrum of the Complete Bipartite Graph $K_{p,q}$

We now determine the spectrum of the complete bipartite graph $K_{p,q}$.

Theorem 11.6.1. $Sp(K_{p,q}) = \begin{pmatrix} 0 & \sqrt{pq} & -\sqrt{pq} \\ p+q-2 & 1 & 1 \end{pmatrix}$.

Proof. Let $V(K_{p,q})$ have the bipartition (X, Y) with $|X| = p$ and $|Y| = q$. Then the adjacency matrix of $K_{p,q}$ is of the form

$$A = \begin{pmatrix} 0 & J_{p,q} \\ J_{q,p} & 0 \end{pmatrix},$$

where $J_{r,s}$ stands for the all-1 matrix of size r by s. Clearly, $\mathrm{rank}(A) = 2$, as the maximum number of independent rows of A is 2. Hence, zero is an eigenvalue of A repeated $p + q - 2$ times (as the null space of A is of dimension $p + q - 2$). Thus, the characteristic polynomial of A is of the form $\lambda^{p+q-2}(\lambda^2 + c_2)$.

[Recall that by Lemma 11.4.1, the coefficient of λ^{p+q-1} in $\chi(G; \lambda)$ is zero.] Further, again by the same lemma, $-c_2 = $ the number of edges of $K_{p,q} = pq$. This proves the result. $\qquad\square$

11.7 The Determinant of the Adjacency Matrix of a Graph

We now present the elegant formula given by Harary for the determinant of the adjacency matrix of a graph in terms of certain of its subgraphs.

Definition 11.7.1. A *linear subgraph* of a graph G is a subgraph of G whose components are single edges or cycles.

Theorem 11.7.2 (Harary [92]). *Let A be the adjacency matrix of a simple graph G. Then*

$$\det A = \sum_{H} (-1)^{e(H)} 2^{c(H)},$$

where the summation is over all the spanning linear subgraphs H of G, and $e(H)$ and $c(H)$ denote, respectively, the number of even components and the number of cycles in H.

Proof. Let G be of order n with $V = \{v_1, \ldots, v_n\}$, and $A = (a_{ij})$. A typical term in the expansion of $\det A$ is

$$sgn(\pi) a_{1\pi(1)} a_{2\pi(2)} \cdots a_{n\pi(n)},$$

where π is a permutation on $\{1, 2, \ldots, n\}$ and $sgn(\pi) = 1$ or -1 according to whether π is an even or odd permutation. This term is zero if and only if for some i, $1 \leq i \leq n$, $a_{i\pi(i)} = 0$, that is, if and only if $\pi(i) = i$ or $\pi(i) = j(\neq i)$ and $v_i v_j \notin E(G)$. Hence, this term is nonzero if and only if the permutation π is a product of disjoint cycles of length *at least* 2, and in this case, the value of the term is $sgn(\pi).1.1 \ldots 1 = sgn(\pi)$. Each cycle (ij) of length 2 in π corresponds to the single edge $v_i v_j$ of G, while each cycle $(ij \ldots p)$ of length $r > 2$ in π corresponds to a cycle of length r of G. Thus, each nonvanishing term in the expansion of $\det A$ gives rise to a linear subgraph H of G and conversely. Now for any cycle C of S_n, $sgn(C) = 1$ or -1 according to whether C is an odd or even cycle. Hence, $sgn(\pi) = (-1)^{e(H)}$, where $e(H)$ is the number of even components of H (that is, components that are either single edges or even cycles of the graph H). Moreover, any cycle of H has two different orientations. Hence, each of the undirected cycles of H of length ≥ 3 yields two distinct even cycles in S_n. [For example, the 4-cycle $(v_{i_1} v_{i_2} v_{i_3} v_{i_4})$ gives rise to two cycles $(v_{i_1} v_{i_2} v_{i_3} v_{i_4})$ and $(v_{i_4} v_{i_3} v_{i_2} v_{i_1})$ in H.] This proves the result.   □

Corollary 11.7.3 (Sachs [167]). *Let $\chi(G; x) = x^n + a_1 x^{n-1} + \cdots + a_n$ be the characteristic polynomial of G. Then*

$$a_i = \sum_{H} (-1)^{\omega(H)} 2^{c(H)},$$

where the summation is over all linear subgraphs H of order i of G, and $\omega(H)$ and $c(H)$ denote, respectively, the number of components and the number of cycle components of H.

Proof. Recall that $a_i = (-1)^i \sum_H \det A$, where H runs through all the induced subgraphs of order i of G. But by Theorem 11.7.2,

$$\det H = \sum_{H_i} (-1)^{e(H_i)} 2^{c(H_i)},$$

where H_i is a spanning linear subgraph of H and $e(H_i)$ stands for the number of even components of H_i, while $c(H_i)$ stands for the number of cycles in H_i. The corollary follows from the fact that i and the number of odd components of H_i have the same parity. □

11.8 Spectra of Product Graphs

We have already defined in Chap. 1 the graph products: Cartesian product, direct product, and strong product. In this section we determine the spectra of these graphs in terms of the spectra of their factor graphs. Our approach is based on Cvetković [46] as described in [47]. Recall that all these three products are associative.

Let $\mathscr{B}$ be a set of binary n-tuples $(\beta_1, \beta_2, \ldots, \beta_n)$ not containing $(0, 0, \ldots, 0)$.

Definition 11.8.1. Given a sequence of graphs $G_1, G_2, \ldots, G_n$, the *NEPS (Noncomplete Extended P-Sum)* of $G_1, G_2, \ldots, G_n$, with respect to $\mathscr{B}$ is the graph G with $V(G) = V(G_1) \times V(G_2) \times \ldots \times V(G_n)$, and in which two vertices $(x_1, x_2, \ldots, x_n)$ and $(y_1, y_2, \ldots, y_n)$ are adjacent if and only if there exists an n-tuple $(\beta_1, \beta_2, \ldots, \beta_n) \in \mathscr{B}$ with the property that if $\beta_i = 1$, then $x_i y_i \in E(G_i)$ and if $\beta_i = 0$, then $x_i = y_i$.

Example 11.8.2. (i) $n = 2$ and $\mathscr{B} = \{(1, 1)\}$. Here the graphs are G_1 and G_2. The vertices (x_1, x_2) and (y_1, y_2) are adjacent in the NEPS of G_1 and G_2 with respect to $\mathscr{B}$ if and only if $x_1 y_1 \in E(G_1)$ and $x_2 y_2 \in E(G_2)$. Hence, $G = G_1 \times G_2$, the *direct product* of G_1 and G_2.

(ii) $n = 2$ and $\mathscr{B} = \{(0, 1), (1, 0)\}$. Here G is the Cartesian product $G_1 \square G_2$.

(iii) $n = 2$ and $\mathscr{B} = \{(0, 1), (1, 0), (1, 1)\}$. Here $G = (G_1 \square G_2) \cup (G_1 \times G_2) = G_1 \boxtimes G_2$, the strong product of G_1 and G_2.

Now, given the adjacency matrices of $G_1, \ldots G_n$, the adjacency matrix of the NEPS graph G with respect to the basis $\mathscr{B}$ is expressible in terms of the Kronecker product of matrices, which we now define:

Definition 11.8.3. Let $A = (a_{ij})$ be an m by n matrix and $B = (b_{ij})$ be a p by q matrix. Then $A \otimes B$, the *Kronecker product of A with B*, is the mp by nq matrix obtained by replacing each entry a_{ij} of A by the double array $a_{ij} B$ (where $a_{ij} B$ is the matrix obtained by multiplying each entry of B by a_{ij}).

It is well known and easy to check that

$$(A \otimes B)(C \otimes D) = (AC \otimes BD), \tag{11.4}$$

whenever the matrix products AC and BD are defined. Clearly, this can be extended to any finite product whenever the products are defined.

Remark 11.8.4. Let us look more closely at the product $A_1 \otimes A_2$, where A_1 and A_2 are the adjacency matrices of the graphs G_1 and G_2 of orders n and t, respectively. To fix any particular entry of $A_1 \otimes A_2$, let us first label $V(G_1) = V_1$ and $V(G_2) = V_2$ as $V_1 = \{u_1, \ldots, u_n\}$, and $V_2 = \{v_1, \ldots, v_t\}$. Then to fix the entry in $A_1 \otimes A_2$ corresponding to $((u_i, u_j), (v_p, v_q))$, we look at the double array $(A_1)_{(u_i u_j)} A_2$ in $A_1 \otimes A_2$, where $(A_1)_{(u_i u_j)} := \alpha$ stands for the (i, j)th entry of A_1. Then the required entry is just $\alpha\{(p, q)$th entry of $A_2\}$. Hence, it is **1** if and only if $(A_1)_{(u_i u_j)} = 1 = (A_2)_{(v_p v_q)}$, that is, if and only if $u_i u_j \in E(G_1)$ and $v_p v_q \in E(G_2)$, and 0 otherwise. In other words, $A_1 \otimes A_2$ is the adjacency matrix of $G_1 \times G_2$. By associativity, $A(G_1) \otimes \ldots \otimes A(G_r)$ is the adjacency matrix of the graph product $G_1 \times \ldots \times G_r$.

Our next theorem determines the adjacency matrix of the NEPS G in terms of the adjacency matrices of G_i, $1 \le i \le n$ for all three products mentioned above.

Theorem 11.8.5 (Cvetković [46]). *Let G be the NEPS of the graphs $G_1, \ldots, G_n$ with respect to the basis $\mathcal{B}$. Let A_i be the adjacency matrix of G_i, $1 \le i \le n$. Then the adjacency matrix A of G is given by*

$$A = \sum_{\beta = (\beta_1, \ldots, \beta_n) \in \mathcal{B}} A_1^{\beta_1} \otimes \ldots \otimes A_n^{\beta_n}.$$

Proof. Label the vertex set of each of the graphs G_i, $1 \le i \le n$, and order the vertices of G lexicographically. Form the adjacency matrix A of G with respect to this ordering. Then (by the description of Kronecker product of matrices given in Remark 11.8.4) we have $(A)_{(x_1, \ldots, x_n)(y_1, \ldots, y_n)} = \sum_{\beta \in \mathcal{B}} (A_1^{\beta_1})_{(x_1, y_1)} \cdots (A_n^{\beta_n})_{(x_n, y_n)}$,

where $(M)_{(x,y)}$ stands for the entry in M corresponding to the vertices x and y. But by lexicographic ordering, $(M)_{(x_1, \ldots, x_n)(y_1, \ldots, y_n)} = 1$ if and only if there exists a $\beta = (\beta_1, \ldots, \beta_n) \in \mathcal{B}$ with $(A_i^{\beta_i})_{(x_i, y_i)} = 1$ for each $i = 1, \ldots, n$. This of course means that $x_i y_i \in E(G_i)$ if $\beta_i = 1$ and $x_i = y_i$ if $\beta_i = 0$ (the latter condition corresponds to $A_i^{\beta_i} = I$). $\square$

We now determine the spectrum of the NEPS graph G with respect to the basis $\mathcal{B}$ in terms of the spectra of the factor graphs G_i.

Theorem 11.8.6 (Cvetković [46]). *Let G be the NEPS of the graphs $G_1, \ldots, G_n$ with respect to the basis $\mathcal{B}$. Let k_i be the order of G_i and A_i, the adjacency matrix of G_i. Let $\{\lambda_{i1}, \ldots, \lambda_{ik_i}\}$ be the spectrum of G_i, $1 \le i \le n$. Then*

$$Sp(G) = \{\Lambda_{i_1 i_2 \ldots i_n} : 1 \le i_j \le k_j \text{ and } 1 \le j \le n\},$$

where $\Lambda_{i_1 i_2 \ldots i_n} = \sum_{\beta = (\beta_1, \ldots, \beta_n) \in \mathcal{B}} \lambda_{1 i_1}^{\beta_1} \ldots, \lambda_{n i_n}^{\beta_n}, 1 \le i_j \le k_j \text{ and } 1 \le j \le n.$

Proof. There exist vectors x_{ij} with $A_i x_{ij} = \lambda_{ij} x_{ij}$, $1 \leq i \leq n$; $1 \leq j \leq k_j$. Now consider the vector $x = x_{1i_1} \otimes \ldots \otimes x_{ni_n}$. Let A be the adjacency matrix of G. Then from Theorem 11.8.5 and (11.4) (rather its extension),

$$
\begin{aligned}
Ax &= \left(\sum_{\beta \in \mathscr{B}} A_1^{\beta_1} \otimes \ldots \otimes A_n^{\beta_n} \right) (x_{1i_1} \otimes \ldots \otimes x_{ni_n}) \\
&= \sum_{\beta \in \mathscr{B}} (A_1^{\beta_1} x_{1i_1} \otimes \ldots \otimes A_n^{\beta_n} x_{ni_n}) \\
&= \sum_{\beta \in \mathscr{B}} (\lambda_{1i_1}^{\beta_1} x_{1i_1} \otimes \ldots \otimes \lambda_{ni_n}^{\beta_n} x_{ni_n}) \\
&= \left(\sum_{\beta \in \mathscr{B}} \lambda_{1i_1}^{\beta_1} \ldots \lambda_{ni_n}^{\beta_n} \right) x \\
&= \Lambda_{i_1 i_2 \ldots i_n} x.
\end{aligned}
$$

Thus, $\Lambda_{i_1 i_2 \ldots i_n}$ is an eigenvalue of G. This yields $k_1 k_2 \ldots k_n$ eigenvalues of G and hence all the eigenvalues of G. $\qquad \square$

Corollary 11.8.7. *Let $Sp(G_1) = \{\lambda_1, \ldots, \lambda_n\}$ and $Sp(G_2) = \{\mu_1, \ldots, \mu_t\}$ and let A_1 and A_2 be the adjacency matrices of G_1 and G_2, respectively. Then*

(i) $A(G_1 \times G_2) = A_1 \otimes A_2$; *and* $Sp(G_1 \times G_2) = \{\lambda_i \mu_j : 1 \leq i \leq n, 1 \leq j \leq t\}$.

(ii) $A(G_1 \square G_2) = (I_n \otimes A_2) + (A_1 \otimes I_t)$; *and* $Sp(G_1 \square G_2) = \{\lambda_i + \mu_j : 1 \leq i \leq n, 1 \leq j \leq t\}$.

(iii) $A(G_1 \boxtimes G_2) = (A_1 \otimes A_2) + (I_n \otimes A_2) + (A_1 \otimes I_t)$; *and* $Sp(G_1 \boxtimes G_2) = \{\lambda_i \mu_j + \lambda_i + \mu_j : 1 \leq i \leq n, 1 \leq j \leq t\}$.

$\qquad \square$

11.9 Cayley Graphs

11.9.1 Introduction

Cayley graphs are a special type of regular graphs constructed out of groups.

Let Γ be a finite group and $S \subset \Gamma$ be such that

(i) $e \notin S$ (e is the identity of Γ),

(ii) If $a \in S$, then $a^{-1} \in S$, and

(iii) S generates Γ.

Construct a graph G with $V(G) = \Gamma$ and in which $ab \in E(G)$ if and only if $b = as$ for some $s \in S$. Since $as = as'$ in Γ implies that $s = s'$, it follows that each vertex a of G is of degree $|S|$; that is, G is a $|S|$-regular graph. Moreover, if a and b are any two vertices of G, there exists c in Γ such that $ac = b$. But as S generates Γ, $c = s_1 s_2 \ldots s_p$, where $s_i \in S$, $1 \le i \le p$. Hence, $b = a s_1 s_2 \ldots s_p$, which implies that b is reachable from a in G by means of the path $a(as_1)(as_1 s_2) \ldots (as_1 s_2 \ldots s_p = b)$. Thus, G is a connected simple graph. [Condition (i) implies that G has no loops.] G is known as the *Cayley graph* $\mathrm{Cay}(\Gamma; S)$ of the group Γ defined by the set S.

We now consider a special family of Cayley graphs. Take $\Gamma = (\mathbb{Z}_n, +)$, the additive group of integers modulo n. If $S = \{s_1, s_2, \ldots, s_p\}$, then $0 \notin S$ and $s_i \in S$ if and only if $n - s_i \in S$. The vertices adjacent to 0 are $s_1, s_2, \ldots, s_p$, while those adjacent to i are $(s_1 + i)(\mathrm{mod}\, n), (s_2 + i)(\mathrm{mod}\, n), \ldots, (s_p + i)(\mathrm{mod}\, n)$. Consequently, the adjacency matrix of $\mathrm{Cay}(\mathbb{Z}_n; S)$ is a circulant and, by Lemma 11.2.3, its eigenvalues are

$$\{\omega^{s_1} + \omega^{s_2} + \cdots + \omega^{s_p} : \omega = \text{an } n\text{th root of unity}\}.$$

11.9.2   Unitary Cayley Graphs

We now take $S(\subset \mathbb{Z}_n)$ to be the set U_n of numbers less than n and prime to n. Note that $(a, n) = 1$ if and only if $(n - a, n) = 1$. The corresponding Cayley graph $\mathrm{Cay}(\mathbb{Z}_n; U_n)$ is denoted by X_n and called the *unitary Cayley graph* mod n (Note that U_n is the set of multiplicative units in $\mathbb{Z}_n$.) Now suppose that $(n_1, n_2) = 1$. What is the Cayley graph $X_{n_1 n_2}$? Given $x \in \mathbb{Z}_{n_1 n_2}$, there exist unique $c \in \mathbb{Z}_{n_1}$ and $d \in \mathbb{Z}_{n_2}$ such that

$$x \equiv c(\mathrm{mod}\, n_1) \quad \text{and} \quad x \equiv d(\mathrm{mod}\, n_2). \tag{11.5}$$

Conversely, given $c \in \mathbb{Z}_{n_1}$ and $d \in \mathbb{Z}_{n_2}$, by the Chinese Remainder Theorem [5], there exists a unique $x \in \mathbb{Z}_{n_1 n_2}$ satisfying (11.5). Define $f : \mathbb{Z}_{n_1 n_2} \to \mathbb{Z}_{n_1} \times \mathbb{Z}_{n_2}$ (the direct product of the groups $\mathbb{Z}_{n_1}$ and $\mathbb{Z}_{n_2}$) by setting $f(x) = (c, d)$. Then f is an additive group isomorphism with $f(U_{n_1 n_2}) = U_{n_1} \times U_{n_2}$.

Two vertices a and b are adjacent in $\mathbb{Z}_{n_1 n_2}$ if and only if $a - b \in U_{n_1 n_2}$. Let

$$a \equiv a_1(\mathrm{mod}\, n_1), \qquad\qquad b \equiv b_1(\mathrm{mod}\, n_1)$$

$$a \equiv a_2(\mathrm{mod}\, n_2) \quad \text{and} \quad b \equiv b_2(\mathrm{mod}\, n_2).$$

Then $(a - b, n_1 n_2) = 1$ if and only if $(a - b, n_1) = 1 = (a - b, n_2)$; equivalently, $(a_1 - b_1, n_1) = 1 = (a_2 - b_2, n_2)$ or, in other words, $a_1 b_1 \in E(X_{n_1})$ and $a_2 b_2 \in E(X_{n_2})$. Thus, we have the following result.

Theorem 11.9.1. *If $(n_1, n_2) = 1$, the unitary Cayley graph $X_{n_1 n_2}$ is isomorphic to $X_{n_1} \times X_{n_2}$ (where $\times$ on the right stands for direct product of graphs).*

11.9.3 Spectrum of the Cayley Graph X_n

The adjacency matrix of X_n, as noted before, is a circulant, and hence the spectrum $(\lambda_1, \lambda_2, \ldots, \lambda_n)$ of X_n (see the proof of Lemma 11.2.3) is given by

$$\lambda_r = \sum_{\substack{1 \leq j \leq n, \\ (j,n)=1}} \omega^{rj}, \quad \text{where } \omega = e^{\frac{2\pi i}{n}}. \tag{11.6}$$

The sum in (11.6) is the well-known Ramanujan sum $c(r, n)$ [123]. It is known that

$$c(r, n) = \mu(t_r)\frac{\phi(n)}{\phi(t_r)}, \quad \text{where } t_r = \frac{n}{(r, n)}, \ 0 \leq r \leq n - 1. \tag{11.7}$$

In relation (11.7), μ stands for the Möebius function. Further, as t_r divides n, $\phi(t_r)$ divides $\phi(n)$, and therefore $c(r, n)$ is an integer for each r. These remarks yield the following theorem.

Theorem 11.9.2. *The eigenvalues of the unitary Cayley graph* $X_n = \mathrm{Cay}(\mathbb{Z}_n, U_n)$ *are all the integers* $c(r, n)$, $0 \leq r \leq n - 1$.

For more spectral properties of Cayley graphs, the reader may consult Klotz and Sander [123].

11.10 Strongly Regular Graphs

Strongly regular graphs form an important class of regular graphs.

Definition 11.10.1. A *strongly regular graph* with parameters (n, k, λ, μ) (for short: $srg(n, k, \lambda, \mu)$) is a k-regular connected graph G of order n with the following properties:

 (i) Any two adjacent vertices of G have exactly λ common neighbors in G.
 (ii) Any two nonadjacent vertices of G have exactly μ common neighbors in G.

Example 11.10.2.   (i) The cycle C_5 is an $srg(5, 2, 0, 1)$.
 (ii) The Petersen graph P is an $srg(10, 3, 0, 1)$.
 (iii) The cycle C_6 is a regular graph that is not strongly regular.
 (iv) The Clebesch graph is an $srg(16, 5, 0, 2)$ (see Fig. 11.2).
 (v) Let $q \equiv 1 \pmod 4$ be an odd prime power. The *Paley graph* $P(q)$ of order q is the graph with vertex set $GF(q)$, the Galois field of order q, with two vertices adjacent in $P(q)$ if and only if their difference is a nonzero square in $GF(q)$. $P(q)$ is an $srg(q, \frac{1}{2}(q - 1), \frac{1}{4}(q - 5), \frac{1}{4}(q - 1))$.
 Our next theorem gives a necessary condition for the existence of a strongly regular graph with parameters (n, k, λ, μ).

Fig. 11.2 Clebsch Graph

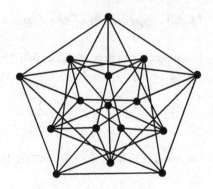

Theorem 11.10.3. *If G is a strongly regular graph with parameters* (n, k, λ, μ), *then*

$$k(k - \lambda - 1) = \mu(n - k - 1).$$

Proof. We prove this theorem by counting the number of induced paths on three vertices in G having the same vertex v of G as an end vertex in two different ways. There are k neighbors w of v in G. For each such w, there are λ vertices that are common neighbors of v and w. Each of the remaining $k - 1 - \lambda$ neighbors of w induces a P_3 with v as an end vertex. As this is true for each neighbor w of v in G, there are $k(k - \lambda - 1)$ paths of length 2 with v as an end vertex.

We now compute this number in a different way. There are $n - 1 - k$ vertices x of G that are nonadjacent to v. Each such pair v, x is commonly adjacent to μ vertices of G. Each one of these μ vertices gives rise to an induced P_3 in G with v as an end vertex. This number is $\mu(n - k - 1)$. $\qquad\square$

Suppose now $\lambda = k - 1$. As G is k-regular, $ab \in E(G)$, implies that every vertex c adjacent to a is adjacent to b, and every vertex d adjacent to b is adjacent to a. Now consider the adjacent pair b, c. Then c must be adjacent to d and so on. Thus, each component of G must be a clique (complete subgraph) of size k.

A similar reasoning applies when $\mu = 0$. Further, if $k = n - 1$, G is K_n, the clique on n vertices. We treat these cases (that is, the cases when $\lambda = k - 1; \mu = 0; k = n - 1$) as degenerate cases.

Theorem 11.10.4. *If G is an* $srg(n, k, \lambda, \mu)$, *then its complement* G^c *of G is an* $srg(n, n - 1 - k, n - 2 - 2k + \mu, n - 2k + \lambda)$.

Proof. Trivially, G^c is $n - 1 - k$ regular. Suppose now $uv \in E(G^c)$. Then $uv \notin E(G)$. There are k vertices adjacent to u in G and k vertices adjacent to v in G. Out of these $2k$ vertices, μ are common vertices adjacent to both u and v in G^c. Hence, there are $(n - 2) - (2k - \mu) = n - 2 - 2k + \mu$ vertices commonly adjacent to u and v in G^c. By a similar argument, if $uv \notin E(G^c)$, u and v are commonly adjacent to $n - 2k + \lambda$ vertices in G^c. Thus, G^c is an $srg(n, n - 1 - k, n - 2 - 2k + \mu, n - 2k + \lambda)$. $\qquad\square$

We now present another necessary condition for a graph to be strongly regular.

Theorem 11.10.5. *If G is an $srg(n, k, \lambda, \mu)$, with $\mu > 0$, then the two numbers*

$$\frac{1}{2}\left(n - 1 \pm \frac{(n-1)(\mu - \lambda) - 2k}{\sqrt{(\mu - \lambda)^2 + 4(k - \mu)}}\right)$$

are both nonnegative integers.

Proof. We prove the theorem by showing that the two numbers are (algebraic) multiplicities of eigenvalues of G. If A is the adjacency matrix of G, $J - I - A$ (where J is the all-1 matrix of order n) is the adjacency matrix of G^c. What is the (i, j)th entry of A^2? If $i = j$, the number of v_i-v_j walks of length 2 is k since they are all of the form $v_i v_p v_i$, and there are k adjacent vertices v_p to v_i in G. Hence, each diagonal entry of A^2 is k.

Now let $i \neq j$. Then $v_i v_j \in E(G)$ if and only if $v_i v_j \notin E(G^c)$. Hence if $i \neq j$, the (i, j)th entry is 1 or 0 in A if and only if the (i, j)th entry is 0 or 1, respectively, in $J - I - A$. Now there are exactly λ vertices commonly adjacent to v_i and v_j in G, and hence there are λ paths of length 2 in G. Hence, the 1's in A will be replaced by λ in A^2, and this is given by the matrix λA. Finally, let $i \neq j$ and $v_i v_j \notin E(G)$. The number of v_i-v_j walks of length 2 in G is μ. The (i, j)th entry 0 in A should now be replaced by μ. This can be done by replacing the (i, j)th entry 1 in $(J - I - A)$ by μ. Thus,

$$A^2 = kI + \lambda A + \mu(J - I - A). \tag{11.8}$$

Hence, if $\mathbf{1}$ is the column vector $(1, 1, \ldots, 1)^T$ of length n,

$$A^2\mathbf{1} = (kI + \lambda A + \mu(J - I - A))\mathbf{1},$$

and this gives (as $A\mathbf{1} = k\mathbf{1}$)

$$k^2\mathbf{1} = k\mathbf{1} + \lambda k\mathbf{1} + \mu(n - 1 - k)\mathbf{1}. \tag{11.9}$$

Note that (11.9) yields another proof for Theorem 11.10.3.

Now as G is connected, $\mathbf{1}$ is a unique eigenvector corresponding to the eigenvalue k of G. Let $\mathbf{x}$ be any eigenvector corresponding to an eigenvalue $\theta \neq k$ of G so that $A\mathbf{x} = \theta\mathbf{x}$. Then $\mathbf{x}$ is orthogonal to $\mathbf{1}$. Taking the product of both sides of (11.8) with $\mathbf{x}$, we get (as $J\mathbf{x} = 0$)

$$\theta^2\mathbf{x} = A^2\mathbf{x} = k\mathbf{x} + \lambda\theta\mathbf{x} + \mu(-\mathbf{x} - \theta\mathbf{x}),$$

and therefore (as $\mathbf{x}$ is a nonzero vector),

$$\theta^2 - (\lambda - \mu)\theta - (k - \mu) = 0. \tag{11.10}$$

This quadratic equation has two real roots (being eigenvalues of G), say, r and s, which are given by

$$\frac{1}{2}\left[(\lambda - \mu) \pm \sqrt{(\lambda - \mu)^2 + 4(k - \mu)}\right]. \tag{11.11}$$

Denote the multiplicities of r and s by a and b. Now

$$k + ra + sb = 0 \text{ (sum of the eigenvalues)},$$

$$\text{and } 1 + a + b = n \text{ (total number of eigenvalues)}.$$

Solution of these equations gives

$$a = -\frac{k + s(n-1)}{r - s},$$

$$b = \frac{k + r(n-1)}{r - s}. \tag{11.12}$$

Substitution of the values of r and s as given in (11.11) into (11.12) yields the two numbers given in the statement of Theorem 11.10.5.   □

11.11   Ramanujan Graphs

Ramanujan graphs constitute yet another family of regular graphs. In recent times, a great deal of interest has been shown in Ramanujan graphs by researchers in diverse fields—graph theory, number theory, and communication theory.

Definition 11.11.1. A *Ramanujan graph* is a k-regular connected graph G, $k \geq 2$, such that if λ is any eigenvalue of G with $|\lambda| \neq k$, then $\lambda \leq 2\sqrt{k - 1}$.

Example 11.11.2. The following graphs are Ramanujan graphs.

1. The complete graphs K_n, $n \geq 3$.
2. Cycles C_n.
3. $K_{n,n}$. Here $k = n$, and $Sp(K_{n,n}) = \begin{pmatrix} n & 0 & -n \\ 1 & 2n-2 & 1 \end{pmatrix}$.
4. The Petersen graph.

The *Möebius ladder* M_h is the cubic graph obtained by joining the opposite vertices of the cycle C_{2h}. By Exercise 14.6, we have

$$\lambda_j = 2\cos\frac{\pi j}{h} + (-1)^j, \quad 0 \leq j \leq 2h - 1.$$

Take $h = 2p$ and $j = 4p - 2$. Then $\lambda_{4p-2} = 2\cos\frac{\pi(4p-2)}{2p} + 1 = 2\cos\frac{\pi}{p} + 1 > 2\sqrt{k-1}$ (when p becomes large) $= 2\sqrt{2}$ (as $k = 3$). Hence, not every regular graph is a Ramanujan graph.

11.11.1 Why Are Ramanujan Graphs Important?

Let G be a k-regular Ramanujan graph of order n, and A its adjacency matrix. As A is symmetric, $\mathbb{R}^n$ has an orthonormal basis $\{u_1, \ldots, u_n\}$ of eigenvectors of A. Since $A\mathbf{1} = k\mathbf{1}$, we can take $u_1 = (\frac{1}{\sqrt{n}}, \ldots, \frac{1}{\sqrt{n}})^T$. Let $Sp(G) = \{\lambda_1 \geq \lambda_2 \geq \ldots \geq \lambda_n\}$, where $\lambda_1 = k$ and $Au_i = \lambda_i u_i$, $1 \leq i \leq n$. We can write $A = \sum_{i=1}^{n} \lambda_i u_i u_i^T$. (This is seen from the fact that the matrices on the two sides when postmultiplied by u_j, $1 \leq j \leq n$ both yield $\lambda_j u_j$.) More generally, as λ_i^p, $1 \leq i \leq n$ are the eigenvalues of A^p, we have

$$A^p = \sum_{i=1}^{n} \lambda_i^p u_i u_i^T. \tag{11.13}$$

Let $u_i = ((u_i)_1, (u_i)_2, \ldots, (u_i)_r, \ldots, (u_i)_n)$. Then the (r, s)th entry of $A^p = \sum_{i=1}^{n} \lambda_i^p (u_i)_r (u_i^T)_s = \sum_{i=1}^{n} \lambda_i^p (u_i)_r (u_i)_s := X_{rs}$. As A is a binary matrix, $X_{r,s}$ is a nonnegative integer. Moreover, as G is k-regular and connected, $\lambda_1 = k$. Set $\lambda(G) := \max_{|\lambda_i| \neq k} |\lambda_i|$. Then,

$$X_{rs} = \sum_{i=1}^{n} \lambda_i^p (u_i)_r (u_i)_s$$

$$= \left| \sum_{i=1}^{n} \lambda_i^p (u_i)_r (u_i)_s \right|$$

$$\geq k^p (u_1)_r (u_1)_s - \left| \sum_{i=2}^{n} \lambda_i^p (u_i)_r (u_i)_s \right|$$

$$= k^p \frac{1}{\sqrt{n}} \frac{1}{\sqrt{n}} - |\Sigma_0|, \text{ where } \Sigma_0 = \sum_{i=2}^{n} \lambda_i^p (u_i)_r (u_i)_s. \tag{11.14}$$

Assume that G is not bipartite, so that $\lambda_n \neq -k$ (see Exercise 14.22). Hence, the eigenvalues $\lambda_2, \ldots, \lambda_n$ all satisfy $|\lambda_i| \leq \lambda(G)$, and therefore

$$|\Sigma_0| \leq \lambda(G)^p \sum_{i=2}^{n} |(u_i)_r||(u_i)_s|$$

$$\leq \lambda(G)^p \left(\sum_{i=2}^{n} |(u_i)_r^2| \right)^{1/2} \left(\sum_{i=2}^{n} |(u_i)_s^2| \right)^{1/2} \quad \text{(by Cauchy–Schwarz inequality)}$$

$$= \lambda(G)^p \left[1 - (u_1)_r^2 \right]^{1/2} \left[1 - (u_1)_s^2 \right]^{1/2} \quad \text{(as the } u_i\text{'s are unit vectors)}$$

$$= \lambda(G)^p \left[1 - \frac{1}{n}\right]^{1/2} \left[1 - \frac{1}{n}\right]^{1/2}$$

$$= \lambda(G)^p \left[1 - \frac{1}{n}\right].$$

Hence, the (r, s)th entry of A^p is positive if $\frac{k^p}{n} > \lambda(G)^p(1 - \frac{1}{n})$, that is, if $\frac{k^p}{\lambda(G)^p} > n - 1$.

This gives, on taking logarithms, that if

$$p > \frac{\log(n - 1)}{\log(k/\lambda(G))},$$

then every entry of A^p is positive. Now the diameter of G is the least positive integer p for which A^p is positive (see Exercise 14.3). Hence, the diameter D of G satisfies the inequality

$$D \leq \frac{\log(n - 1)}{\log(k/\lambda(G))} + 1.$$

Thus, we have proved the following theorem.

Theorem 11.11.3 (Chung [37]). *Let G be a k-regular connected nonbipartite graph with n vertices and diameter D. Then*

$$D \leq \frac{\log(n - 1)}{\log\left(k/\lambda(G)\right)} + 1. \tag{11.15}$$

Now assume that G is bipartite and k-regular. Then $\lambda_1 = k$ and $\lambda_n = -k$ so that $|\lambda_1| = k = |\lambda_n|$. Working as above, one gets the following inequality in this case:

$$D \leq \frac{\log(n - 2)/2}{\log\left(k/\lambda(G)\right)} + 2. \tag{11.16}$$

The last two inequalities show that to minimize D, one has to minimize $\lambda(G) = \max_{|\lambda_i| \neq k} |\lambda_i|$. Such graphs are useful in communication theory—the smaller the diameter, better the communication. In the case of Ramanujan graphs, we demand that $\lambda(G) \leq 2\sqrt{k - 1}$. Hence, Ramanujan graphs with sufficiently small values for $\lambda(G)$ could be used in cost-efficient communication networks.

We mention that in constructing k-regular Ramanujan graphs G for a fixed $k \geq 2$, it is not possible to bring down the upper bound $2\sqrt{k - 1}$ for $\lambda(G)$ since as per a result of Serre [132], for $\epsilon > 0$, there exists a positive constant $c = c(k, \epsilon)$, which depends only on k and ϵ such that the adjacency matrix of every k-regular graph on n vertices has at least cn eigenvalues larger than $(2 - \epsilon)\sqrt{k - 1}$ (see also [137]). Again, not every unitary Cayley graph is a Ramanujan graph. Unitary Cayley graphs that are Ramanujan graphs have been completely characterized by Droll [58].

We conclude this section with a remark that explains the name "Ramanujan graph" (see also [162]). The question that one may ask is the following: Does there exist a sequence $\{G_i\}$ of graphs with an increasing number of vertices, satisfying the bound $\lambda(G_i) \leq 2\sqrt{k-1}$? That is, can we give an explicit construction for such a sequence of Ramanujan graphs? The only case known for which such sequences have been constructed is when $k-1$ equals a prime power. In all these cases, the proof that the eigenvalues satisfy the required bound is by means of the Ramanujan's conjecture in the theory of modular forms, proved by Deligne [48] in 1974 in the case when $k-1$ is a prime, and by the work of Drinfeld [57] in the case when $k-1$ is a prime power. This explains how Ramanujan's name has entered into the definition of these graphs.

We have only touched upon the periphery of Ramanujan graphs. A deeper study of Ramanujan graphs requires expertise in number theory. Interested readers can refer to the two survey articles by Ram Murty ([162, 163]) and the relevant references contained therein.

11.12 The Energy of a Graph

11.12.1 Introduction

In this section, we discuss another application of eigenvalues of graphs. "The energy of a graph" is a concept borrowed from chemistry. Every chemical molecule can be represented by means of its corresponding *molecular graph*: Each vertex of the graph corresponds to an atom of the molecule, and two vertices of the graph are adjacent if and only if there is a bond between the corresponding molecules (the number of bonds being immaterial). The π-electron energy of a conjugated hydrocarbon, as calculated with the "Hückel molecular orbital (HMO) method," coincides with the energy (as we are going to define below) of the corresponding graph. Conjugated hydrocarbons are of great importance for science and technology. A *conjugated hydrocarbon* can be characterized as a molecule composed entirely of carbon and hydrogen atoms in which every carbon atom has exactly three neighbors (which are either carbon or hydrogen atoms). See Fig. 11.3, which gives the structure of butadiene.

The graph corresponding to a conjugated hydrocarbon is taken as follows (which is somewhat different from our description of molecular graphs given earlier in Chap. 1, and there are valid theoretical reasons for doing this): Every carbon atom is represented by a vertex and every carbon–carbon bond by an edge. Hydrogen atoms are ignored. Thus, the graph for the molecule of Fig. 11.3 is the path P_4 on four vertices. These graphs of the conjugated hydrocarbons are connected and their vertex degrees are at most 3. However, for our general definition of graph energy, there is no such restriction and our graphs are quite general—only, they are simple.

Fig. 11.3 Butadiene

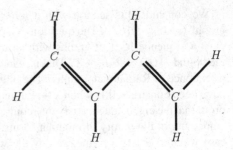

The pioneer in this area is Ivan Gutman. Readers interested in doing research in graph energy should consult [82, 83, 84].

Definition 11.12.1. The *energy* of a graph G is the sum of the absolute values of its eigenvalues.

Hence if $\lambda_1, \lambda_2, \ldots, \lambda_n$ are the eigenvalues of a graph G of order n, the energy $\mathcal{E}(G)$ of G is given by

$$\mathcal{E}(G) = \sum_{i=1}^{n} |\lambda_i|.$$

Example 11.12.2. 1. The energy of the complete graph $K_n = \mathcal{E}(K_n) = (n-1) + (n-1)|-1| = 2(n-1)$.

2. The spectrum of the Petersen graph is $\begin{pmatrix} 3 & 1 & -2 \\ 1 & 5 & 4 \end{pmatrix}$. (See Exercise 14.14). Hence, its energy is $1.3 + 5.1 + 4.2 = 16$.

11.12.2 Maximum Energy of k-Regular Graphs

Theorem 11.12.3. *The maximum energy of a k-regular graph G of order n is $k + \sqrt{k(n-1)(n-k)}$.*

Proof. Let $Sp(G) = \{\lambda_1 = k, \lambda_2, \ldots, \lambda_n\}$. Apply Cauchy–Schwarz's inequality to the two vectors $(|\lambda_2|, |\lambda_3|, \ldots, |\lambda_n|)$ and $(1, 1, \ldots, 1)$. This gives

$$\left(\sum_{i=2}^{n} |\lambda_i|.1\right)^2 \leq \left(\sum_{i=2}^{n} |\lambda_i|^2\right)\left(\sum_{i=2}^{n} 1^2\right)$$

so that (as $\sum_{i=1}^{n} \lambda_i^2 = 2m$)

$$(\mathcal{E}(G) - \lambda_1) \leq \sqrt{(2m - \lambda_1^2)(n-1)}.$$

Therefore, $\mathcal{E}(G) \leq F(\lambda_1)$, where

$$F(\lambda_1) := \lambda_1 + \sqrt{(2m - \lambda_1^2)(n - 1)}. \tag{11.17}$$

If G is k-regular, by Theorem 11.5.1, $\lambda_1 = k$, and hence

$$\mathcal{E}(G) \leq k + \sqrt{(n - 1)(2m - k^2)}$$
$$= k + \sqrt{k(n - 1)(n - k)} \quad \text{(as } 2m = nk\text{).} \tag{11.18}$$

$\square$

The same method yields for a general (not necessarily regular) graph the following theorem of Gutman [84].

Theorem 11.12.4. *Let G be a graph of order n and size m. Then*

$$\mathcal{E}(G) \leq \frac{2m}{n} + \sqrt{(n - 1)\left[2m - \left(\frac{2m}{n}\right)^2\right]}. \tag{11.19}$$

Proof. As usual, let A stand for the adjacency matrix of G. We have

$$\sqrt{\frac{2m}{n}} \leq \frac{2m}{n} \leq \lambda_1 \leq \sqrt{2m}. \tag{11.20}$$

Relation (11.20) is a consequence of the fact that $0 < \lambda_1^2 \leq \lambda_1^2 + \cdots + \lambda_n^2 = 2m$ and Exercise 14.10. Moreover, by (11.17), $\mathcal{E}(G) \leq F(\lambda_1)$. Now the function $F(x) = x + \sqrt{(n - 1)(2m - x^2)}$ is strictly decreasing in the interval $\sqrt{\frac{2m}{n}} < x \leq \sqrt{2m}$. Hence, by inequality (11.20),

$$\mathcal{E}(G) \leq F(\lambda_1) \leq F\left(\frac{2m}{n}\right) = \frac{2m}{n} + \sqrt{(n - 1)\left[2m - \left(\frac{2m}{n}\right)^2\right]}. \tag{11.21}$$

$\square$

Theorem 11.12.5. *If $2m > n$ and G is a graph on n vertices and m edges, then $\mathcal{E}(G) = \frac{2m}{n} + \sqrt{(n - 1)\left[2m - \left(\frac{2m}{n}\right)^2\right]}$ if and only if G is either $\frac{n}{2}K_2$, K_n or a non-complete connected strongly regular graph for which the spectrum $\{\lambda_1 > \lambda_2 \geq \cdots \geq \lambda_n\}$ has the property that $|\lambda_2| = |\lambda_3| = \ldots = |\lambda_n| = \sqrt{\frac{2m - \left(\frac{2m}{n}\right)^2}{n - 1}}$.*

Proof. We have already seen in Theorem 11.12.4 that $\mathcal{E}(G) \leq F\left(\frac{2m}{n}\right) = \frac{2m}{n} + \sqrt{(n - 1)\left(2m - \left(\frac{2m}{n}\right)^2\right)}$. If $G \simeq \frac{n}{2}K_2$, then $Sp(G) = \begin{pmatrix} 1 & -1 \\ \frac{n}{2} & \frac{n}{2} \end{pmatrix}$, and if $G \simeq K_n$, then $Sp(K_n) = \begin{pmatrix} 1 & -1 \\ 1 & n - 1 \end{pmatrix}$. In both cases, it is routine to check that $\mathcal{E}(G) = F(\frac{2m}{n})$.

If the third alternative holds, G is regular of degree (see Theorem 11.5.1) $\frac{2m}{n} = \lambda_1$,

and hence $\mathcal{E}(G)$, which, by hypothesis, is equal to $\lambda_1 + (n-1)\sqrt{\frac{2m-\left[\frac{2m}{n}\right]^2}{n-1}} =$

$\frac{2m}{n} + \sqrt{(n-1)\left[2m - \left(\frac{2m}{n}\right)^2\right]} = F\left(\frac{2m}{n}\right)$, and therefore equality holds in (11.19).

Conversely, assume that equality is attained in (11.19) for some graph G. Then (11.20) and (11.21) together with the strict decreasing nature of $F(x)$ imply that λ_1 (maximum degree of G) $= \frac{2m}{n}$ (average degree of G). Hence, G must be regular of degree $\frac{2m}{n}$. Moreover, in this case, equality must be attained in the Cauchy–Schwarz inequality applied in the proof of Theorem 11.12.3 and hence $(|\lambda_2|, \ldots, |\lambda_n|)$ is a scalar multiple of $(1, 1, \ldots, 1)$. In other words, $|\lambda_2| = \ldots = |\lambda_n|$. Consequently, $|\lambda_2| + \cdots + |\lambda_n| = (n-1)|\lambda_i| = \sqrt{(n-1)\left[2m - \left(\frac{2m}{n}\right)^2\right]}$, and

therefore, $|\lambda_i| = \sqrt{\frac{\left[2m - \left(\frac{2m}{n}\right)^2\right]}{n-1}}$, $2 \le i \le n$. This results in three cases.

Case (i). *The graph G has two distinct eigenvalues λ_1 and λ_2 with $|\lambda_1| = |\lambda_2|$.* [If all the eigenvalues of G are equal, they must all be zero (as $\sum \lambda_i = 0$) and hence G must be the totally disconnected graph K_n^c.] Let m_1 and m_2 be the multiplicities of λ_1 and λ_2, respectively. Then $0 = tr(A) = m_1\lambda_1 + m_2\lambda_2$ and

hence $m_1 = m_2$. This gives $\lambda_2 = -\lambda_1$, and therefore $\lambda_2 = -\sqrt{\frac{\left[2m - \left(\frac{2m}{n}\right)^2\right]}{n-1}} = -\lambda_1 = -\frac{2m}{n}$, which reduces to $n = 2m$. If G has ω components, $\frac{n}{2} = m \ge n-\omega$, and hence $\omega \ge \frac{n}{2}$. As G has no isolated vertices, $\omega = \frac{n}{2}$, and hence $G \simeq \frac{n}{2}K_2$.

Case (ii). *G has two eigenvalues with distinct absolute values.* Then $|\lambda_1| \ne |\lambda_2| (=$

$|\lambda_3| = \ldots = |\lambda_n|)$. Now $tr(A) = 0$ gives $\frac{2m}{n} + (n-1)\sqrt{\frac{\left[2m - \left(\frac{2m}{n}\right)^2\right]}{n-1}} = 0$, which gives, on simplification, $m = \frac{n(n-1)}{2}$. Hence, $G \simeq K_n$.

Case (iii). *G has exactly three distinct eigenvalues with distinct absolute values*

equal to $\frac{2m}{n}$ and $\sqrt{\frac{\left[2m - \left(\frac{2m}{n}\right)^2\right]}{n-1}}$.

The conclusion in this case is an immediate consequence of a result of Shrikhande and Bhagwandas [172], which states that a k-regular connected graph G of order n is strongly regular if and only if it has three distinct eigenvalues $\alpha_1 = k, \alpha_2, \alpha_3$ and if G is an $srg(n, k, \lambda, \mu)$, then $\lambda = k + \alpha_2\alpha_3 + \alpha_2 + \alpha_3$, and $\mu = k + \alpha_2\alpha_3$. Hence, G must be connected (as the multiplicity of k is 1) and strongly regular. $\square$

Remark 11.12.6. Theorems 11.12.4 and 11.12.5 deal with the case $2m \ge n$. If $2m < n$, the graph G is disconnected with $n - 2m$ isolated vertices. Removal of these isolated vertices from G results in a subgraph G' of G with $n' = 2m$ vertices and hence $G' \simeq mK_2$ by Theorem 11.12.5. Therefore, in this case, $G \simeq mK_2 \cup K_{n-2m}^c$.

We conclude this section with the theorem of Koolen and Moulton [124] on the characterization of maximum energy graphs with a given number n of vertices.

Theorem 11.12.7. *Let G be a graph on n vertices. Then $\mathcal{E}(G) \leq \frac{n}{2}(1 + \sqrt{n})$ with equality holding if and only if G is a strongly regular graph with parameters $(n, \frac{n+\sqrt{n}}{2}, \frac{n+2\sqrt{n}}{4}, \frac{n+2\sqrt{n}}{4})$.*

Proof. First, assume that $2m \geq n$. By Theorem 11.12.4,

$$\mathcal{E}(G) \leq F(m) = \frac{2m}{n} + \sqrt{(n-1)\left[2m - \left(\frac{2m}{n}\right)^2\right]}. \qquad (11.22)$$

We are interested in finding the values of m for which $F(m)$ attains the maximum. This is done by the methods of elementary calculus: $F'(m) = 0$ gives $16nm^2 - 8n^3m + (n^5 - n^4) = 0$, and this is the case when $m = \frac{n^2 \pm n\sqrt{n}}{4}$. Further, $F''(m) < 0$ only for $m = \frac{n^2 + n\sqrt{n}}{4}$. Hence, $f(m)$ attains its maximum when $m = \frac{n^2 + n\sqrt{n}}{4}$, and for this value of m, $F(\frac{2m}{n}) = F(\frac{n+\sqrt{n}}{2}) = \frac{n^3 + \sqrt{n}}{2}$.

Next, assume that G is a strongly regular graph with parameters given in Theorem 11.12.5. As G is k-regular with $k = \frac{n+\sqrt{n}}{2}$, by Theorem 11.12.3,

$$\mathcal{E}(G) = k + \sqrt{k(n-1)(n-k)}$$

$$= \frac{n + \sqrt{n}}{2} + \sqrt{\left(\frac{n+\sqrt{n}}{2}\right)(n-1)\left(n - \frac{n+\sqrt{n}}{2}\right)}$$

$$= \frac{n^{3/2} + n}{2}, \text{ the maximum possible energy.}$$

Finally, assume that G attains the maximum possible energy. We have to show that G is an $srg(n, \frac{n+\sqrt{n}}{2}, \frac{n+2\sqrt{n}}{4}, \frac{n+2\sqrt{n}}{4})$. By Theorem 11.12.5, three cases arise:

Case (i). $\frac{n(1+\sqrt{n})}{2} = \mathcal{E}(G) = \mathcal{E}(\frac{n}{2}K_2) = n$. This case cannot arise (as $n > 1$).

Case (ii). $\frac{n(1+\sqrt{n})}{2} = \mathcal{E}(G) = \mathcal{E}(K_n) = 2(n-1)$. This gives $(n-1)(n-4)^2 = 0$ and hence $G \simeq K_4$, and corresponds to a degenerate strongly regular graph.

Case (iii). G is a strongly regular graph with two eigenvalues λ_2 and λ_3 (distinct from $\lambda_1 = \frac{2m}{n}$) both having the same absolute value $\sqrt{\frac{2m - \left(\frac{2m}{n}\right)^2}{n-1}}$.

Recall that λ_2 and λ_3 are the roots of the quadratic equation (see (11.10) of Theorem 11.10.5) $x^2 + (\mu - \lambda)x + (\mu - k) = 0$, where $k = \lambda_1 = \frac{2m}{n}$. Now $\lambda_2 + \lambda_3 = 0$ gives $\lambda = \mu$. Moreover, $\lambda_2 \lambda_3 = \mu - k$ gives $\mu - \frac{2m}{n} = \lambda_2 \lambda_3 = -\left(\frac{2m - (\frac{2m}{n})^2}{n-1}\right)$. Substituting $m = \frac{nk}{2} = n\frac{n(n+\sqrt{n})}{2}$ and simplifying, we get $\mu = \frac{n+2\sqrt{n}}{4} = \lambda$. Hence, G is strongly regular with parameters $(n, \frac{n+\sqrt{n}}{2}, \frac{n+2\sqrt{n}}{4}, \frac{n+2\sqrt{n}}{4})$. $\qquad\square$

Finally, if $2m < n$, by Remark 11.12.6, $\mathcal{E}(G) \leq 2m < n < \frac{n(1+\sqrt{n})}{2}$, $n \geq 1$. Hence, the case $2m = \frac{n(1+\sqrt{n})}{2}$ cannot arise.

11.12.3 Hyperenergetic Graphs

The complete graph K_n, as seen in Example 11.12.2, has energy $2(n - 1)$. It was conjectured by Gutman [82] that if G is any graph of order n, then $\mathcal{E}(G) \leq 2(n - 1) = \mathcal{E}(K_n)$. This was disproved by many: Gutman et al. in [86] and Walikar et al. in [189]. Graphs G of order n for which $\mathcal{E}(G) > 2(n - 1)$ have been called *hyperenergetic graphs*.

Example 11.12.8.   (i) K_n is nonhyperenergetic.
(ii) All cubic graphs are nonhyperenergetic. If G is cubic, $k = 3$. Hence, by Theorem 11.12.3, $\mathcal{E}(G) \leq 3 + \sqrt{3(n - 1)(n - 3)}$. Now, $3 + \sqrt{3(n - 1)(n - 3)} \leq 2(n - 1)$ if and only if $n^2 - 8n + 16 \geq 0$, which is true as $n \geq 4$ for any cubic graph.
(iii) If $n \geq 5$, $L(K_n)$ is hyperenergetic.
(iv) If $n \geq 4$, $L(K_{n,n})$ is hyperenergetic.
    We leave the proofs of (iii) and (iv) as exercises (See [188]).

Theorem 11.12.9 (Stevanović and Stanković [173]). *The graph $K_n - H$, where H is a Hamilton cycle of K_n, $n \geq 10$ is hyperenergetic.*

Proof. Let $H = (v_1 v_2 \ldots v_n)$. Then the adjacency matrix A of $K_n - H$ is a circulant with first row $(0\ 0\ 1\ 1\ \ldots\ 1\ 0)$. Hence, by Lemma 11.2.3,

$$Sp(K_n - H) = \{\omega^{2r} + \omega^{3r} + \cdots + \omega^{(n-2)r} : \omega = \text{a primitive } n\text{th root of unity}\}.$$

$$= \begin{cases} n - 3, & \text{if } \omega = 1 \\ -1 - \omega^r - \omega^{r(n-1)}, & \text{if } \omega \neq 1, \ 1 \leq r \leq n - 1. \end{cases}$$

$$= \begin{cases} n - 3, & \text{if } \omega = 1 \\ -1 - 2\cos\frac{2\pi r}{n}, & \text{if } \omega \neq 1, \ 1 \leq r \leq n - 1. \end{cases}$$

Hence, $\mathcal{E}(K_n - H) = (n - 3) + \sum_{r=1}^{n-1} \left| -1 - 2\cos\frac{2\pi r}{n} \right|$, which shows that (by the definition of Riemann integral) $\mathcal{E}(K_n - H) \to (n - 3) + \frac{n-1}{2\pi} \int_0^{2\pi} \left| -1 - 2\cos x \right| dx$ as $n \to \infty$.
    Hence,

$$\lim_{n \to \infty} \frac{\mathcal{E}(K_n - H)}{n - 1} = 1 + \frac{1}{2\pi} \int_0^{2\pi} \left| -1 - 2\cos x \right| dx$$

$$= \frac{4\sqrt{3}}{2\pi} + \frac{1}{3}$$

$$\approx 2.43599,$$

which shows that there exists n_0 such that for $n \geq n_0$, $\mathcal{E}(K_n - H) > 2(n - 1)$. Computations show that $n_0 = 10$. $\square$

Theorem 11.12.9 disproves a conjecture given in [10].

Graph $\overline{C}(n; k_1, \ldots, k_r)$, $k_1 < k_2 < \ldots < k_r < \frac{n}{2}$, $k_i \in \mathbb{N}$ for each i is the circulant graph with vertex set $V = \{0, 1, \ldots, n-1\}$ such that a vertex u is adjacent to all the vertices of $V \setminus \{u\}$, except $u \pm k_i \pmod{n}$, $i = 1, 2, \ldots, n$. Note that $K_n - H$ is just $\overline{C}(n; 1)$. Stevanović and Stanković also generalized Theorem 11.12.9 as follows:

Theorem 11.12.10 (Stevanović and Stanković [173]). *Given $k_1 < k_2 < \ldots < k_r$, there exists $n_0 \in \mathbb{N}$ such that for $n \geq n_0$, the graph $\overline{C}(n; k_1, \ldots, k_r)$ is hyperenergetic.*

11.12.4 Energy of Cayley Graphs

We now compute the energy of the unitary Cayley graph X_{p^α}, where p is a prime. This can be done in two ways: In [10], Balakrishnan computed it using cyclotomic polynomials, while in [161], Ramaswamy and Veena computed it using Ramanujan sums. We present the latter method.

Theorem 11.12.11. *If p is a prime and $r \geq 1$, $\mathcal{E}(X_{p^\alpha}) = 2\phi(p^\alpha)$.*

Proof. If $\alpha = 1$, $X_{p^\alpha} = X^p = K_p$, and $\mathcal{E}(K_p) = 2(p - 1) = 2\phi(p)$. So assume that $\alpha \geq 2$. By Theorem 11.9.2, $Sp(X_{p^\alpha}) = \{\lambda_0, \ldots, \lambda_{n-1}\}$, where

$$\lambda_{r+1} = c(r, p^\alpha) = \mu(t_r) \frac{\phi(p^\alpha)}{\phi(t_r)}, \text{ and } t_r = \frac{p^\alpha}{(r, p^\alpha)}, \ 0 \leq r < p^\alpha.$$

Here μ is the Möebius function and ϕ is Euler's totient function.

We consider three cases:

(i) $(r, p^\alpha) = p^\alpha$. Then as $r < p^\alpha$, $r = 0$, $t_r = 1$, and $\mu(t_r) = 1$, and therefore $\lambda_1 = \frac{\phi(p^\alpha)}{\phi(1)} = p^\alpha - p^{\alpha-1}$.

(ii) $(r, p^\alpha) = 1$. Then $t_r = p^\alpha$ and therefore $\mu(t_r) = 0$. Hence, $\lambda_{r+1} = 0$ in this case.

(iii) If $1 < (r, p^\alpha) < p^\alpha$, then $(r, p^\alpha) = p^m$, $1 \leq m \leq \alpha - 1$. If $(r, p^\alpha) = p^{\alpha-1}$, then $r \in \{1.p^{\alpha-1}, 2.p^{\alpha-1}, \ldots, (p-1).p^{\alpha-1}\}$, and hence $t_r = p$ and therefore $\mu(t_r) = -1$. Hence, $\lambda_{r+1} = -\frac{\phi(p^\alpha)}{\phi(p)} = -p^{\alpha-1}$.

In all the other cases, $\mu(t_r) = 0$, and therefore $\lambda_{r+1} = 0$.

Hence, $Sp(X_{p^\alpha}) = \begin{pmatrix} p^\alpha - p^{\alpha-1} & -p^{\alpha-1} & 0 \\ 1 & p - 1 & p^\alpha - p \end{pmatrix}$, and therefore

$$\mathcal{E}(X_{p^\alpha}) = (p^\alpha - p^{\alpha-1}) + (p - 1)p^{\alpha-1}$$

$$= 2(p^\alpha - p^{\alpha-1})$$

$$= 2\phi(p^\alpha).$$

$\square$

In [10], Theorem 11.12.11 has been applied to prove the next result, which in essence constructs for each n, a k-regular graph of order n for a suitable k whose energy is small compared to the maximum possible energy among the class of k-regular graphs.

Theorem 11.12.12 ([10]). *For each $\epsilon > 0$, there exist infinitely many n for each of which there exists a k-regular graph G of order n with $k < n - 1$ and $\frac{\mathcal{E}(G)}{B} < \epsilon$, where B is the maximum energy that a k-regular graph on n vertices can attain.*

Proof. The Cayley graph X_{p^α} is $\phi(p^\alpha)$-regular. The maximum energy that a k-regular graph can attain is (by Theorem 11.12.3) $B = k + \sqrt{k(n-1)(n-k)}$. Taking $n = p^\alpha$ and $k = \phi(p^\alpha)$, this bound becomes

$$B = \phi(p^\alpha) + \sqrt{\phi(p^\alpha)(p^\alpha - 1)(p^\alpha - \phi(p^\alpha))}$$

$$= (p^\alpha - p^{\alpha-1}) + \sqrt{(p^\alpha - p^{\alpha-1})(p^\alpha - 1)p^{\alpha-1}}$$

$$= (p^\alpha - p^{\alpha-1}) + p^{\alpha-1}\sqrt{(p-1)(p^\alpha - 1)}.$$

Hence, $\dfrac{\mathcal{E}(X_{p^\alpha})}{B} = \dfrac{2(p^\alpha - p^{\alpha-1})}{(p^\alpha - p^{\alpha-1}) + p^{\alpha-1}\sqrt{(p-1)(p^\alpha - 1)}}$

$$= \frac{2}{1 + \sqrt{1 + p + \cdots + p^{\alpha-1}}} \longrightarrow 0 \text{ as either } p \to \infty \text{ or } \alpha \to \infty.$$

$\square$

In [10], the following dual question was also raised: Given a positive integer $n \geq 3$, and $\epsilon > 0$, does there exist a k-regular graph G of order n such that $\frac{\mathcal{E}(G)}{B} > 1 - \epsilon$ for some $k < (n-1)$? An affirmative answer to this question is given in [133] and [190], not for general n, but when n is a prime power $p^m \equiv 1 \pmod{4}$. Interestingly, both papers give the same example, namely, the Paley graph [see Example 11.10.2(v)].

Lemma 11.12.13. $\mathcal{E}(G_1 \times G_2) = \mathcal{E}(G_1)\mathcal{E}(G_2)$.

Proof. From Corollary 11.8.7, it follows that if $Sp(G_1) = \{\lambda_1, \ldots, \lambda_p\}$ and $Sp(G_2) = \{\mu_1, \ldots, \mu_q\}$, then $Sp(G_1 \times G_2) = \{\lambda_i \mu_j : 1 \leq i \leq p, 1 \leq j \leq q\}$. Hence,

$$\mathcal{E}(G_1 \times G_2) = \sum_{i=1}^{p} \sum_{j=1}^{q} |\lambda_i \mu_j| = \left(\sum_{i=1}^{p} |\lambda_i|\right)\left(\sum_{j=1}^{q} |\mu_j|\right) = \mathcal{E}(G_1)\mathcal{E}(G_2). \qquad \square$$

Recall Theorem 11.9.2, which states that if $(n_1, n_2) = 1$, $X_{n_1 n_2} \simeq X_{n_1} \times X_{n_2}$.

Corollary 11.12.14 ([161]). $\mathcal{E}(X_n) = 2^k \phi(n)$, *where k is the number of distinct prime factors of n.*

Proof. If p_1 and p_2 are distinct primes, then by Lemma 11.12.13, $\mathcal{E}(X_{p_1^{\alpha_1} p_2^{\alpha_2}}) = \mathcal{E}(X_{p_1^{\alpha_1}})\mathcal{E}(X_{p_2^{\alpha_2}}) = 2\phi(p_1^{\alpha_1}).2\phi(p_2^{\alpha_2}) = 2^2\phi(p_1^{\alpha_1} p_2^{\alpha_2})$. By induction, it follows that if $n = p_1^{\alpha_1} \ldots p_k^{\alpha_k}$, where the p_i's are the distinct prime factors of n, $\mathcal{E}(X_n) = \prod_{i=1}^{k} \mathcal{E}(p_i^{\alpha_i}) = \prod_{i=1}^{k} 2\phi(p_i^{\alpha_i}) = 2^k\phi(n)$. $\qquad\square$

The last result has also been proved in [108] using different techniques.

11.13   Energy of the Mycielskian of a Regular Graph

In this section we determine the energy of the Mycielskian of a k-regular graph G in terms of the energy of G. [We defined the Mycielskian $\mu(G)$ of a graph G in Chap. 7.] Some papers that deal with the energy of regular graphs are [4, 87, 109].

Theorem 11.13.1 ([13]). *Let G be a k-regular graph on n vertices. Then the energy $\mathcal{E}(\mu(G))$ of $\mu(G)$ is given by*

$$\mathcal{E}(\mu(G)) = \sqrt{5}\mathcal{E}(G) - (\sqrt{5} - 1)k + 2|t_3|, \qquad (11.23)$$

where $\mathcal{E}(G)$ is the energy of G and t_3 is the unique negative root of the cubic

$$t^3 - kt^2 - (n + k^2)t + kn. \qquad (11.24)$$

Proof. Denote by A the adjacency matrix of G. As A is real symmetric, $A = PDP^T$, where D is the diagonal matrix $\mathrm{diag}(\lambda_1, \ldots, \lambda_n)$, and P is an orthogonal matrix with orthonormal eigenvectors p_i's, that is, $Ap_i = \lambda_i p_i$ for each i. In particular, if e denotes the n-vector with all entries equal to 1, then $p_1 = \frac{e}{\sqrt{n}}$ (G being regular). Hence, $e^T p_i = 0$ for $i = 2, \ldots, n$, and consequently $e^T P = [\sqrt{n}, 0, \ldots, 0]$.

Now, by the definition of $\mu(G)$, the adjacency matrix of $\mu(G)$ is the matrix of order $(2n + 1)$ given by

$$\mu(A) = \begin{bmatrix} A & A & 0 \\ A & 0 & e \\ 0 & 0e^T & 0 \end{bmatrix}.$$

Since $A = PDP^T$, we have

$$\mu(A) = \begin{bmatrix} PDP^T & PDP^T & 0 \\ PDP^T & 0 & e \\ 0 & e^T & 0 \end{bmatrix}$$

$$= \begin{bmatrix} P & 0 & 0 \\ 0 & P & 0 \\ 0 & 0 & 1 \end{bmatrix} \begin{bmatrix} D & D & 0 \\ D & 0 & P^T e \\ 0 & e^T P & 0 \end{bmatrix} \begin{bmatrix} P^T & 0 & 0 \\ 0 & P^T & 0 \\ 0 & 0 & 1 \end{bmatrix}.$$

As P is an orthogonal matrix, the spectrum of $\mu(A)$ is the same as the spectrum of

$$
B = \begin{bmatrix} D & D & 0 \\ D & 0 & [\sqrt{n},0,\ldots,0]^{\mathrm{T}} \\ 0 & [\sqrt{n},0,\ldots,0] & 0 \end{bmatrix}.
$$

The determinant of the characteristic matrix of B is given by

$$
\begin{vmatrix}
x-\lambda_1 & & 0 & -\lambda_1 & & 0 & 0 \\
 & x-\lambda_2 & & & -\lambda_2 & & 0 \\
0 & & x-\lambda_n & 0 & & -\lambda_n & 0 \\
\hline
-\lambda_1 & & 0 & x & & 0 & -\sqrt{n} \\
 & -\lambda_2 & & & x & & 0 \\
0 & & -\lambda_n & 0 & & x & 0 \\
\hline
0 & 0 & \cdots & 0 & -\sqrt{n} & 0 & \cdots & 0 & x
\end{vmatrix}.
$$

We now expand $\det(xI - B)$ along its first, $(n + 1)$th, and $(2n + 1)$th columns by Laplace's method [61]. [Recall that Laplace's method of expansion of a determinant T along any chosen set $\{C_{j_1}, \ldots, C_{j_k}\}$ of k columns (or rows) is a natural generalization of the usual expansion of a determinant along any column (or row). This expansion of T is given by $\sum_{(i_1,\ldots,i_k)} (-1)^{(i_1+\cdots+i_k)+(j_1+\cdots+j_k)} U_k V_{2n+1-k}$, where U_k is the determinant minor of T of order k common to the k rows $R_{i_1}, \ldots, R_{i_k}$, and the k columns $C_{j_1}, \ldots, C_{j_k}$, and V_{2n+1-k} is the complementary determinant minor of order $(2n + 1 - k)$ obtained by deleting these sets of k rows and k columns from T.] In this expansion, only the 3×3 minor, which is common to the above three columns, and the 1-st, $(n + 1)$th, and $(2n + 1)$th rows are nonzero. This minor is

$$
M_1 = \begin{vmatrix} x-\lambda_1 & -\lambda_1 & 0 \\ -\lambda_1 & 0 & -\sqrt{n} \\ 0 & -\sqrt{n} & 0 \end{vmatrix}.
$$

We now expand the complementary minor of M_1 along the 2nd and $(n + 2)$th columns, and then the resulting complementary minor by the 3rd and $(n + 3)$th columns, and so on. These expansions give the spectrum of $\mu(A)$ to be the union of the spectra of the matrices

$$
\begin{bmatrix} \lambda_1 & \lambda_1 & 0 \\ \lambda_1 & 0 & \sqrt{n} \\ 0 & \sqrt{n} & 0 \end{bmatrix}, \begin{bmatrix} \lambda_2 & \lambda_2 \\ \lambda_2 & 0 \end{bmatrix}, \ldots, \begin{bmatrix} \lambda_n & \lambda_n \\ \lambda_n & 0 \end{bmatrix}.
$$

(Another way to see this is to observe that B is orthogonally similar to the direct sum of the above n matrices.) Thus, the spectrum of $\mu(A)$ is

$$\left\{ t_1,\, t_2,\, t_3,\, \lambda_2 \left(\frac{1 \pm \sqrt{5}}{2} \right),\, \ldots,\, \lambda_n \left(\frac{1 \pm \sqrt{5}}{2} \right) \right\},$$

where t_1, t_2, t_3 are the roots of the cubic polynomial $t^3 - kt^2 - (n + k^2)t + kn$ (note that as G is k-regular, $\lambda_1 = k$), which has two positive roots and one negative root, say, $t_1 > t_2 > 0 > t_3$.

Let $\mathcal{E}(G)$ denote the energy of G. Then $\mathcal{E}(G) = \sum_i |\lambda_i| = k + |\lambda_2| + \cdots + |\lambda_n|$. Hence, the energy of the Mycielskian $\mu(G)$ of G, when G is k-regular, is

$$\mathcal{E}(\mu(G)) = |t_1| + |t_2| + |t_3| + \left(\left| \frac{1 + \sqrt{5}}{2} \right| + \left| \frac{1 - \sqrt{5}}{2} \right| \right)(|\lambda_2| + \cdots + |\lambda_n|)$$

$$= t_1 + t_2 + |t_3| + \sqrt{5}(|\lambda_2| + \cdots + |\lambda_n|)$$

$$= k - t_3 + |t_3| + \sqrt{5}(\mathcal{E}(G) - k)$$

$$= \sqrt{5}\mathcal{E}(G) - (\sqrt{5} - 1)k + 2|t_3|.$$

□

11.13.1 An Application of Theorem 11.13.1

We have shown in Theorem 11.12.7 that the maximum energy that a graph G of order n can have is $\frac{n^{\frac{3}{2}} + n}{2}$ and that G has maximum energy if and only if it is a strongly regular graph with parameters $\left(n, \frac{n + \sqrt{n}}{2}, \frac{n + 2\sqrt{n}}{4}, \frac{n + 2\sqrt{n}}{4} \right)$. If $n > 25$ and G has maximum energy, $\mathcal{E}(G) > 3n > 2(n - 1)$ and hence G is hyperenergetic. Also, from (11.23), $\mathcal{E}(\mu(G)) > 3n\sqrt{5} - (\sqrt{5} - 1)k + 2|t_3| > 3n\sqrt{5} - (\sqrt{5} - 1)(n - 1) > 2\sqrt{5}n + \sqrt{5} + 1 > 4n = 2[(2n + 1) - 1]$. Hence, the Mycielskians of maximal energy graphs of order $n > 25$ are all hyperenergetic. More generally, if G is a regular graph of order n and $\mathcal{E}(G) > 3n$, then $\mu(G)$ is hyperenergetic.

Example 11.13.2 ([13]). We now present two examples. Consider two familiar regular graphs, namely, the Petersen graph P and a unitary Cayley graph.

1. For P, $n = 10$, $m = 15$, and $k = 3$. The spectrum of P is $\left(\begin{smallmatrix} 3 & 1 & -2 \\ 1 & 5 & 4 \end{smallmatrix} \right)$. Hence, $\mathcal{E}(P) = 16$. Now for P, the polynomial (3.2) is $t^3 - 3t^2 - 19t + 30$, and its unique negative root t_3 is ≈ -3.8829. Hence, from (11.23), $\mathcal{E}(\mu(P)) \approx 16\sqrt{5} - (\sqrt{5} - 1)3 + 2(3.8829) = 39.8347 < 40 = 2(21 - 1)$, where 21 is the order of $\mu(P)$. Hence, $\mu(P)$ is non-hyperenergetic.
2. Consider the unitary Cayley graph $G = \mathrm{Cay}(\mathbb{Z}_{210}, U)$, U being the group of multiplicative units of the additive group $(\mathbb{Z}_{210}, +)$. Then $\mathcal{E}(G) = 2^4 \phi(210) = 768 > 3 \times 210 = 3n$. Here $n = 210$, $k = \phi(210) = 48$. Hence, the polynomial

(11.24) becomes $t^3 - 48t^2 - 2514t + 10,080$, for which the unique negative root t_3 is ≈ -34.18. Hence, from (11.23), $\mathcal{E}(\mu(G)) \approx 768\sqrt{5} - (\sqrt{5} - 1)$ $48 + 2(34.18) = 1,726.352 > 2((2n + 1) - 1) = 840$. Thus, $\mu(\text{Cay}(\mathbb{Z}_{210}, U))$ is hyperenergetic, as expected.

11.14   Exercises

14.1. Prove: For any nontrivial graph of order n with spectrum $= \{\lambda_1 \geq \lambda_2 \geq \ldots \geq \lambda_n\}$, $\lambda_n \leq 0$.

14.2. If G is k-regular, show that the number of connected components of G is the multiplicity of k as an eigenvalue of G.

14.3. Prove by induction on l that the number of v_i-v_j walks of length l in the connected graph G with $V(G) = \{v_1, \ldots, v_n\}$ is the (i, j)th entry of A^l. Hence show that the diameter of a connected graph G is the least positive integer d with $A^d > 0$ (that is, all the entries of A^d are positive).

14.4. Let G be a complete r-partite graph of order n with all parts of the same size. Find $Sp(G)$. In particular, find $Sp(K_{p,p})$, and compare it with Theorem 11.6.1.

14.5. If G and H are disjoint graphs, prove that $Sp(G \cup H) = Sp(G) \cup Sp(H)$, and that $\phi(G \cup H; \lambda) = \phi(G; \lambda)\phi(H; \lambda)$.

14.6. The Möbius ladder M_h is the cubic graph obtained by joining the opposite vertices of the cycle C_{2h}. Show that the spectrum of the Möbius ladder is given by $\lambda_j = 2\cos\frac{\pi j}{h} + (-1)^j$, $0 \leq j \leq 2h - 1$.

14.7. Find the spectrum of

(i) $C_4 \times C_3$,
(ii) $K_4 \square K_3$,
(iii) $L(K_{4,3})$ using Sachs' theorem and hence comment on the results in (ii) and (iii).

14.8. If G is a bipartite graph of odd order, prove that $\det A$ is zero.

14.9. If G is bipartite and A is nonsingular, show that G has a perfect matching.

14.10. Let G be a graph of order n with adjacency matrix A. Prove that for any nonzero vector $\mathbf{x} \in \mathbb{R}^n$, $\lambda_1 = \lambda_{\max}(G) \geq \frac{(A\mathbf{x}, \mathbf{x})}{(\mathbf{x}, \mathbf{x})} \geq \lambda_{\min}(G) = \lambda_n$. In particular, show that $\lambda_1 \geq \frac{2m}{n}$. (Hint: Use the fact that $\mathbb{R}^n$ has an orthonormal basis of eigenvectors of A.)

14.11. Find the spectrum of $K_4 \boxtimes K_3$.

14.12. Show that a graph G on n vertices is connected if and only if each entry of $(A + I)^{n-1}$ is positive.

14.13. Let $\lambda_1, \ldots, \lambda_r$ be the distinct eigenvalues of A, the adjacency matrix of a graph G of order n. Prove that the minimal polynomial of A is $(x - \lambda_1) \ldots (x - \lambda_r)$. (Hint: Same as for Exercise 14.10.)

14.14. Using the fact that the Petersen graph $P \simeq (L(K_5))^c$, show that $Sp(P) = \begin{pmatrix} 3 & 1 & -2 \\ 1 & 5 & 4 \end{pmatrix}$.

14.15. Is the Petersen graph a Cayley graph?

14.16. When n is a prime, show that the unitary Cayley graph X_n is the complete graph K_n.

14.17. Let G be a finite abelian group of order n. Prove that $Sp\big(\text{Cay}(G; S)\big) = \{\lambda_\chi = \sum_{s \in S} \chi(s) : \chi = $ an irreducible character of $G\}$. [Hint: Show that if $G = \{g_1, \ldots, g_n\}$, then $\big(\chi(g_1), \ldots, \chi(g_n)\big)^{\mathrm{T}}$ is an eigenvector of the adjacency matrix of the Cayley graph.]

14.18. If λ is any eigenvalue of a graph G of order n and size m, prove that $\lambda \le \sqrt{\frac{2m}{n}(n-1)}$. (Hint: Imitate the relevant steps given in the proof of Theorem 11.12.3.)

14.19. Show that the Petersen graph is a Ramanujan graph.

14.20. Show that the graphs $K_n \times K_n \times \ldots \times K_n$ p times, where $p \ge 4$, $n \ge 3$, form a family of regular graphs none of which is a Ramanujan graph.

14.21. Prove that if G is bipartite, then $\chi(G; x)$ is of the form $\phi(x^2)$ or $x\phi(x^2)$. Hence, prove that if λ is an eigenvalue of a bipartite graph with multiplicity $m(\lambda)$, then so is $-\lambda$.

14.22. Show that a connected k-regular graph is bipartite if and only if $-k \in Sp(G)$. [Hint: Suppose $-k \in Sp(G)$ and that $(x_1 = 1, x_2, \ldots, x_n)^{\mathrm{T}}$ is an eigenvector corresponding to $-k$. Then if $v_{i_1}, \ldots, v_{i_k}$ are the neighbors of v_1, $x_{i_1} = x_{i_2} = \ldots = x_{i_k} = -1$. If v_j is a neighbor of v_{i_1}, $x_j = 1$, etc.]

14.23. Establish the inequality (11.16).

14.24. Prove: If $n \ge 5$, $L(K_n)$ is hyperenergetic.

14.25. Prove: If $n \ge 4$, $L(K_{n,n})$ is hyperenergetic.

Notes

Standard references for algebraic graph theory are [22, 47, 75].

Many variations of graph energy have been studied recently. Two types of energy of digraphs have been considered—skew energy of digraphs by taking the adjacency matrix of the digraph to be skew-symmetric [1, 171] and another by taking it to be the nonsymmetric $(0, 1)$-matrix [157]. The Laplacian matrix of a simple graph G is the matrix $L = \Delta - A$, where Δ is the diagonal matrix whose diagonal entries are the degrees of the vertices of G, and A is the adjacency matrix of G. The energy of the matrix L is the Laplacian energy of G [85]. Yet another graph energy that has been studied is the distance energy of a graph [110]. It is the energy of the distance matrix $D = (d_{ij})$ of G, where d_{ij} is the distance between the vertices v_i and v_j of G. More generally, Nikiforov has introduced and studied [147] the concept of the energy of any p by q matrix.

List of Symbols

$(d_1, d_2, \ldots, d_n)$	The degree sequence of a graph, page 11
$(G_1)_y$	The G_1-fiber or G_1-layer at the vertex y of G_2, page 28
$(G_2)_x$	The G_2-fiber or G_2-layer at the vertex x of G_1, page 28
$(S(v_1), S(v_2), \ldots, S(v_n))$	The score vector of a tournament with vertex set $\{v_1, v_2, \ldots, v_n\}$, page 44
$[S, \bar{S}]$	An edge cut of graph G, page 50
$\alpha(G)$	The stability or the independence number of graph G, page 98
$\beta(G)$	The covering number of graph G, page 98
$\chi(G)$	The chromatic number of graph G, page 144
$\chi(G; \lambda)$	The characteristic polynomial of graph G, page 242
χ'	The edge-chromatic number of a graph, page 159
$\chi'(G)$	The edge-chromatic number of graph G, page 159
$\Delta(G)$	The maximum degree of graph G, page 10
$\delta(G)$	The minimum degree of graph G, page 10
γ-set	A minimum dominating set of a graph, page 221
$\Gamma(G)$	The upper domination number of graph G, page 228
$\gamma(G)$	The domination number of graph G, page 221
$\kappa(G)$	The vertex connectivity of graph G, page 53
$\lambda(G)$	The edge connectivity of graph G, page 53
$\lambda_c(G)$	The cyclical edge connectivity of graph G, page 61
$\mathcal{E}(G)$	The energy of graph G, page 271
$\mathcal{F}$	The set of faces of the plane graph G, page 178
$\mu(G)$	The Mycielskian of graph G, page 156
$\omega(G)$	The number of components of graph G, page 14
$\phi_1 \circ \phi_2$	The composition of the mappings ϕ_1 and ϕ_2 (ϕ_2 followed ϕ_1), page 18
$\psi(G)$	The pseudoachromatic number of graph G, page 151
$\begin{pmatrix} \lambda_1 & \lambda_2 & \cdots & \lambda_s \\ m_1 & m_2 & \cdots & m_s \end{pmatrix}$	The spectrum of a graph in which the λ_i is repeated m_i times, $1 \le i \le s$, page 242

R. Balakrishnan and K. Ranganathan, *A Textbook of Graph Theory*,
Universitext, DOI 10.1007/978-1-4614-4529-6,
© Springer Science+Business Media New York 2012

$\rho(G)$	The 2-packing number of graph G, page 230
$\varphi(G)$	The number of faces of a plane graph G, page 180
$\simeq$	Is isomorphic to, page 5
$\tau(G)$	The number of spanning trees of graph G, page 81
$A(D)$	The set of arcs of digraph D, page 37
A^{T}	The transpose of the matrix A, page 242
$A_L(G)$	The adjacency matrix of the line graph of graph G, page 246
$b(f)$	The boundary of a face f in a plane graph, page 178
$B(G)$	The bipartite graph of graph G, page 214
$b(v)$	The number of blocks of G containing the vertex v, page 70
c	The capacity function of a network, page 61
$c(a)$	The capacity of arc a, page 61
$c(B)$	The number of cut vertices of G belonging to the block B, page 70
$c(H)$	The number of cycle components of H, page 250
C_k	The cycle of length k, page 13
D	A directed graph, or digraph, page 37
$d(f)$	The degree of the face f in a plane graph, page 178
$d(u,v)$	The length of a shortest u-v path in a graph, page 14
$d(v)$	The degree or valency of the vertex v in a graph, page 10
$d_D^+(v)$	The outdegree of v in digraph D, page 38
$d_D^-(v)$	The indegree of v in digraph D, page 38
$d_G(v)$	The degree or valency of the vertex v in G, page 10
D_n	The dihedral group of order $2n$, page 35
$d_D(v)$	The degree of v in D, page 38
$E(G)$	The edge set of graph G, page 2
$e(H)$	The number of even components in H, page 250
$e(v)$	The eccentricity of vertex v, page 77
f	The flow function in a network, page 62
$f(G;\lambda)$	The chromatic polynomial of graph G, page 170
$f : G \to H$	A homomorphism f of graphs from G to H, page 153
$f^+(S)$	$f([S,\bar{S}])$, where $S \subseteq V(D)$, page 62
$f^-(S)$	$f([\bar{S},S])$, where $S \subseteq V(D)$, page 62
f_{uv}	$f((u,v))$, the flow on the arc (u,v), page 62
$G(D)$	The underlying graph of digraph D, page 37
$G(X,Y)$	A bipartite graph G with bipartition (X,Y), page 6
$G * H$	A general graph product of the two graphs G and H, page 26
$G + uv$	The supergraph of G obtained by adding the new edge uv, page 8
$G - e$	The subgraph of G obtained by deleting the edge e, page 9

$G - E'$	The subgraph of G obtained by the deletion of the edges in $E' \subset E(G)$, page 8
$G - S$	The subgraph of G obtained by the deletion of the vertices in $S \subset V(G)$, page 8
$G - v$	The subgraph of G obtained by deleting the vertex v, page 8
$G[E']$	The subgraph of G induced by the subset E' of $E(G)$, page 8
$G[S]$	The subgraph of G induced by the subset S of $V(G)$, page 8
$G \circ e$	The graph obtained from G by contracting the edge e, page 81
G^*	The canonical embedding of the plane graph G, page 186
G^c	The complement of a simple graph G, page 7
G^k	The kth power of graph G, page 30
$G_1 + G_2$	The sum of the two graphs G_1 and G_2, page 25
$G_1 \square G_2$	The Cartesian product of the graph G_1 with the graph G_2, page 27
$G_1 \boxtimes G_2$	The normal or strong product of the graph G_1 with the graph G_2, page 28
$G_1 \cap G_2$	The intersection of the two graphs G_1 and G_2, page 25
$G_1 \cup G_2$	The union of the two graphs G_1 and G_2, page 24
$G_1 \times G_2$	The Kronecker or direct or tensor product of the graph G_1 with the graph G_2, page 27
$G_1 \vee G_2$	The join of the two graphs G_1 and G_2, page 25
$G_1[G_2]$	The lexicographic or composition or wreath product of the graph G_1 with the graph G_2, page 27
$i(G)$	The independence domination number of graph G, page 227
I_D	The incidence relation of digraph D, page 37
I_G	The incidence relation of graph G, page 2
$I_G(e)$	The incidence relation of the edge e in graph G, page 2
$IR(G)$	The upper irredundance number of graph G, page 228
$ir(G)$	The irredundance number of graph G, page 228
$K(G)$	The clique graph of graph G, page 215
K_n	The complete graph on n vertices, page 6
$K_{1,q}$	The star of size q, page 6
$K_{p,q}$	The complete bipartite graph with part sizes p and q, page 6
$L(G)$	The line graph or the edge graph of graph G, page 20
$m(G)$	The size of G = the number of edges in graph G, page 3
N	A network, page 61

$n(G)$	The order of G = the number of vertices in graph G, page 3
$N_G(v)$	The open neighborhood of the vertex v in graph G, page 3
$N_G[v]$	The closed neighborhood of the vertex v in graph G, page 3
$NEPS$	The Non-complete Extended P-Sum (of graphs), page 251
$O(G)$	The number of odd components of graph G, page 107
$O^+(v)$	The number of outneighbors of v in a digraph, page 46
P	The Petersen graph, page 5
P^{-1}	The inverse of the path P, page 13
P_k	The path of length k, page 13
$r(G)$	The radius of graph G, page 77
s	The source of a network, page 61
$s(G)$	The number of cycle decompositions of graph G, page 121
$S(v)$	The score of the vertex v in a tournament, page 44
S_n	The symmetric group of degree n, page 19
$Sp(G)$	The spectrum of graph G, page 241
t	The sink of a network, page 61
$V(D)$	The set of vertices of digraph D, page 37
$V(G)$	The vertex set of graph G, page 2
X_n	$\mathrm{Cay}(\mathbb{Z}_n; U_n)$, the unitary Cayley graph, page 254
$a(G)$	The achromatic number of graph G, page 151
$\mathrm{Aut}(G)$	The group of automorphisms of the graph G, page 18
$b(G)$	The b-chromatic number of graph G, page 152
$\mathrm{cap}\ K$	The sum of capacities of the arcs in K, page 63
$\mathrm{cl}(G)$	The closure of graph G, page 127
$\mathrm{diam}(G)$	The diameter of graph G, page 35
$\mathrm{ext}\ J$	The exterior of a closed Jordan curve, page 176
$\mathrm{int}\ J$	The interior of a closed Jordan curve, page 176
$\mathrm{val}\ f$	The value of the flow f in a network, page 63

References

1. Adiga, C., Balakrishnan, R., So, W.: The skew energy of a digraph. Linear Algebra Appl. **432**, 1825–1835 (2010)
2. Aharoni, R., Szabó, T.: Vizing's conjecture for chordal graphs. Discrete Math. **309**(6), 1766–1768 (2009)
3. Aho, A.V., Hopcroft, J.E., Ullman, J.D.: The Design and Analysis of Computer Algorithms. Addison-Wesley, Reading, MA (1974)
4. Akbari, S., Moazami, F., Zare, S.: Kneser graphs and their complements are hyperenergetic. MATCH Commun. Math. Comput. Chem. **61**, 361–368 (2009)
5. Apostol, T.M.: Introduction to Analytic Number Theory. Springer-Verlag, Berlin (1989)
6. Appel, K., Haken, W.: Every planar map is four colorable: Part I-discharging. Illinois J. Math. **21**, 429–490 (1977)
7. Appel, K., Haken, W.: The solution of the four-color-map problem, Sci. Amer. **237**(4) 108–121 (1977)
8. Appel, K., Haken, W., Koch, J.: Every planar map is four colorable: Part II—reducibility. Illinois J. Math. **21**, 491–567 (1977)
9. Aravamudhan, R., Rajendran, B.: Personal communication
10. Balakrishnan, R.: The energy of a graph. Linear Algebra Appl. **387**, 287–295 (2004)
11. Balakrishnan, R., Paulraja, P.: Powers of chordal graphs. J. Austral. Math. Soc. Ser. A **35**, 211–217 (1983)
12. Balakrishnan, R., Paulraja, P.: Chordal graphs and some of their derived graphs. Congressus Numerantium **53**, 71–74 (1986)
13. Balakrishnan, R., Kavaskar, T., So, W.: The energy of the Mycielskian of a regular graph. Australasian J. Combin. **52**, 163–171 (2012)
14. Bapat, R.B.: Graphs and Matrices, Universitext Series, Springer. (2011)
15. Barcalkin, A.M., German, L.F.: The external stability number of the Cartesian product of graphs. Bul. Akad. Štiince RSS Moldoven **94**(1), 5–8 (1979)
16. Behzad, M., Chartrand, G., Lesniak-Foster, L.: Graphs and Digraphs. Prindle, Weber & Schmidt International Series, Boston, MA (1979)
17. Beineke, L.W.: On derived graphs and digraphs. In: Sachs, H., Voss, H.J., Walther, H. (eds.) Beiträge zur Graphentheorie, pp. 17–33. Teubner, Leipzig (1968)
18. Benzer, S.: On the topology of the genetic fine structure. Proc. Nat. Acad. Sci. USA **45**, 1607–1620 (1959)
19. Berge, C.: Graphs and Hypergraphs, North-Holland Mathematical Library, Elsevier, **6**, (1973)
20. Berge, C.: Graphes, Third Edition. Dunod, Paris, 1983 (English, Second and revised edition of part 1 of the 1973 English version, North-Holland, 1985)

R. Balakrishnan and K. Ranganathan, *A Textbook of Graph Theory,*
Universitext, DOI 10.1007/978-1-4614-4529-6,
© Springer Science+Business Media New York 2012

21. Berge, C., Chvátal, V.: Topics on perfect graphs. Annals of Discrete Mathematics, **21**. North Holland, Amsterdam (1984)

22. Biggs, N.: Algebraic Graph Theory, 2nd ed. Cambridge University Press, Cambridge (1993)

23. Birkhoff, G., Lewis, D.: Chromatic polynomials. Trans. Amer. Math. Soc. **60**, 355–451 (1946)

24. Bondy, J.A.: Pancyclic graphs. J. Combin. Theory Ser. B **11**, 80–84 (1971)

25. Bondy, J.A., Chvátal, V.: A method in graph theory. Discrete Math. **15**, 111–135 (1976)

26. Bondy, J.A., Halberstam, F.Y.: Parity theorems for paths and cycles in graphs. J. Graph Theory **10**, 107–115 (1986)

27. Bondy, J.A., Murty, U.S.R.: Graph Theory with Applications. MacMillan, India (1976)

28. Bollobás, B.: Modern Graph Theory, Graduate Texts in Mathematics. Springer (1998)

29. Brešar, B., Rall, D.F.: Fair reception and Vizing's conjecture. J. Graph Theory **61**(1), 45–54 (2009)

30. Brešsar, B., Dorbec, P., Goddard, W., Hartnell, B.L., Henning, M.A., Klavžar, S., Rall, D.F.: Vizing's conjecture: A survey and recent results. J. Graph Theory, **69**, 46–76 (2012)

31. Brooks, R.L.: On coloring the nodes of a network. Proc. Cambridge Philos. Soc. **37**, 194–197 (1941)

32. Bryant, D.: Another quick proof that $K_{10} \neq P + P + P$. Bull. ICA Appl. **34**, 86 (2002)

33. Cayley, A.: On the theory of analytical forms called trees. Philos. Mag. **13**, 172–176 (1857); Mathematical Papers, Cambridge **3**, 242–246 (1891)

34. Chartrand, G., Ollermann, O.R.: Applied and algorithmic graph theory. International Series in Pure and Applied Mathematics. McGraw-Hill, New York (1993)

35. Chartrand, G., Wall, C.E.: On the Hamiltonian index of a graph. Studia Sci. Math. Hungar. **8**, 43–48 (1973)

36. Chudnovsky, M., Robertson, N., Seymour, P., Thomas, R.: The strong perfect graph theorem. Ann. Math. **164**, 51–229 (2006)

37. Chung, F.R.K.: Diameters and eigenvalues. J. Amer. Math. Soc. **2**, 187–196 (1989)

38. Chvátal, V.: On Hamilton's ideals. J. Combin. Theory Ser. B **12**, 163–168 (1972)

39. Chvátal, V., Erdős, P.: A note on Hamiltonian circuits. Discrete Math. **2**, 111–113 (1972)

40. Chvátal, V., Sbihi, N.: Bull-free Berge graphs are perfect. Graphs Combin. **3**, 127–139 (1987)

41. Clark, J., Holton, D.A.: A First Look at Graph Theory. World Scientific, Teaneck, NJ (1991)

42. Clark, W.E., Suen, S.: An inequality related to Vizings conjecture. Electron. J. Combin. **7**(1), Note 4, (electronic) (2000)

43. Clark, W.E., Ismail, M.E.H., Suen, S.: Application of upper and lower bounds for the domination number to Vizing's conjecture. Ars Combin. **69**, 97–108 (2003)

44. Cockayne, E.J.: Domination of undirected graphs—a survey, In: Alavi, Y., Lick, D.R. (eds.) Theory and Application of Graphs in America's Bicentennial Year. Springer-Verlag, New York (1978)

45. Cockayne, E.J., Hedetniemi, S.T., Miller, D.J.: Properties of hereditary hypergraphs and middle graphs. Networks **7**, 247–261 (1977)

46. Cvetković, D.M.: Graphs and their spectra. Thesis, Univ. Beograd Publ. Elektrotehn. Fak., Ser. Mat. Fiz., No. 354–356, pp. 01–50 (1971)

47. Cvetković, D.M., Doob, M., Sachs, H.: Spectra of Graphs—Theory and Application, Third revised and enlarged edition. Johann Ambrosius Barth Verlag, Heidelberg/Leipzig (1995)

48. Deligne, P.: La conjecture de Weil I. Publ. Math. IHES **43**, 273–307 (1974)

49. Demoucron, G., Malgrange, Y., Pertuiset, R.: Graphes planaires: Reconnaissance et construction de représentaitons planaires topologiques. Rev. Française Recherche Opérationnelle **8**, 33–47 (1964)

50. Descartes, B.: Solution to advanced problem no. 4526. Am. Math. Mon. **61**, 269–271 (1954)

51. Diestel, R.: Graph theory. In: Graduate Texts in Mathematics, vol. 173, 4th edn. Springer, Heidelberg (2010)

52. Dijkstra, E.W.: A note on two problems in connexion with graphs. Numerische Math. **1**, 269–271 (1959)

53. Dirac, G.A.: A property of 4-chromatic graphs and some remarks on critical graphs. J. London Math. Soc. **27**, 85–92 (1952)

54. Dirac, G.A.: Some theorems on abstract graphs. Proc. London Math. Soc. **2**, 69–81 (1952)
55. Dirac, G.A.: Généralisations du théoréme de Menger. C.R. Acad. Sci. Paris **250**, 4252–4253 (1960)
56. Dirac, G.A.: On rigid circuit graphs-cut sets-coloring. Abh. Math. Sem. Univ. Hamburg **25**, 71–76 (1961)
57. Drinfeld, V.: The proof of Peterson's conjecture for $GL(2)$ over a global field of characteristic p. Funct. Anal. Appl. **22**, 28–43 (1988)
58. Droll, A.: A classification of Ramanujan Cayley graphs. Electron. J. Cominator. **17**, #N29 (2010)
59. El-Zahar, M., Pareek, C.M.: Domination number of products of graphs. Ars Combin. **31** (1991) 223–227
60. Fáry, I.: On straight line representation of planar graphs. Acta Sci. Math. Szeged **11**, 229–233 (1948)
61. Ferrar, W.L.: A Text-Book of Determinants, Matrices and Algebraic Forms. Oxford University Press (1953)
62. Fiorini, S., Wilson, R.J.: Edge-colourings of graphs. Research Notes in Mathematics, vol. 16. Pitman, London (1971)
63. Fleischner, H.: Elementary proofs of (relatively) recent characterizations of Eulerian graphs. Discrete Appl. Math. **24**, 115–119 (1989)
64. Fleischner, H.: Eulerian graphs and realted topics. Ann. Disc. Math. **45** (1990)
65. Ford, L.R. Jr., Fulkerson, D.R.: Maximal flow through a network. Canad. J. Math. **8**, 399–404 (1956)
66. Ford, L.R. Jr., Fulkerson, D.R.: Flows in Networks. Princeton University Press, Princeton (1962)
67. Fournier, J.-C.: Colorations des arétes d'un graphe. Cahiers du CERO **15**, 311–314 (1973)
68. Fournier, J.-C.: Demonstration simple du theoreme de Kuratowski et de sa forme duale. Discrete Math. **31**, 329–332 (1980)
69. Fulkerson, D.R.: Blocking and anti-blocking pairs of polyhedra. Math. Programming **1**, 168–194 (1971)
70. Fulkerson, D.R., Gross, O.A.: Incidence matrices and interval graphs. Pacific J. Math. **15**, 835–855 (1965)
71. Garey, M.R., Johnson, D.S.: Computers and Intractability: A Guide to the Theory of NP-Completeness, W.H. Freeman & Co., San Francisco (1979)
72. Gibbons, A.: Algorithmic Graph Theory. Cambridge University Press, Cambridge (1985)
73. Gilmore, P.C., Hoffman, A.J.: A characterization of comparability graphs and interval graphs. Canad. J. Math. **16**, 539–548 (1964)
74. Goddard, W.D., Kubicki, G., Ollermann, O.R., Tian, S.L.: On multipartite tournaments. J. Combin. Theory Ser. B **52**, 284–300 (1991)
75. Godsil, C.D., Royle, G.: Algebraic graph theory. Graduate Texts in Mathematics, vol. 207. Springer-Verlag, Berlin (2001)
76. Golumbic, M.C.: Algorithmic Graph Theory and Perfect Graphs. Academic Press, New York (1980)
77. Gould, R.J.: Graph Theory. Benjamin/Cummings Publishing Company, Menlo Park, CA (1988)
78. Grinberg, É.Ja.: Plane homogeneous graphs of degree three without Hamiltonian circuits (Russian). Latvian Math. Yearbook **4**, 51–58 (1968)
79. Gross, J.L., Tucker, T.W.: Topological Graph Theory. Wiley, New York (1987)
80. Gross, J.L., Yellen, J.: Handbook of graph theory. Discrete Mathematics and Its Applications, vol. 25. CRC Press, Boca Raton, FL (2004)
81. Gupta, R.P.: The chromatic index and the degree of a graph. Notices Amer. Math. Soc. **13**, Abstract 66T-429 (1966)
82. Gutman, I.: The energy of a graph. Ber. Math. Stat. Sekt. Forschungszent. Graz. **103**, 1–22 (1978)

83. Gutman, I.: The energy of a graph: Old and new results. In: Betten, A., Kohnert, A., Laue, R., Wassermann, A. (eds.) Algebraic Combinatorics and Applications, pp. 196–211, Springer, Berlin (2001)

84. Gutman, I., Polansky, O.: Mathematical Concepts in Organic Chemistry. Springer-Verlag, Berlin (1986)

85. Gutman, I., Zhou, B.: Laplacian energy of a graph. Linear Algebra Appl. **414**, 29–37 (2006)

86. Gutman, I., Soldatović, T., Vidović, D.: The energy of a graph and its size dependence. A Monte Carlo approach. Chem. Phys. Lett. **297**, 428–432 (1998)

87. Gutman, I., Firoozabadi, S.Z., de la Pěna, J.A., Rada, J.: On the energy of regular graphs. MATCH Commun. Math. Comput. Chem. **57**, 435–442 (2007)

88. Gyárfás, A., Lehel, J., Nešetril, J., Rödl, V., Schelp, R.H., Tuza, Z.: Local k-colorings of graphs and hypergraphs. J. Combin. Theory Ser. B **43**, 127–139 (1987)

89. Hajnal, A., Surányi, J.: Über die Auflösung von Graphen in Vollständige Teilgraphen. Ann. Univ. Sci. Budapest, Eötvös Sect. Math. **1**, 113–121 (1958)

90. Hall, P.: On representatives of subsets. J. London Math. Soc. **10**, 26–30 (1935)

91. Hall, M. Jr.: Combinatorial Theory. Blaisdell, Waltham, MA (1967)

92. Harary, F.: The determinant of the adjacency matrix of a graph. SIAM Rev **4**, 202–210 (1962)

93. Harary, F.: Graph Theory. Addison-Wesley, Reading, MA (1969)

94. Harary, F., Nash-Williams, C.St.J.A.: On Eulerian and Hamiltonian graphs and line graphs. Canad. Math. Bull. **8**, 701–710 (1965)

95. Harary, F., Palmer, E.M.: Graphical Enumeration, Academic Press, New York (1973)

96. Harary, F., Tutte, W.T.: A dual form of Kuratowski's theorem. Canad. Math. Bull. **8**, 17–20 (1965)

97. Harary, F., Norman, R.Z., Cartwright, D.: Structural Models: An Introduction to the Theory of Directed Graphs. Wiley, New York (1965)

98. Hartnell, B., Rall, D.F.: On Vizing's conjecture. Congr. Numer. **82**, 87–96 (1991)

99. Hartnell, B., Rall, D.F.: Domination in Cartesian products: Vizing's conjecture. In: Domination in Graphs, Advanced Topics, vol. 209. Monographs and Textbooks in Pure and Applied Mathematics, pp. 163–189. Marcel Dekker, New York (1998)

100. Haynes, T.W., Hedetniemi, S.T., Slater, P.J.: Fundamentals of Domination in Graphs. Marcel Dekker, New York (1998)

101. Haynes, T.W., Hedetniemi, S.T., Slater, P.J.: Domination in Graphs: Advanced Topics. Marcel Dekker, New York (1998)

102. Hayward, R.B.: Weakly triangulated graphs. J. Combin. Theory Ser. B **39**, 200–208 (1985)

103. Heawood, P.J.: Map colour theorems. Quart. J. Math. **24**, 332–338 (1890)

104. Hedetniemi, S.T.: Homomorphisms of graphs and automata. Tech. Report 03105-44-T, University of Michigan (1966)

105. Hell, P., Nešetril, J.: Graphs and homorphisms. Oxford Lecture Series in Mathematics and its Applications, vol. 28. Oxford University Press, Oxford (2004)

106. Holton, D.A., Sheehan, J.: The Petersen graph. Australian Mathematical Society Lecture Series, vol. 7, Cambridge University Press, Cambridge (1993)

107. Hougardy, S., Le, V.B., Wagler, A.: Wing-triangulated graphs are perfect. J. Graph Theory **24**, 25–31 (1997)

108. Ilić, A.: The energy of unitary Cayley graphs. Linear Algebra Appl. **431**, 1881–1889 (2009)

109. Ilić, A.: Distance spectra and distance energy of integral circulant graphs. Linear Algebra Appl. **433**, 1005–1014 (2010)

110. Indulal, G., Gutman, I., Vijayakumar, A.: On distance energy of graphs. MATCH Commun. Math. Comput. Chem. **60**, 461–472 (2008)

111. Irving, R.W., Manlove, D.F.: The b-chromatic number of a graph. Discrete Appl. Math. **91**, 127–141 (1999)

112. Jacobson, M.S., Kinch, L.F.: On the domination number of product graphs: I. Ars. Combin. **18**, 33–44 (1984)

113. Jaeger, F.: A note on sub-Eulerian graphs. J. Graph Theory **3**, 91–93 (1979)

114. Jaeger, F.: Nowhere-zero flow problems. In: Beineke, L.W., Wilson, R.J. (eds.) Selected Topics in Graph Theory III, pp. 71–95. Academic Press, London (1988)

115. Jaeger, F., Payan, C.: Relations du type Nordhaus–Gaddum pour le nombre d'absorption d'un graphe simple. C. R. Acad. Sci. Paris A **274**, 728–730 (1972)

116. Jensen, T.R., Toft, B.: Graph coloring problems. Wiley-Interscience Series in Discrete Mathematics and Optimization. Wiley, New York (1995)

117. Jordan, C.: Sur les assemblages de lignes. J. Reine Agnew. Math. **70**, 185–190 (1869)

118. Jüng, H.: Zu einem Isomorphiesatz von Whitney für Graphen. Math. Ann. **164**, 270–271 (1966)

119. Jünger, M., Pulleyblank, W.R., Reinelt, G.: On partitioning the edges of graphs into connected subgraphs. J. Graph Theory **9**, 539–549 (1985)

120. Kainen, P.C., Saaty, T.L.: The Four-Color Problem (Assaults and Conquest). Dover Publications, New York (1977)

121. Kempe, A.: On the geographical problem of four colours. Amer. J. Math. **2**, 193–200 (1879)

122. Kilpatrick, P.A.: Tutte's first colour-cycle conjecture. Ph.D. thesis, Cape Town, (1975)

123. Klotz, W., Sander, T.: Some properties of unitary Cayley graphs. Electron. J. Combinator. **14**, 1–12 (2007)

124. Koolen, J.H., Moulton, V.: Maximal energy graphs. Adv. Appl. Math. **26**, 47–52 (2001)

125. Kouider, M., Mahéo, M.: Some bounds for the b-chromatic number of a graph. Discrete Math. **256**, 267–277 (2002)

126. Kratochvíl, J., Tuza, Z., Voigt, M.: On the b-chromatic number of graphs. Lecture Notes Comput. Sci. **2573**, 310–320 (2002)

127. Kruskal, J.B. Jr.: On the shortest spanning subtree of a graph and the travelling salesman problem. Proc. Amer. Math. Soc. **7**, 48–50 (1956)

128. Kundu, S.: Bounds on the number of disjoint spanning trees. J. Combin. Theory Ser. B **17**, 199–203 (1974)

129. Kuratowski, C.: Sur le problème des courbes gauches en topologie. Fund. Math. **15**, 271–283 (1930)

130. Laskar, R., Shier, D.: On powers and centers of chordal graphs. Discrete Applied Math. **6**, 139–147 (1983)

131. Lesniak, L.M.: Neighborhood unions and graphical properties. In: Alavi, Y., Chartrand, G., Ollermann, O.R., Schwenk, A.J. (eds.) Proceedings of the Sixth Quadrennial International Conference on the Theory and Applications of Graphs: Graph Theory, Combinatorics and Applications. Western Michigan University, pp. 783–800. Wiley, New York (1991)

132. Li, W.C.W.: Number theory with applications. Series of University Mathematics, vol. 7. World Scientific, Singapore (1996)

133. Li, X., Li, Y., Shi, Y.: Note on the energy of regular graphs. Linear Algebra Appl. **432**, 1144–1146 (2010)

134. Lovász, L.: Normal hypergraphs and the perfect graph conjecture. Discrete Math. **2**, 253–267 (1972)

135. Lovász, L.: Three short proofs in graph theory. J. Combin Theory Ser. B **19**, 111–113 (1975)

136. Lovász, L., Plummer, M.D.: Matching theory. Annals of Discrete Mathematics, vol. 29. North-Holland Mathematical Studies, vol. 121 (1986)

137. Lubotzky, A., Phillips, R., Sarnak, P.: Ramanujan graphs. Combinatorica **8**, 261–277 (1988)

138. McKee, T.A.: Recharacterizing Eulerian: Intimations of new duality. Discrete Math. **51**, 237–242 (1984)

139. Meir, A., Moon, J.W.: Relations between packing and covering numbers of a tree. Pacific J. Math. **61**, 225–233 (1975)

140. Menger, K.: Zur allgemeinen Kurventheorie. Fund. Math. **10**, 96–115 (1927)

141. Moon, J.W.: On subtournaments of a tournament. Canad. Math. Bull. **9**, 297–301 (1966)

142. Moon, J.W.: Various proofs of Cayley's formula for counting trees. In: Harary, F. (eds.) A Seminar on Graph Theory, pp. 70–78. Holt, Rinehart and Winston, Inc., New York (1967)

143. Moon, J.W.: Topics on Tournaments. Holt, Rinehart and Winston Inc., New York (1968)

144. Mycielski, J.: Sur le coloriage des graphs. Colloq. Math. **3**, 161–162 (1955)

145. Nash-Williams, C.St.J.A.: Edge-disjoint spanning trees of finite graphs. J. London Math. Soc. **36**, 445–450 (1961)

146. Nebesky, L.: On the line graph of the square and the square of the line graph of a connected graph, Casopis. Pset. Mat. **98**, 285–287 (1973)

147. Nikiforov, V.: The energy of graphs and matrices. J. Math. Anal. Appl. **326**, 1472–1475 (2007)

148. Nordhaus, E.A., Gaddum, J.W.: On complementary graphs. Amer. Math. Monthly **63**, 175–177 (1956)

149. Oberly, D.J., Sumner, D.P.: Every connected, locally connected nontrivial graph with no induced claw is Hamiltonian. J. Graph Theory **3**, 351–356 (1979)

150. Ore, O.: Note on Hamilton circuits. Amer. Math. Monthly **67**, 55 (1960)

151. Ore, O.: Theory of graphs. Amer. Math. Soc. Transl. **38**, 206–212 (1962)

152. Ore, O.: The Four-Color Problem. Academic Press, New York (1967)

153. Papadimitriou, C.H., Steiglitz, K.: Combinatorial Optimization: Algorithms and Complexity. Prentice-Hall, Upper Saddle River, NJ (1982)

154. Parthasarathy, K.R., Ravindra, G.: The strong perfect-graph conjecture is true for $K_{1,3}$-free graphs. J. Combin. Theory Ser. B **21**, 212–223 (1976)

155. Parthasarathy, K.R.: Basic Graph Theory. Tata McGraw-Hill Publishing Company Limited, New Delhi (1994)

156. Parthasarathy, K.R., Ravindra, G.: The validity of the strong perfect-graph conjecture for $(K_4 - e)$-free graphs. J. Combin. Theory Ser. B **26**, 98–100 (1979)

157. Peña, I., Rada, J.: Energy of digraphs. Linear and Multilinear Algebra **56**(5), 565–579 (2008)

158. Petersen, J.: Die Theorie der regulären Graphen. Acta Math. **15**, 193–220 (1891)

159. Prim, R.C.: Shortest connection networks and some generalizations. Bell System Techn. J. **36**, 1389–1401 (1957)

160. Rall, D.F.: Total domination in categorical products of graphs. Discussiones Mathematicae Graph Theory **25**, 35–44 (2005)

161. Ramaswamy, H.N., Veena, C.R.: On the energy of unitary Cayley graphs. Electron. J. Combinator. **16** (2009)

162. Ram Murty, M.: Ramanujan graphs. J. Ramanujan Math. Soc. **18**(1), 1–20 (2003)

163. Ram Murty, M.: Ramanujan graphs and zeta functions. Jeffery-Williams Prize Lecture. Canadian Mathematical Society, Canada (2003)

164. Ray-Chaudhuri, D.K., Wilson, R.J.: Solution of Kirkman's schoolgirl problem. In: Proceedings of the Symposium on Mathematics, vol. 19, pp. 187–203. American Mathematical Society, Providence, RI (1971)

165. Rédei, L.: Ein kombinatorischer satz. Acta. Litt. Sci. Szeged **7**, 39–43 (1934)

166. Roberts, F.S.: Graph theory and its applications to problems in society. CBMS-NSF Regional Conference Series in Mathematics. SIAM, Philadelphia (1978)

167. Sachs, H.: Über teiler, faktoren und charakteristische polynome von graphen II. Wiss. Z. Techn. Hochsch. Ilmenau **13**, 405–412 (1967)

168. Sampathkumar, E.: A characterization of trees. J. Karnatak Univ. Sci. **32**, 192–193 (1987)

169. Schwenk, A.J., Lossers, O.P.: Solutions of advanced problems. Am. Math. Mon. **94**, 885–887 (1987)

170. Serre, J.-P.: Trees. Springer-Verlag, New York (1980)

171. Shader, B., So, W.: Skew spectra of oriented graphs. Electron. J. Combinator. **16**, 1–6 (2009)

172. Shrikhande, S.S., Bhagwandas: Duals of incomplete block designs. J. Indian Stat. Assoc. **3**, 30–37 (1965)

173. Stevanović, D., Stanković, I.: Remarks on hyperenergetic circulant graphs. Linear Algebra Appl. **400**, 345–348 (2005)

174. Sumner, D.P.: Graphs with 1-factors. Proc. Amer. Math. Soc. **42**, 8–12 (1974)

175. Toida, S.: Properties of an Euler graph. J. Franklin. Inst. **295**, 343–346 (1973)

176. Trinajstic, N.: Chemical Graph Theory—Volume I. CRC Press, Boca Raton, FL (1983)

177. Trinajstic, N.: Chemical Graph Theory—Volume II. CRC Press, Boca Raton, FL (1983)

178. Tucker, A.: The validity of perfect graph conjecture for K_4-free graphs. In: Berge, C., Chvátal, V. (eds.) Topics on Perfect Graphs, vol. 21, pp. 149–157 (1984)

179. Tutte, W.T.: The factorization of linear graphs. J. London Math. Soc. **22**, 107–111 (1947)
180. Tutte, W.T.: A theorem on planar graphs. Trans. Amer. Math. Soc. **82**, 570–590 (1956)
181. Tutte, W.T.: On the problem of decomposing a graph into n connected factors. J. London Math. Soc. **36**, 221–230 (1961)
182. Vizing, V.G.: The Cartesian product of graphs. Vycisl/Sistemy **9**, 30–43 (1963)
183. Vizing, V.G.: On an estimate of the chromatic class of a p-graph (in Russian). Diskret. Analiz. **3**, 25–30 (1964)
184. Vizing, V.G.: A bound on the external stability number of a graph. Dokl. Akad. Nauk. SSSR **164**, 729–731 (1965)
185. Vizing, V.G.: Some unsolved problems in graph theory. Uspekhhi Mat. Nauk. **23**(6), 117–134 (1968)
186. Wagner, K.: Über eine eigenschaft der ebenen komplexe. Math. Ann. **114**, 570–590 (1937)
187. Walikar, H.B., Acharya, B.D., Sampathkumar, E.: Recent developments in the theory of domination in graphs. MRI Lecture Notes in Mathematics, vol. 1. Mehta Research Institue, Allahabad (1979)
188. Walikar, H.B., Ramane, H.S., Hampiholi, P.R.: On the energy of a graph. In: Mulder, H.M., Vijayakumar, A., Balakrishnan, R. (eds.) Graph Connections, pp. 120–123. Allied Publishers, New Delhi (1999)
189. Walikar, H.B., Gutman, I., Hampiholi, P.R., Ramane, H.S.: Non-hyperenergetic graphs. Graph Theory Notes New York **41**, 14–16 (2001)
190. Walikar, H.B., Ramane, H.S., Jog, S.R.: On an open problem of R. Balakrishnan and the energy of products of graphs. Graph Theory Notes New York **55**, 41–44 (2008)
191. Welsh, D.J.A.: Matroid Theory. Academic Press, London (1976)
192. West, D.B.: Introduction to Graph Theory, 2nd ed. Prentice Hall, New Jersey (2001)
193. Whitney, H.: Congruent graphs and the connectivity of graphs. Amer. J. Math. **54**, 150–168 (1932)
194. Yap, H.P.: Some topics in graph theory. London Mathematical Society Lecture Notes Series, vol. 108, Cambridge University Press, Cambridge (1986)
195. Zykov, A.A.: On some properties of linear complexes (in Russian). Math. Sbornik N. S. **24**, 163–188 (1949); Amer. Math Soc. Trans. **79** (1952)

Index

R. Balakrishnan and K. Ranganathan, *A Textbook of Graph Theory*,
Universitext, DOI 10.1007/978-1-4614-4529-6,
© Springer Science+Business Media New York 2012